Robust Adaptive Control for Fractional-Order Systems with Disturbance and Saturation

WILEY-ASME PRESS SERIES LIST

Robust Adaptive Control for Fractional-Order Systems with Disturbance and Saturation	Chen	November 2017
Robot Manipulator Redundancy Resolution	Zhang	October 2017
Combined Cooling, Heating, and Power Systems: Modeling, Optimization, and Operation	Shi	August 2017
Applications of Mathematical Heat Transfer and Fluid Flow Models in Engineering and Medicine	Dorfman	February 2017
Bioprocessing Piping and Equipment Design: A Companion Guide for the ASME BPE Standard	Huitt	December 2016
Nonlinear Regression Modeling for Engineering Applications	Rhinehart	September 2016
Fundamentals of Mechanical Vibrations	Cai	May 2016
Introduction to Dynamics and Control of Mechanical Engineering Systems	To	March 2016

Robust Adaptive Control for Fractional-Order Systems with Disturbance and Saturation

Mou Chen
Nanjing University of Aeronautics and Astronautics, Jiangsu, Nanjing, China

Shuyi Shao
Nanjing University of Aeronautics and Astronautics, Jiangsu, Nanjing, China

Peng Shi
University of Adelaide, Adelaide, South Australia

This Work is a co-publication between ASME Press and John Wiley & Sons Ltd.

WILEY

ASME PRESS

This edition first published 2018

Registered Offices
John Wiley & Sons, Inc., 111 River Street, Hoboken, NJ 07030, USA
John Wiley & Sons Ltd, The Atrium, Southern Gate, Chichester, West Sussex, PO19 8SQ, UK

Editorial Office
The Atrium, Southern Gate, Chichester, West Sussex, PO19 8SQ, UK

For details of our global editorial offices, customer services, and more information about Wiley products, visit us at www.wiley.com.

Wiley also publishes its books in a variety of electronic formats and by print-on-demand. Some content that appears in standard print versions of this book may not be available in other formats.

Library of Congress Cataloging-in-Publication Data

Names: Chen, Mou, author. | Shao, Shuyi, author | Shi, Peng, 1958- author.
Title: Robust adaptive control for fractional-order systems with disturbance and saturation / by Professor Mou Chen, Doctor Shuyi Shao, Professor Peng Shi.
Description: Hoboken, NJ : John Wiley & Sons, 2017. | Includes bibliographical references and index. |
Identifiers: LCCN 2017027175 (print) | LCCN 2017041115 (ebook) | ISBN 9781119393337 (pdf) | ISBN 9781119393313 (epub) | ISBN 9781119393276 (cloth)
Subjects: LCSH: Adaptive control systems.
Classification: LCC TJ217 (ebook) | LCC TJ217 .S53 2017 (print) | DDC 629.8/36–dc23
LC record available at https://lccn.loc.gov/2017027175

Cover Design: Wiley
Cover Image: © CaryllN/Gettyimages

Set in 10/12pt WarnockPro by SPi Global, Chennai, India
Printed and bound in Malaysia by Vivar Printing Sdn Bhd

10 9 8 7 6 5 4 3 2 1

To our families, for their love and support

Contents

Preface

This book is devoted to an investigation of some issues of tracking control and synchronization control for fractional-order nonlinear systems in the presence of system uncertainty, external disturbance, and input saturation. On the basis of definitions of the fractional integral and fractional derivatives, lemmas of stability analysis for fractional-order systems, design techniques of disturbance observers, approximation methods of system uncertainty, and handling methods of input saturation, the main research motives of this book are given as follows:

1) In the modeling process, there exist a vast amount of uncertainties caused by modeling error, which may not only degrade the performance of the control system but also even lead to instability of the dynamics system. Therefore, the uncertainty should be considered in the control design to improve the closed-loop system performance of fractional-order systems. Furthermore, neural networks can approximate any continuous uncertain dynamics with an arbitrary accuracy. Many adaptive neural control schemes have been reported for uncertain integer-order nonlinear systems. However, the neural network approximation technique has rarely been considered in uncertain fractional-order nonlinear systems in past decades.
2) A practical system is often subjected to external unknown disturbances. The disturbance may lead to oscillations and even increase the instability of the system. In the field of traditional control, it is well known that feedforward control provides an effective disturbance compensation method which can achieve prompt disturbance attenuation. However, the disturbance has to be measured by sensors for the implementation of traditional feedforward control. Unfortunately, disturbances are usually difficult or even impossible to be measured physically using sensors. Since disturbance observers can estimate external disturbances from known information of the controlled plants, the output of disturbance observers can be used to design the control law. As a result, disturbance rejection is guaranteed to improve the performance and robustness of the closed-loop system. Therefore, disturbance estimation techniques could be used to alleviate the restriction faced by traditional feedforward control and reject the effect of external disturbances. However, disturbance observers have seldom been reported for uncertain fractional-order nonlinear systems subject to external disturbances in the existing literatures. In addition, neural-network-based fractional-order disturbance observers need to be further designed for uncertain fractional-order nonlinear systems.

3) Since the interactive design is rendered more difficult by incorporating the neural network and the disturbance observer, the neural network approximation technique and the disturbance observer have rarely been considered together for integer-order nonlinear systems, although disturbance observers have been widely developed for integer-order nonlinear systems. Conversely, neural-network-based fractional-order disturbance observers have not been designed for uncertain fractional-order nonlinear systems by the interactive design method in the existing literature. For fractional-order systems, the disturbance-observer-based adaptive neural control schemes need to be further developed for uncertain fractional-order nonlinear systems with unknown disturbances.
4) Saturation nonlinearity is a common problem for actuators in a wide range of practical systems. Input saturation can degrade system control performance and even lead to system instability if it is ignored in the control design. Furthermore, control design under consideration of input saturation is a challenging problem for any uncertain nonlinear system. So far, many control design schemes for integer-order nonlinear systems with input saturation have been studied. However, the issues of input saturation and disturbance have rarely been considered together in the control of fractional-order nonlinear systems, although a number of studies have considered input saturation. Therefore, new control schemes need to be further studied for fractional-order nonlinear systems in the presence of system uncertainty, external disturbance, and input saturation.

Based on these research motives, the main contributions of this book are contained in 12 chapters. Chapter 1 introduces some background knowledge. Chapter 2 provides definitions of the fractional integral and fractional derivatives and corresponding lemmas for the stability analysis of fractional-order systems, and introduces some typical fractional-order systems. Chapter 3 gives a fractional-order PID controller and a frequency-domain fractional-order disturbance observer. In Chapter 4, two fractional-order controllers are designed, for integer-order and fractional-order systems, respectively. Chapter 5 develops a disturbance-observer-based sliding-mode control scheme for fractional-order nonlinear systems with external disturbances. In Chapter 6, an adaptive neural control issue is investigated for a fractional-order rotational mechanical system subject to system uncertainties and external disturbance. Chapter 7 considers system uncertainties, external disturbance, and input saturation in the tracking control of fractional-order chaotic systems. In Chapter 8, a stabilization issue is studied for continuous-time fractional-order positive systems based on disturbance observers. Chapter 9 investigates an adaptive sliding-mode synchronization control of fractional-order chaotic systems with disturbances. In Chapter 10, the problem of anti-synchronization control is investigated for fractional-order nonlinear systems based on a disturbance observer and the neural network. In Chapter 11, the input saturation issue is considered for the synchronization of fractional-order systems, while Chapter 12 which considers the synchronization controller design for fractional-order chaotic systems with disturbance and input saturation.

This book intends to provide readers with a good understanding of how to achieve tracking control and synchronization control of fractional-order nonlinear systems with system uncertainties, external disturbance, and input saturation. The book can be used as a reference for the academic research on fractional-order nonlinear systems or used in Ph.D. study of control theory and engineering.

We would like to acknowledge support from the National Natural Science Foundation of China (grant no. 61573184) and the Fundamental Research Funds for the Central Universities (grant no. NE2016101).

Nanjing, China
January, 2017

Mou Chen
Shuyi Shao
Peng Shi

Series Preface

The Wiley-ASME Press Series in Mechanical Engineering brings together two established leaders in mechanical engineering publishing to deliver high-quality, peer-reviewed books covering topics of current interest to engineers and researchers worldwide.

The series publishes across the breadth of mechanical engineering, comprising research, design and development, and manufacturing. It includes monographs, references and course texts.

Prospective topics include emerging and advanced technologies in Engineering Design; Computer-Aided Design; Energy Conversion & Resources; Heat Transfer; Manufacturing & Processing; Systems & Devices; Renewable Energy; Robotics; and Biotechnology.

Symbols and Acronyms

D^{α}	Caputo fractional derivative
$D_{0,t}^{-\alpha}$	fractional-order integral
${}_{\mathrm{GL}}D_{0,t}^{\alpha}$	Grünwald–Letnikov fractional derivative
${}_{\mathrm{RL}}D_{0,t}^{\alpha}$	Riemann–Liouville fractional derivative
$E_{\alpha,\beta}(z)/\mathscr{E}_{\alpha,\beta}(z)$	Mittag–Leffler function
eig	eigenvalue
$\mathbf{I}$	identity matrix
$\mathscr{L}(\cdot)$	Laplace transform
$\Re$	field of real numbers
R^r	r-dimensional real vector space
R^+	positive real numbers
sign	signum function
$\lambda_{\max}(\cdot)$	maximum eigenvalue
$\lambda_{\min}(\cdot)$	minimum eigenvalue
$\lvert\cdot\rvert$	absolute value
$\lVert\cdot\rVert$	2-norm
$(\cdot)^{\mathrm{T}}$	transposition
$\forall$	for all
$\in$	belongs to
$\sum$	sum

FOCS	fractional-order chaotic system
FODO	fractional-order disturbance observer
FONS	fractional-order nonlinear system
FOPS	fractional-order positive system
GL	Grünwald–Letnikov definition
MIMO	multi-input and multi-output
ML	Mittag–Leffler
RL	Riemann–Liouville definition
SISO	single-input and single-output
SMDO	sliding-mode disturbance observer
SMFODO	sliding-mode fractional-order disturbance observer
T–S	Takagi–Sugeno

1

Introduction

Over the past decades, fractional calculus has attracted increasing interest from researchers, and has been widely applied in fields in engineering and physics, such as system control [1], electromechanics [2], and signal processing [3]. Since the mathematical model of a real plant can be accurately described via the fractional-order differential method [4, 5], many systems can be expressed as fractional differential equations, for example, fractional-order economic systems [6], fractional-order biological population models [7], fractional-order financial systems [8], and fractional-order chaotic and hyperchaotic systems [9–15]. With the development of fractional calculus, problems of control and synchronization control for fractional-order systems have been extensively investigated. So far, some important control schemes have been reported for fractional-order systems as follows.

Sliding-model control It is well known that sliding-mode control is an effective robust control scheme and has the features of fast global convergence and high robustness to external disturbances [16]. In recent years, sliding-mode control has been investigated for integer-order linear and nonlinear systems [17–25] and many important results have been reported for the control of fractional-order systems using the sliding-mode technique. Dadras and Momeni [6] studied the control for a fractional-order economical system using the sliding-mode method. Chaos control was investigated for a class of fractional-order chaotic systems based on the sliding-mode approach by Chen *et al.* [26]. Yin *et al.* [27] designed a sliding-mode controller for a class of fractional-order chaotic systems. A no-chattering sliding-mode control strategy was proposed for a class of fractional-order chaotic systems by Chen *et al.* [28]. Yin *et al.* [29] realized chaos control for a class of fractional-order chaotic systems via an adaptive sliding-mode controller. Tavazoei and Haeri [30] and Wang *et al.* [31] achieved synchronization for two fractional-order chaotic systems, using the stability theory of fractional-order systems and the active sliding-mode control method. Sliding-mode synchronization control was realized for uncertain fractional-order Duffing–Holmes systems by Hosseinnia *et al.* [32]. Aghababa [33] investigated stabilization and synchronization for a class of chaotic fractional order systems via a novel fractional-order sliding-mode method. A robust fractional-order sliding-mode scheme was proposed and synchronization was realized for uncertain fractional-order chaotic systems by Zhang and Yan [34]. Li *et al.* [35] presented a new three-dimensional fractional-order chaotic system and studied its adaptive sliding-mode synchronization. Synchronization was studied for a class of fractional-order arbitrary dimensional hyperchaotic systems based on the

Robust Adaptive Control for Fractional-Order Systems with Disturbance and Saturation, First Edition.
Mou Chen, Shuyi Shao, and Peng Shi.

sliding-mode control method by Liu *et al.* [36]. These works focused on the control of fractional-order systems via the sliding-mode approach.

Adaptive control As is well known, the adaptive control method is a valid control technique for linear and nonlinear systems with parameter uncertainty. So far, integer-order linear and nonlinear systems have been extensively investigated based on the adaptive control method [37–47]. Furthermore, many important results have also been reported for the control of fractional-order systems by using adaptive control strategies. Vinagre *et al.* [48] investigated the use of fractional calculus in conventional model reference adaptive control systems. Odibat [49] proposed an adaptive feedback control scheme for the synchronization of fractional-order chaotic systems with different fractional orders. An adaptive sliding-mode controller was designed for uncertain fractional-order chaotic systems with external disturbance by Yuan *et al.* [50]. Synchronization control was investigated for fractional-order chaotic systems with uncertain parameters by the adaptive function projective control method by Zhou and Ding [51]. Yin *et al.* [52] explored an adaptive fractional-order switching-type control method for three-dimensional fractional-order nonlinear systems. The synchronization of two different uncertain fractional-order time-delay chaotic systems was studied using an adaptive fuzzy sliding-mode control by Lin and Lee [53]. Lin *et al.* [54] studied a chaos synchronization between two different uncertain fractional-order chaotic systems based on adaptive fuzzy sliding-mode control. Adaptive control and synchronization control were investigated for a fractional-order chaotic system by Li and Tong [55]. Lin and Kuo [56] proposed a novel adaptive fuzzy H^{∞} control to deal with chaos synchronization between two different uncertain fractional-order chaotic systems. The adaptive control scheme was reported for the control of fractional-order systems in the literature.

Active control Control problems for fractional-order systems using active control schemes have been widely studied. Recently, some important results have been reported. Bhalekar and Daftardar-Gejji [57] realized synchronization between two different fractional-order chaotic systems via active control. Phase and anti-phase synchronization were investigated between two identical and non-identical fractional-order chaotic systems, based on an active control technique, by Taghvafard and Erjaee [58]. Agrawal *et al.* [59] used the active control method to realize the synchronization of two different pairs of fractional-order systems. An active control methodology was presented for controlling the chaotic behavior of a fractional-order version of the Rössler system by Razminia *et al.* [60]. Radwan *et al.* [61] explored control and switching synchronization for fractional-order chaotic systems using an active control technique. A simple integer-order control scheme was proposed for fractional-order systems, based on an active disturbance rejection method, by Li *et al.* [62]. Senejohnny and Delavari [63] developed a novel observer scheme for synchronization of fractional-order chaotic systems. Anti-synchronization was investigated between two identical chaotic fractional-order Qi or Genesio–Tesi systems, as well as between two different fractional-order Genesio–Tesi and Qi systems, using active control method by Srivastava *et al.* [64]. Bhalekar [65] studied synchronization for non-identical fractional-order hyperchaotic systems using active control. An active control scheme was developed for the fractional-order chaotic economic system by Baskonus *et al.*

[66]. These works focused on the control of fractional-order systems using the active control method.

Pinning control As we know, real-world complex networks normally have a large number of nodes. Therefore, it is usually difficult to add controllers to all nodes to control a complex network. To reduce the number of controllers, the method of pinning some of the nodes is employed to control a complex network. Some important results have been proposed for integer-order complex networks [67–73]. With the development of fractional calculus, control problems for fractional-order systems have also been widely investigated. An adaptive pinning synchronization control was studied for fractional-order complex dynamical networks by Chai *et al.* [74]. Sun *et al.* [75] studied multi-group consensus of heterogeneous fractional-order nonlinear agents via pinning control. The pinning control problem of fractional-order weighted complex dynamical networks was investigated by Tang *et al.* [76]. Yu *et al.* [77] explored the leader-following consensus problem of fractional-order multi-agent systems using adaptive pinning control. Adaptive pinning synchronization was developed for fractional-order uncertain complex dynamical networks with delay by Liang *et al.* [78]. Liu *et al.* [79] proposed an adaptive synchronization scheme for a class of fractional-order complex networks via pinning control. Cluster projective synchronization was investigated for complex networks with fractional-order nodes by pinning control by Yang *et al.* [80]. Wang *et al.* [81] developed a pinning impulsive control scheme to study the synchronization of fractional-order complex dynamical networks. The problem of pinning synchronization was explored for fractional-order complex networks with Lipschitz-type nonlinear dynamics by Wang *et al.* [82]. Wang *et al.* [83] studied a projective cluster synchronization scheme for fractional-order coupled-delay complex networks via adaptive pinning control. Synchronization and anti-synchronization were investigated for a new uncertain fractional-order modified unified chaotic systems using a novel active pinning control by Pan *et al.* [84]. In the literature, the pinning control method is effective for the control of fractional-order nonlinear systems.

The aforementioned works focused on control and synchronization control for fractional-order systems via sliding-mode control, adaptive control, and pinning control strategies. In the literature, the issue of adaptive tracking control for fractional-order systems with system uncertainty, unknown disturbance, and input saturation has rarely been considered. Therefore, it is significative to investigate effective robust adaptive control techniques for fractional-order chaotic systems subjected to model uncertainty, unknown disturbance, and input saturation.

Since the uncertain items in nonlinear systems cannot be precisely ascertained, radial basis function neural networks are employed to approximate the uncertainty; neural-network-based control schemes have been proposed to control integer-order uncertain nonlinear systems. Zhang and Ge [85] studied an adaptive neural control scheme for a class of uncertain multi-input and multi-output (MIMO) nonlinear state time-varying delay systems. Robust adaptive neural network control was explored for uncertain MIMO nonlinear systems with input nonlinearities by Chen *et al.* [86]. An adaptive neural output feedback tracking control was investigated for uncertain nonlinear MIMO systems with the discrete-time form by Liu *et al.* [87]. An adaptive neural control scheme was developed for non-strict-feedback stochastic nonlinear systems with unknown backlash-like hysteresis by Wang *et al.* [88]. Liu *et al.* [89]

investigated adaptive output feedback control of uncertain nonlinear single-input and single-output (SISO) systems. An adaptive neural network controller was designed for a class of SISO uncertain nonlinear systems in pure-feedback form by Wang and Huang [90]. Hovakimyan *et al.* [91] developed the adaptive output feedback control scheme for uncertain nonlinear systems via single-hidden-layer neural networks. A robust adaptive neural control scheme was presented for a class of perturbed strict feedback nonlinear systems with both completely unknown virtual control coefficients and unknown nonlinearities by Ge and Wang [92]. Wang and Huang [93] proposed a neural-network-based adaptive dynamic surface control scheme for a class of uncertain nonlinear systems in strict-feedback form. A direct adaptive output feedback control scheme was developed for highly uncertain nonlinear systems using neural networks by Calise *et al.* [94]. Ge *et al.* [95] presented an adaptive neural control scheme for a class of strict-feedback nonlinear systems with unknown time delays. An adaptive neural control scheme was proposed for a class of nonlinear time-delay systems with unknown virtual control coefficients by Ge *et al.* [96]. Yu *et al.* [97] explored neural-network-based adaptive dynamic surface control for permanent magnet synchronous motors. Optimal control was studied for nonlinear discrete time-varying systems based on a new neural network approximation structure by Kiumarsi *et al.* [98]. Chen *et al.* [99] presented an adaptive L_2-gain control scheme for a class of uncertain nonlinear systems with a backstepping method and neural networks. A robust adaptive neural network synchronization controller was designed for a class of chaotic systems with uncertain time delay by Chen and Chen [100]. Chen *et al.* [101] presented a sliding-mode adaptive synchronization controller for two chaotic systems using the radial-basis function neural network. The aforementioned works conclude that the neural network approximation is effective in tackling uncertainties in nonlinear systems. However, the neural network has rarely been considered in the control of uncertain fractional-order chaotic systems (FOCSs); this must be further investigated for uncertain FOCS in the presence of unknown disturbance and input saturation.

It is well known that the control performance of nonlinear plants is characterized by time-varying external unknown disturbances. Meanwhile, external disturbances in nonlinear systems cannot be efficiently handled using the neural network approximation. To compensate for the influence of unknown disturbances, nonlinear disturbance observers have been designed to estimate the unknown external disturbance, and disturbance-observer-based control schemes have been widely proposed by Li *et al.* [102]. Chen *et al.* [103] studied a nonlinear disturbance observer for robot manipulators. A control method based on a disturbance observer was proposed for a nonlinear system with disturbance by Chen [104]. Chen *et al.* [105] explored an adaptive fuzzy tracking control scheme using a disturbance observer for MIMO nonlinear systems. An overview of disturbance-observer-based control and related methods is given by Chen *et al.* [106]. Yang *et al.* [107] introduced a sliding-mode control method for systems with mismatched uncertainties based on a nonlinear disturbance observer. On the basis of the designed disturbance observer, an adaptive dynamic surface control scheme was investigated for near-space vehicles by Chen and Yu [108]. Chen and Yu [109] developed a disturbance-observer-based adaptive sliding-mode control scheme for near-space vehicles. A robust control scheme was proposed for a class of systems with uncertainty and time delay based on the disturbance-observer technique by Chen and Chen [110]. Chen *et al.* [111] proposed a robust adaptive tracking control for an underwater robot in the presence of parametric uncertainties and unknown external disturbances. A robust synchronization control scheme was investigated for chaotic systems subjected to system uncertainties and unknown external disturbances by Chen *et al.* [112]. Chen and Ge [113] proposed an adaptive neural output feedback control scheme for uncertain nonlinear systems in the presence of unknown hysteresis, external disturbances, and unmeasured states. A robust attitude control scheme based on the backstepping technique was developed for near-space vehicles with time-varying disturbances by Chen and Jiang [114]. Chen and Chen [115] proposed a sliding-mode control scheme for a class of nonlinear systems based on disturbance observers. The sliding-mode control scheme was developed for near-space vehicles with strong nonlinearity, high coupling, parameter uncertainty, and unknown time-varying disturbance, based on radial basis function neural networks and the nonlinear disturbance observer, by Zhou and Chen [116]. Chen [117] proposed a nonlinear disturbance-observer-enhanced dynamic inversion control scheme for missiles. Robust control of nonlinear systems with disturbances and uncertainties was studied using a disturbance-observer-based control technique by Yang *et al.* [118]. Yang *et al.* [119] presented a nonlinear disturbance-observer-based control scheme for MIMO nonlinear systems subjected to a mismatch condition. According to these discussions, the anti-disturbance ability of control systems can be improved by employing a disturbance observer in the robust control design for the uncertain nonlinear system with external disturbance. However, fractional-order disturbance observers (FODOs) have rarely been reported for the control of uncertain fractional-order systems with unknown disturbance and input saturation. Therefore, the development of the disturbance-observer-based neural control scheme is significant for the control of uncertain fractional-order systems.

Conversely, system control performance can be degraded under input saturation, which can even lead to closed-loop system instability [120, 121]. In recent years, some schemes have been proposed for the control of nonlinear systems with input saturation. Adaptive fuzzy output feedback control was investigated for uncertain nonlinear systems subjected to input saturation by Li *et al.* [122, 123]. Cao *et al.* [124] analyzed the stability of linear systems with state delay and input saturation. Adaptive control was studied for single-input uncertain nonlinear systems in the presence of input saturation and unknown external disturbance by Wen *et al.* [125]. Lan and Huang [126] investigated semiglobal stabilization and output regulation problems of singular linear systems subjected to input saturation. A neural network tracking control scheme was proposed for ocean surface vessels with input saturation by Chen *et al.* [127]. Chen *et al.* [128] proposed an adaptive tracking control for a class of uncertain MIMO nonlinear systems with asymmetric input constraints. An adaptive sliding-mode control scheme was presented for Lorenz chaos in the presence of input saturation by Yau and Chen [129]. Choi [130] studied the stabilization problem for linear discrete time systems subjected to input saturation. The stabilization problem was investigated for a linear system with distributed input delay and input saturation by Zhou *et al.* [131]. Li *et al.* [132] presented a novel decentralized adaptive neural control scheme for a class of interconnected large-scale uncertain nonlinear time-delay systems in the presence of input saturation. The problem of global tracking and stabilization control was studied for internally damped mobile robots with unknown parameters, and subjected to input torque saturation and external disturbances by Huang *et al.* [133]. Yang *et al.* [134]

proposed a decentralized adaptive neural output feedback control scheme for a class of large-scale time-delay systems with saturating inputs. Sector-condition-based results were proposed for adaptive control and synchronization of chaotic systems with input saturation by Iqbal *et al.* [135]. In the literature, input saturation, unknown disturbance, and uncertainty in the robust tracking control design for fractional-order systems have rarely been considered, although a number of studies have considered input saturation. The asymptotic stabilization method was studied for linear fractional-order systems with input saturation by Lim *et al.* [136]. Luo [137] derived a sufficient condition for the asymptotic stability of nonlinear fractional-order systems subjected to the effects of input saturation. Therefore, the development of a disturbance-observer-based neural control scheme is significant for the control of uncertain fractional-order systems in the presence of input saturation and unknown disturbance.

Many physical systems, called positive systems, have the peculiarity of having only non-negative states, such as absolute temperature, population level, the height of the human body, or concentration of a substance in a chemical process [138, 139]. In the past decade, positive system control has received much attention and many important results have been proposed. On the basis of Gersgorin's theorem, a stabilization method was proposed for positive linear continuous-time and discrete-time systems by Kaczorek [140]. Gao *et al.* [141] presented a sufficient and necessary linear matrix inequality condition for the stabilization of positive linear systems. A static feedback controller was designed to stabilize positive linear continuous-time systems by Wang and Huang [142]. Benzaouia *et al.* [143] reported on the sufficient conditions of asymptotic stability for positive Takagi–Sugeno (T–S) fuzzy systems based on multiple Lyapunov functions. The problem of ℓ_1-induced controller design was investigated for discrete-time positive systems with the use of linear Lyapunov functions by Chen *et al.* [144]. Chen *et al.* [145] studied the stabilization problem for continuous-time positive systems with interval uncertainties based on a designed ℓ_1-induced output-feedback controller. An ℓ_1-induced sparse controller was designed for continuous-time positive systems with interval uncertainties by Chen *et al.* [146]. Chen and colleagues [147, 148] explored positive filtering problems for positive continuous-time systems and positive T–S fuzzy systems based on the ℓ_1-induced performance. The ℓ_∞-gain analysis problem was studied for positive linear systems with unbounded delays by Shen and Lam [149]. Furthermore, the stabilization problems of switched positive linear systems were investigated by Liu and colleagues [150–153]. The aforementioned works focus on investigations of stabilization problems for positive linear continuous-time systems, discrete-time positive systems, switched positive linear systems, and positive T–S fuzzy systems. In addition, many results have been reported for the control of fractional-order positive systems (FOPSs).

On the basis of the definitions of fractional derivatives [154], many significant conclusions have been proposed for FOPSs [155]. Kaczorek [156] introduced a new class of continuous-time FOPSs and gave sufficient conditions for the reachability of FOPS. Stability and stabilization problems were studied for the linear FOPS via a state feedback method by Kaczorek [157]. Hmamed *et al.* [158] presented necessary and sufficient conditions for the boundedness of continuous-time FOPSs. A minimum-energy control problem was investigated for continuous-time FOPSs with bounded inputs by Kaczorek [159]. Mesquine *et al.* [160] explored the robust stabilization problem for continuous-time FOPSs with bounded control. The stabilization problem was studied for continuous-time FOPSs using a Lyapunov function by Benzaouia *et al.* [161]. Kaczorek [162] proposed a minimum-energy control method for fractional positive electrical circuits with bounded inputs. In the literature, the issue of external disturbances for FOPSs has not been considered. Therefore, it is significative to investigate the stabilization problem for FOPSs in the presence of external disturbances.

Motivated by these discussions, tracking control and synchronization control are investigated for fractional-order systems in the presence of system uncertainties, unknown disturbances, and input saturation. The rest of this book is organized as follows. Chapter 2 discusses the principle definitions of the fractional integral and fractional derivatives. Corresponding lemmas are introduced for the stability analysis of fractional-order systems. Furthermore, some fractional-order systems are introduced. In Chapter 3, a fractional-order PID controller and a frequency-domain FODO are introduced. In Chapter 4, two control schemes are studied for nonlinear chaotic systems based on the fractional-order controllers. A sliding-mode control scheme is developed in Chapter 5 for a class of fractional-order nonlinear systems in the presence of external unknown disturbances. In Chapter 6, a robust adaptive neural control is proposed for a fractional-order rotational mechanical system with system uncertainties and external unknown disturbances. An adaptive neural tracking control scheme is designed for uncertain fractional-order chaotic systems subjected to unknown disturbance and input saturation, as described in Chapter 7. In Chapter 8, the problem of stabilization is investigated for continuous-time fractional systems in the presence of unknown constant disturbances with closed-loop positivity, based on a FODO. The FODO-based adaptive sliding-mode synchronization control is proposed for a class of fractional-order chaotic systems with unknown bounded disturbances in Chapter 9. In Chapter 10, the problem of anti-synchronization control is investigated for fractional-order nonlinear systems based on a disturbance observer and the neural network. A synchronization control scheme is developed in Chapter 11 for a class of nonlinear fractional-order systems subjected to input saturation based on the state-feedback control method. In Chapter 12, a synchronization control scheme is studied for fractional-order chaotic systems subjected to input saturation and external unknown disturbances.

2

Fractional Calculus and Fractional-Order Systems

2.1 Fractional Calculus

In this section, several important functions of fractional calculus, the definition of the fractional integral, three main fractional derivatives, and some important lemmas are introduced, which will be used throughout the book.

2.1.1 Several Important Functions of Fractional Calculus

In this subsection, three important functions of fractional calculus will be introduced: the gamma function, the beta function, and the Mittag–Leffler (ML) function.

Definition 2.1 (Gamma Function [154, 163–165]) *The gamma function $\Gamma(z)$ is defined by the following integral form:*

$$\Gamma(z) = \int_0^{\infty} e^{-t} t^{z-1} dt, \qquad \mathrm{Re}(z) > 0, \tag{2.1}$$

where z belongs to the field of complex numbers.

According to the definition of the gamma function (2.1), it has the following property:

$$\Gamma(z+1) = z\Gamma(z). \tag{2.2}$$

On the basis of Equation (2.2), we have

$$\begin{aligned}
\Gamma(2) &= 1 \cdot \Gamma(1) = 1 = 1! \\
\Gamma(3) &= 2 \cdot \Gamma(2) = 2 \cdot 1 = 2! \\
\Gamma(4) &= 3 \cdot \Gamma(3) = 3 \cdot 2 \cdot 1 = 3! \\
\Gamma(5) &= 4 \cdot \Gamma(4) = 4 \cdot 3 \cdot 2 \cdot 1 = 4! \\
\Gamma(6) &= 5 \cdot \Gamma(5) = 5 \cdot 4 \cdot 3 \cdot 2 \cdot 1 = 5! \\
&\vdots \\
\Gamma(n+1) &= n \cdot \Gamma(n) = n(n-1)! = n!
\end{aligned} \tag{2.3}$$

Robust Adaptive Control for Fractional-Order Systems with Disturbance and Saturation, First Edition.
Mou Chen, Shuyi Shao, and Peng Shi.

Definition 2.2 (Beta Function [165]) *The beta function is defined as*

$$B(z_1, z_2) = \int_0^1 \tau^{z_1-1}(1-\tau)^{z_2-1}\mathrm{d}\tau, \qquad \mathrm{Re}(z_1) > 0, \qquad \mathrm{Re}(z_2) > 0, \tag{2.4}$$

where z_1 and z_2 belong to the field of complex numbers.

Combining Equations (2.1) and (2.4), the relationship between the gamma and beta functions is given by

$$B(z_1, z_2) = \frac{\Gamma(z_1)\Gamma(z_2)}{\Gamma(z_1+z_2)} = B(z_2, z_1). \tag{2.5}$$

Definition 2.3 (Mittag–Leffler Function [154, 163–165]) *The ML function plays a very important role in the theory of fractional-order differential equations. The one-parameter ML function is defined as*

$$E_{\alpha_1}(z_3) = \sum_{k=0}^{\infty} \frac{z_3^k}{\Gamma(k\alpha_1+1)}, \tag{2.6}$$

where $\alpha_1 > 0$, z_3 belongs to the field of complex numbers. When $\alpha_1 = 1$, the ML function is regarded as an exponential function. The ML function with two parameters is defined as

$$E_{\alpha_1,\alpha_2}(z_3) = \sum_{k=0}^{\infty} \frac{z_3^k}{\Gamma(k\alpha_1+\alpha_2)}, \tag{2.7}$$

where $\alpha_2 > 0$.

The relationship of the ML function is as follows:

$$E_{\alpha_1,1}(z_3) = \sum_{k=0}^{\infty} \frac{z_3^k}{\Gamma(k\alpha_1+1)} = E_{\alpha_1}(z_3). \tag{2.8}$$

On the basis of Equation (2.7), we have

$$E_{1,1}(z_3) = \sum_{k=0}^{\infty} \frac{z_3^k}{\Gamma(k+1)} = \sum_{k=0}^{\infty} \frac{z_3^k}{k!} = \mathrm{e}^{z_3}, \tag{2.9}$$

$$E_{1,2}(z_3) = \sum_{k=0}^{\infty} \frac{z_3^k}{\Gamma(k+2)} = \sum_{k=0}^{\infty} \frac{z_3^k}{(k+1)!} = \frac{1}{z_3}\sum_{k=0}^{\infty} \frac{z_3^{k+1}}{(k+1)!} = \frac{\mathrm{e}^{z_3}-1}{z_3}, \tag{2.10}$$

$$E_{1,3}(z_3) = \sum_{k=0}^{\infty} \frac{z_3^k}{\Gamma(k+3)} = \sum_{k=0}^{\infty} \frac{z_3^k}{(k+2)!} = \frac{1}{z_3^2}\sum_{k=0}^{\infty} \frac{z_3^{k+2}}{(k+2)!} = \frac{\mathrm{e}^{z_3}-1-z_3}{z_3}. \tag{2.11}$$

For the generalized case, one has

$$E_{1,m_0}(z_3) = \frac{1}{z_3^{m_0-1}}\left(\mathrm{e}^{z_3} - \sum_{k=0}^{m_0-2} \frac{z_3^k}{k!}\right), \tag{2.12}$$

where m_0 is a positive integer.

Furthermore, using the Laplace transform for the ML function (2.7) with two parameters, we have

$$\mathscr{L}\left\{t^{\alpha_2-1}E_{\alpha_1,\alpha_2}(-\lambda t^{\alpha_1}), s\right\} = \frac{s^{\alpha_1-\alpha_2}}{s^{\alpha_1}+\lambda}, \qquad \Re(s) > |\lambda|^{\frac{1}{\alpha_1}}, \tag{2.13}$$

where t and s are the variables in the time domain and Laplace domain, respectively, $\mathscr{L}(\cdot)$ stands for the Laplace transform, $\Re(s)$ denotes the real part of s, and $\lambda \in R$.

2.1.2 Fractional Integral and Derivatives

In this subsection, we will introduce the definitions of the fractional integral and the three main fractional derivatives, i.e., the Grünwald–Letnikov (GL), Riemann–Liouville (RL), and Caputo definitions [154, 163–165].

Definition 2.4 (Fractional Integral [154, 163–165]) *The fractional integral (or the RL integral) $D_{0,t}^{-\alpha_3}$ with fractional order $\alpha_3 \in R^+$ of function $g(t)$ is defined as*

$$D_{0,t}^{-\alpha_3} g(t) = \frac{1}{\Gamma(\alpha_3)} \int_0^t (t-\tau)^{\alpha_3 - 1} g(\tau) \mathrm{d}\tau. \tag{2.14}$$

In this definition, the initial time is set to zero. The same is true for the following definitions.

Definition 2.5 (Grünwald–Letnikov Definition [154, 163–165]) *The GL fractional derivative with fractional order α is given by*

$${}_{\mathrm{GL}}D_{0,t}^{\alpha} g(t) = \sum_{k_1=0}^{m-1} \frac{g^{(k_1)}(0) t^{-\alpha+k_1}}{\Gamma(-\alpha+k_1+1)} + \frac{1}{\Gamma(m-\alpha)} \int_0^t (t-\tau)^{m-\alpha-1} g^{(m)}(\tau) \mathrm{d}\tau, \tag{2.15}$$

where $m-1 \leq \alpha < m$ and $m \in Z^+$.

Property 2.1 ([166])

A) According to the definition of the GL fractional derivative (2.15), we obtain that the Laplace transform of ${}_{\mathrm{GL}}D_{0,t}^{\alpha} g(t)$ is $\mathscr{L}\{{}_{\mathrm{GL}}D_{0,t}^{\alpha} g(t); s\} = s^{\alpha} G(s)$, where $G(s) = \mathscr{L}(g(t))$ and $0 \leq \alpha < 1$.

B) If $\alpha > 1$, the Laplace transform of the GL fractional derivative does not exist in the classical sense. If we take the sense of the finite-part integral, then its Laplace transform is $\mathscr{L}\{{}_{\mathrm{GL}}D_{0,t}^{\alpha} g(t); s\} = s^{\alpha} G(s)$, where $G(s) = \mathscr{L}(g(t))$.

C) For a constant g_0, we obtain

$${}_{\mathrm{GL}}D_{0,t}^{\alpha} g_0 = \frac{g_0 t^{-\alpha}}{\Gamma(1-\alpha)}.$$

Definition 2.6 (Riemann–Liouville Definition [154, 163–165]) *The RL fractional derivative of fractional order α of function $g(t)$ is given by*

$${}_{\mathrm{RL}}D_{0,t}^{\alpha} g(t) = \frac{\mathrm{d}^m}{\mathrm{d}t^m} D_{0,t}^{-(m-\alpha)} g(t) = \frac{1}{\Gamma(m-\alpha)} \frac{\mathrm{d}^m}{\mathrm{d}t^m} \int_0^t (t-\tau)^{m-\alpha-1} g(\tau) \mathrm{d}\tau, \tag{2.16}$$

where $m-1 \leq \alpha < m$ and $m \in Z^+$.

Property 2.2 ([154, 163–165])

A) On the basis of the definition of the RL fractional derivative (2.16), we have

$$\mathscr{L}\left\{{}_{\mathrm{RL}}D_{0,t}^{\alpha} g(t); s\right\} = s^{\alpha} G(s) - \sum_{k_1=0}^{m-1} s^{k_1} \left[{}_{\mathrm{RL}}D_{0,t}^{\alpha-k_1-1} g(t)\right]_{t=0},$$

where $G(s) = \mathscr{L}(g(t))$.

B) For a constant g_1, we obtain

$$ {}_{\mathrm{RL}}D_{0,t}^{\alpha}g_1 = \frac{g_1 t^{-\alpha}}{\Gamma(1-\alpha)}. $$

Definition 2.7 (Caputo Definition [154, 163–165]) *The Caputo fractional derivative of fractional order α of function $g(t)$ is defined by*

$$ {}_{\mathrm{C}}D_{0,t}^{\alpha}g(t) = D_{0,t}^{-(m-\alpha)}\frac{\mathrm{d}^m}{\mathrm{d}t^m}g(t) = \frac{1}{\Gamma(m-\alpha)}\int_0^t (t-\tau)^{m-\alpha-1}g^{(m)}(\tau)\mathrm{d}\tau, \tag{2.17} $$

where $m-1 \leq \alpha < m$ and $m \in Z^+$.

Property 2.3 ([166])

A) From the definition of the Caputo fractional derivative (2.17), we have

$$ \mathscr{L}\left\{{}_{\mathrm{C}}D_{0,t}^{\alpha}g(t); s\right\} = s^{\alpha}G(s) - \sum_{k_1=0}^{m-1} s^{\alpha-k_1-1}g^{(k_1)}(0). $$

B) For a constant g_2, we obtain ${}_{\mathrm{C}}D_{0,t}^{\alpha}g_2 = 0$.

Property 2.4 ([166]) *Considering the RL fractional derivative (2.16) and the Caputo fractional derivative (2.17) for the case of $m-1 \leq \alpha < m$ with $m \in Z^+$, we obtain the following:*

A)

$$ {}_{\mathrm{C}}D_{0,t}^{\alpha}g(t) = {}_{\mathrm{RL}}D_{0,t}^{\alpha}\left(g(t) - \sum_{k_1=0}^{m-1}\frac{t^{k_1}}{k_1!}g^{(k_1)}(0)\right). $$

B) ${}_{\mathrm{C}}D_{0,t}^{\alpha}D_{0,t}^{-\alpha}g(t) = {}_{\mathrm{RL}}D_{0,t}^{\alpha}D_{0,t}^{-\alpha}g(t) = g(t)$ holds for $m = 1$.

C)

$$ D_{0,t}^{-\alpha}{}_{\mathrm{C}}D_{0,t}^{\alpha}g(t) = g(t) - \sum_{k_1=0}^{m-1}\frac{t^{k_1}}{k_1!}g^{(k_1)}(0). $$

D)

$$ D_{0,t}^{-\alpha}{}_{\mathrm{RL}}D_{0,t}^{\alpha}g(t) = g(t) - \sum_{k_1=0}^{m-1}\left[{}_{\mathrm{RL}}D_{0,t}^{\alpha-k_1}g(t)\right]_{t=0}\frac{t^{\alpha-k_1}}{\Gamma(\alpha-k_1+1)}. $$

E)

$$ {}_{\mathrm{RL}}D_{0,t}^{m}D_{0,t}^{-m}g(t) = g(t), $$

$$ D_{0,t}^{-m}{}_{\mathrm{RL}}D_{0,t}^{m}g(t) = g(t) - \sum_{k_1=0}^{m-1}\frac{t^{k_1}}{k_1!}g^{(k_1)}(0). $$

In this book, we use the definition of the Caputo fractional derivative to analyze the problems of control and synchronization control for fractional-order systems. Furthermore, we will use D^{α} instead of ${}_{\mathrm{C}}D_{0,t}^{\alpha}$.

2.1.3 Some Important Lemmas

In this subsection, some lemmas are introduced, which will be used to analyze the control for fractional-order systems in the book.

Lemma 2.1 **([167])** *Let $p(t) \in R$ be a continuous and derivable function. Then, for any time instant $t \geq 0$, the following relationship holds:*

$$\frac{1}{2}D^{\beta}p^2(t) \leq p(t)D^{\beta}p(t), \tag{2.18}$$

where $\beta \in (0,1)$.

Lemma 2.2 **([167])** *For $0 < \beta < 1$, the fractional-order system is described as*

$$D^{\beta}\varphi(t) = h(\varphi(t)). \tag{2.19}$$

We assume that $\varphi(t) = 0$ is the equilibrium point and $\varphi(t) \in R$. If the following condition is satisfied:

$$\varphi(t)h(\varphi(t)) \leq 0, \qquad \forall\varphi(t), \tag{2.20}$$

then the origin of the system (2.19) is stable. Moreover, if

$$\varphi(t)h(\varphi(t)) < 0, \qquad \forall\varphi(t) \neq 0, \tag{2.21}$$

then the origin of the system (2.19) is asymptotically stable.

Lemma 2.3 **([168, 169])** *Consider the following inequality with fractional order $\beta \in (0,1)$:*

$$D^{\beta}q(t) \leq -b_0q(t) + b_1; \tag{2.22}$$

then there exists a constant $t_1 > 0$, such that for all $t \in (t_1, \infty)$, we have

$$\|q(t)\| \leq \frac{2b_1}{b_0}, \tag{2.23}$$

where $q(t)$ is the state variable and b_0 and b_1 are two positive constants.

Lemma 2.4 **([170])** *As a class of linearly parameterized neural networks, the radial basis function neural network is adopted to approximate the continuous function $f(Z) : R^q \to R$, which can be expressed as follows:*

$$\hat{f}(Z) = \hat{\theta}_1^{\top}\phi(Z) + \varepsilon_1, \tag{2.24}$$

where $Z = [z_1, z_2, \cdots, z_q]^{\top} \in R^q$ is the input vector of the neural network, $\hat{\theta}_1 \in R^p$ is a weight vector of the neural network, $\phi(Z) = [\phi_1(Z), \phi_2(Z), \cdots, \phi_p(Z)]^{\top} \in R^p$ is the basis function vector, and ε_1 is the approximation error of the neural network. The optimal weight value θ_1^* of the radial basis function neural network is given by

$$\theta_1^* = \arg\min_{\theta_l \in \Omega_f}[\sup_{z \in S_Z}|\hat{f}(Z|\hat{\theta}_1) - f(Z)|], \tag{2.25}$$

where $\Omega_f = \{\hat{\theta}_1 : \|\hat{\theta}_1\| \leq \overline{N}\}$ is a valid field of the estimation parameter $\hat{\theta}_1$, $\overline{N}$ is a design parameter, and $S_Z \subset R^n$ is an allowable set of the state vector. Using the optimal weight value yields

$$\begin{aligned} f(Z) &= \theta_1^{*\top}\phi(Z) + \varepsilon_1^*, \\ |\varepsilon_1^*| &\leq \overline{\varepsilon}_1, \end{aligned} \tag{2.26}$$

where $\phi(Z)$ satisfies $\|\phi(Z)\| \leq \overline{\phi}$ with $\overline{\phi} > 0$, θ_1^* satisfies $\left\|\theta_1^*\right\| \leq \overline{\theta}_1$ with $\overline{\theta}_1 > 0$, ε_1^* is the optimal approximation error, and $\overline{\varepsilon}_1 > 0$ is the upper bound of the approximation error.

Lemma 2.5 ([171]) *Let $w(t) \in R^n$ be a differentiable function vector. Then, for any time instant $t \geq 0$, the following relationship holds:*

$$\frac{1}{2}D^{\beta_1}(w^\top(t)Pw(t)) \leq w^\top(t)PD^{\beta_1}w(t), \qquad \forall \beta_1 \in (0,1], \qquad \forall t \geq 0, \tag{2.27}$$

where $P \in R^{n\times n}$ is a symmetric and positive definite constant matrix.

Lemma 2.6 ([172]) *Consider the following system:*

$$\begin{aligned} D^\alpha w_1(t) &= w_2(t), \\ D^\alpha w_2(t) &= -R_0 \tanh\left(w_1(t) - v(t) + \frac{w_2(t)\left|w_2(t)\right|}{R_0}\left[1 - \frac{\Gamma^2(\alpha+1)}{\Gamma(2\alpha+1)}\right]\right), \end{aligned} \tag{2.28}$$

where the design parameter R_0 is a positive constant to be chosen, $v(t)$ is the input signal of the system (2.28), $\tanh(\cdot)$ is the hyperbolic tangent function, and $w_1(t)$ and $w_2(t)$ are estimates of $v(t)$ and $D^\alpha v(t)$, respectively.

Define the estimation errors as

$$\begin{aligned} \xi_1(t) &= v(t) - w_1(t), \\ \xi_2(t) &= D^\alpha v(t) - w_2(t). \end{aligned} \tag{2.29}$$

On the basis of the conclusions presented by Wei et al. [172], we obtain that the errors $\xi_1(t)$ and $\xi_2(t)$ are uniformly convergent as $t \to \infty$. Thus, we have that ξ_2 satisfies $\left|\xi_2(t)\right| < \overline{\xi}_2$ with $\overline{\xi}_2 > 0$.

Lemma 2.7 ([173]) *If $A \in R^{n\times n}$, $0 < \alpha_4 \leq 1$, $N > 0$, and μ is such that $\pi\alpha_4/2 < \mu < \min\{\pi, \pi\alpha_4\}$ and $\mu \leq \left|\arg(\mathrm{spec}(A))\right| \leq \pi$, we have*

$$\left\|E_{\alpha_4,\beta_1}(A)\right\| \leq \frac{N}{1+\|A\|}, \tag{2.30}$$

where β_1 stands for set of complex numbers, $\mathrm{spec}(A)$ denotes the eigenvalue of the maximum of the matrix A, and $\arg(\cdot)$ denotes the argument.

Lemma 2.8 Gronwall–Bellman Lemma [174, 175]) *Assume that the function $\rho(t)$ satisfies*

$$\rho(t) \leq \sigma(t) + \int_0^t \varsigma(\tau)\rho(\tau)\mathrm{d}\tau, \tag{2.31}$$

where $\varsigma(\tau)$ and $\sigma(t)$ are known real functions. Then we have

$$\rho(t) \leq \sigma(t) + \int_0^t \varsigma(\tau)\sigma(\tau)\exp\left[\int_\tau^t \varsigma(r)\mathrm{d}r\right]\mathrm{d}\tau. \tag{2.32}$$

If $\sigma(t)$ is differentiable, we obtain

$$\rho(t) \leq \sigma(0)\exp\left(\int_0^t \varsigma(\tau)\mathrm{d}\tau\right) + \int_0^t \dot{\sigma}(\tau)\exp\left(\int_\tau^t \varsigma(\upsilon)\mathrm{d}\upsilon\right)\mathrm{d}\tau. \tag{2.33}$$

In particular, if $\sigma(t)$ is a constant, one has

$$\rho(t) \leq \sigma(0) \exp\left(\int_0^t \varsigma(\tau)\mathrm{d}\tau\right), \tag{2.34}$$

where $\exp(\cdot)$ *denotes the exponential operation.*

Lemma 2.9 **([136])** *Let*

$$S(LK, u_0) = \left\{x(t) \in R^n | -u_0 \leq LKx(t) \leq u_0\right\}, \tag{2.35}$$

where $u_0 = [u_{01}, u_{02}, \cdots, u_{0n}]^\top \in R^n$, $L = \mathrm{diag}[l_1, l_2, l_3, \cdots, l_n]$, *with* $0 < l_i \leq 1$ *for all* $\forall i = 1, 2, \ldots, n$, *and*

$$\varphi(t, x(t)) = \mathrm{sat}(Kx(t)) - LKx(t), \tag{2.36}$$

where K is a real matrix. If $x(t) \in S(LK, u_0)$, the following conclusions are equivalent:

A) $(\mathrm{sat}(Kx(t)) - LKx(t))^\top(\mathrm{sat}(Kx(t)) - Kx(t)) \leq 0.$
B) $\varphi(t, x(t))^\top(\varphi(t, x(t)) - (K - LK)x(t)) \leq 0.$
C) $\|\varphi(t, x(t))\| \leq \|K - LK\| \, \|x(t)\|.$

Lemma 2.10 **([154])** *If $0 < \alpha_5 < 2$, μ_1 satisfies $\pi\alpha_5/2 < \mu_1 < \min\left\{\pi, \pi\alpha_5\right\}$ and $\left|\arg(\sigma)\right| \leq \mu_1$, β_2 is an arbitrary real number, and C_1 and C_2 are real constants, we have*

$$\left|E_{\alpha_5,\beta_2}(\sigma)\right| \leq C_1(1 + |\sigma|)^{(1-\beta_2)/\alpha_5} \exp\left(\mathrm{Re}(\sigma^{1/\alpha_5})\right) + \frac{C_2}{1 + |\sigma|}. \tag{2.37}$$

Lemma 2.11 **([176])** *For the ML function $E_{\alpha_6,\beta_3}(A_0 t^{\alpha_6})$, the following property holds. If $\alpha_6 \geq 1$, then, for $\beta_3 = 1, 2, \alpha_6$, one has*

$$\left\|E_{\alpha_6,\beta_3}(A_0 t^{\alpha_6})\right\| \leq \left\|\mathrm{e}^{A_0 t^{\alpha_6}}\right\|, \qquad t \geq 0, \tag{2.38}$$

where A_0 denotes a matrix and $\|\cdot\|$ denotes any vector or induced matrix norm.
Moreover, if A_0 is a stable matrix, we obtain

$$\left\|\mathrm{e}^{A_0 t^{\alpha_6}}\right\| \leq M\mathrm{e}^{-\eta t}, \qquad t \geq 0, \tag{2.39}$$

where the constant $M \geq 1$, $-\eta(\eta > 0)$ is the largest eigenvalue of the stable matrix A_0.

Lemma 2.12 **([174])** *Suppose $\beta_5 > 0$, $b(t)$ is a non-negative function locally integrable on $0 \leq t < T(T \leq +\infty)$, and $\chi(t)$ is a non-negative, non-decreasing continuous function defined on $0 \leq t < T$, $\chi(t) \leq N$ (constant), and suppose that $\eta(t)$ is non-negative and locally integrable on $0 \leq t < T$, with*

$$\eta(t) \leq b(t) + \chi(t)\int_0^t (t - s)^{\beta_5 - 1}\eta(s)\mathrm{d}s. \tag{2.40}$$

Furthermore, we have

$$\eta(t) \leq b(t) + \int_0^t \left[\sum_{n=1}^{\infty} \frac{(\chi(t)\Gamma(\beta_5))^n}{\Gamma(n\beta_5)}(t - s)^{n\beta_5 - 1}b(s)\right]\mathrm{d}s. \tag{2.41}$$

If $b(t)$ is a non-negative function on $[0, T)$, one obtains

$$\eta(t) \le b(t)E_{\beta_5,1}\left(\chi(t)\Gamma(\beta_5)t^{\beta_5}\right). \tag{2.42}$$

Lemma 2.13 ([177]) *If $0 < \alpha_7 < 1$, $C_0 > 0$, and A_1 is a stable matrix, we have*

$$\lim_{t\to\infty}\left\{-\sigma_0 t + \overline{C}_0\right\} = -\infty, \tag{2.43}$$

where

$$\overline{C}_0 = C_0 \sum_{k_2=0}^{\infty} \frac{\left\|A_1^{(\alpha_7-1)k_2/\alpha_7}\right\|\Gamma(k_2+1)}{\Gamma(k_2\alpha_7+\alpha_7)} \frac{t^{(\alpha_7-1)(k_2+1)+1}}{(\alpha_7-1)(k_2+1)+1}$$

and $\sigma_0 > 0$.

Lemma 2.14 ([177]) *For the ML function $E_{\alpha_8,\beta_4}(A_2t^{\alpha_8})$, the following property holds:*

$$\left\|E_{\alpha_8,\beta_4}(A_2t^{\alpha_8})\right\| \le M_1\left\|e^{A_2^{1/\alpha_8}t}\right\|, \tag{2.44}$$

where $\alpha_8 > 0$ and $\beta_4 > 0$, A_2 denotes a matrix, $\|\cdot\|$ denotes the vector norm or the induced matrix norm, and $M_1 > 0$ is a constant. Note that, from Equation (2.44), if $|\arg(\mathrm{eig}(A_2))| > \alpha_8\pi/2$, $\beta_4 = 1, \alpha_8 > 0$, we have

$$\left\|e^{A_2^{1/\alpha_8}t}\right\| \le M_2e^{-\eta t}, \tag{2.45}$$

where $M_2 > 0$ is a constant, $-\eta(\eta > 0)$ is the largest eigenvalue of the stable matrix A_2^{1/α_8} with $\{\mathrm{Re}(\mathrm{eig}(A_2^{1/\alpha_8}))\} < 0$, and $\mathrm{eig}(A_2^{1/\alpha_3})$ denotes all the eigenvalues of A_2^{1/α_8}.

2.2 Some Typical Fractional-Order Systems

In this section, some fractional-order models are given; the dynamic behaviors of fractional-order systems are also shown by numerical simulation.

2.2.1 Fractional-Order Lorenz System

From the simplified equation of convective rolls in the equations of the atmosphere, the first three-dimensional chaotic system was derived by Lorenz in 1963 [178]. Furthermore, the developed fractional-order Lorenz system [179] is given as follows:

$$\begin{aligned}
D^{\alpha_{01}}x_1(t) &= \hbar_1(x_2(t) - x_1(t)),\\
D^{\alpha_{02}}x_2(t) &= x_1(t)(\hbar_2 - x_3(t)) - x_2(t),\\
D^{\alpha_{03}}x_3(t) &= x_1(t)x_2(t) - \hbar_3x_3(t),
\end{aligned} \tag{2.46}$$

where α_{01}, α_{02}, and α_{03} are fractional orders, $x_1(t)$, $x_2(t)$, and $x_3(t)$ are system state variables, and $\hbar_1$, $\hbar_2$, and $\hbar_3$ are system parameters. For fractional orders chosen as $\alpha_{01} = \alpha_{02} = \alpha_{03} = 0.995$, system parameters given by $\hbar_1 = 10$, $\hbar_2 = 28$, and $\hbar_3 = 8/3$, and initial conditions chosen as $x_1(0) = x_2(0) = x_3(0) = 0.1$, the simulation results of the fractional-order Lorenz system are shown in Figure 2.1. Furthermore, the bifurcation diagram for $\alpha_{02} \in [1.2, 1.4]$ is presented in Figure 2.2. Following the bifurcation diagram (Figure 2.2), the dynamic characteristic of two cycles for $\alpha_{02} = 1.3$ is shown in Figure 2.3.

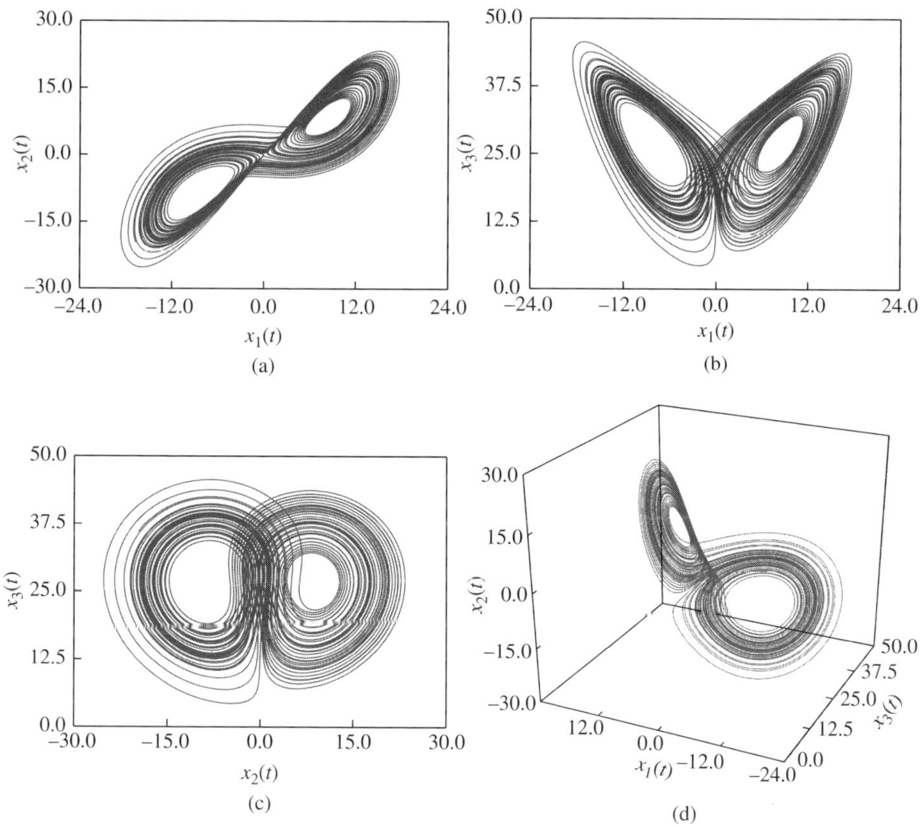

Figure 2.1 Chaotic behaviors of fractional-order Lorenz system (2.46): (a) $x_1(t)$–$x_2(t)$ plane; (b) $x_1(t)$–$x_3(t)$ plane; (c) $x_2(t)$–$x_3(t)$ plane; (d) $x_3(t)$–$x_1(t)$–$x_2(t)$ space.

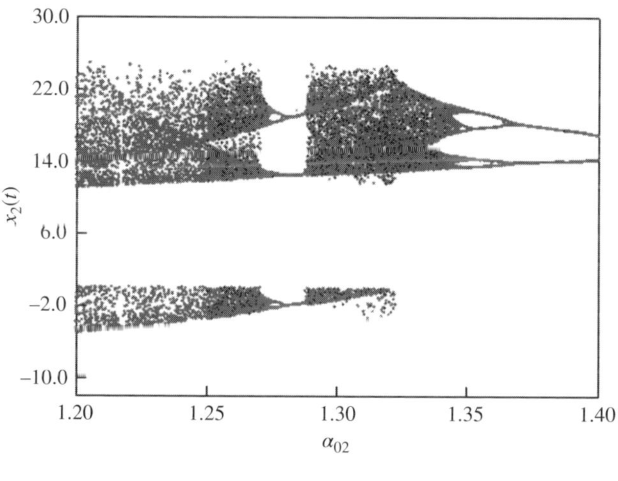

Figure 2.2 Bifurcation diagram of fractional-order Lorenz system (2.46) for $\alpha_{02} \in [1.2, 1.4]$.

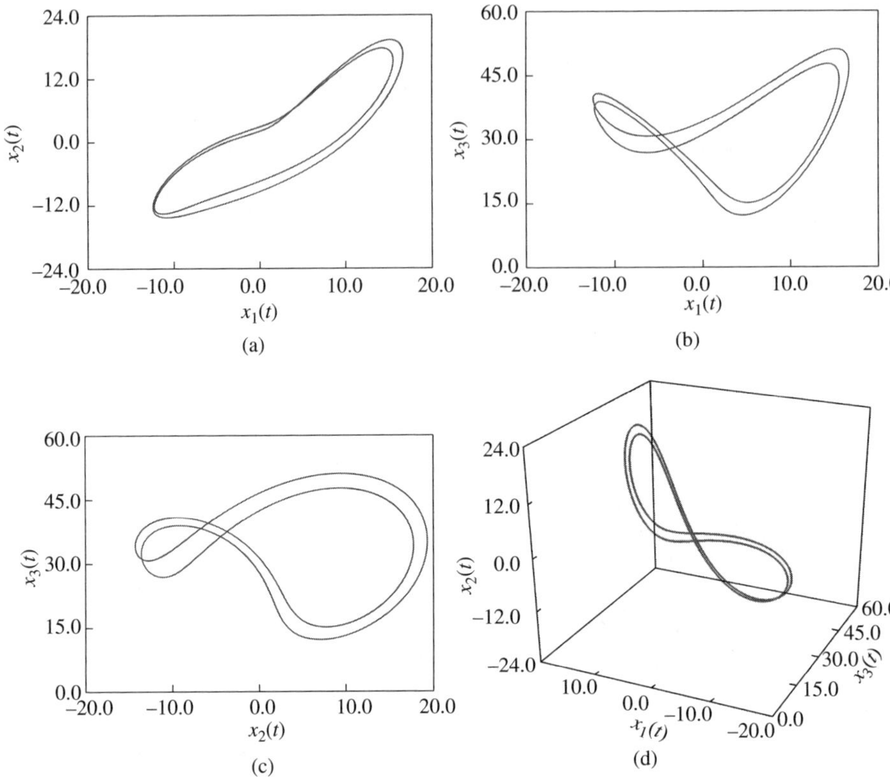

Figure 2.3 Dynamic characteristic of two cycles for fractional-order Lorenz system (2.46): (a) $x_1(t)$–$x_2(t)$ plane; (b) $x_1(t)$–$x_3(t)$ plane; (c) $x_2(t)$–$x_3(t)$ plane; (d) $x_3(t)$–$x_1(t)$–$x_2(t)$ space.

2.2.2 Fractional-Order Van Der Pol Oscillator

The van der Pol oscillator is a non-conservative oscillator with nonlinear damping. The fractional-order van der Pol oscillator is given in the following form [164]:

$$\begin{aligned} D^{\alpha_{01}} x_1(t) &= x_2(t), \\ D^{\alpha_{02}} x_2(t) &= -x_1(t) - \hbar(x_1^2(t) - 1)x_2(t), \end{aligned} \tag{2.47}$$

where α_{01} and α_{02} are fractional orders, $x_1(t)$ and $x_2(t)$ are system state variables, and $\hbar$ is a system parameter. For fractional orders chosen as $\alpha_{01} = 1.2$ and $\alpha_{02} = 0.8$, the system parameter given by $\hbar = 1$, and initial conditions chosen as $x_1(0) = 0.2$ and $x_2(0) = -0.2$, the simulation results of the fractional-order van der Pol oscillator are given in Figure 2.4.

2.2.3 Fractional-Order Genesio–Tesi System

The Genesio-Tesi system was first described using mathematical equations by Petráš [164]. In addition, the fractional-order Genesio–Tesi system is defined as follows [180]:

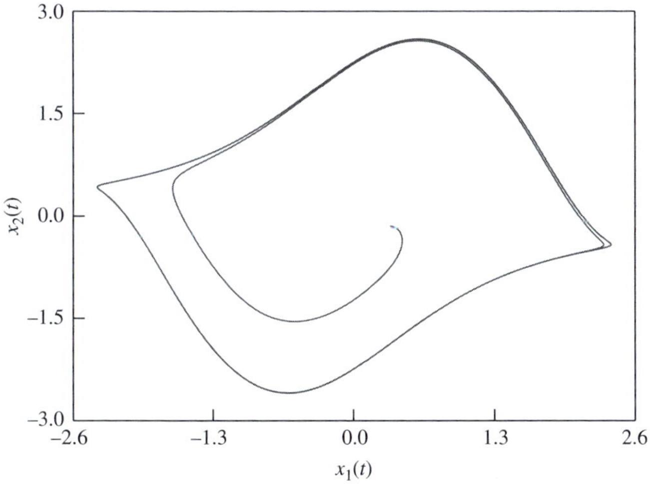

Figure 2.4 Simulation result of fractional-order van der Pol oscillator (2.47).

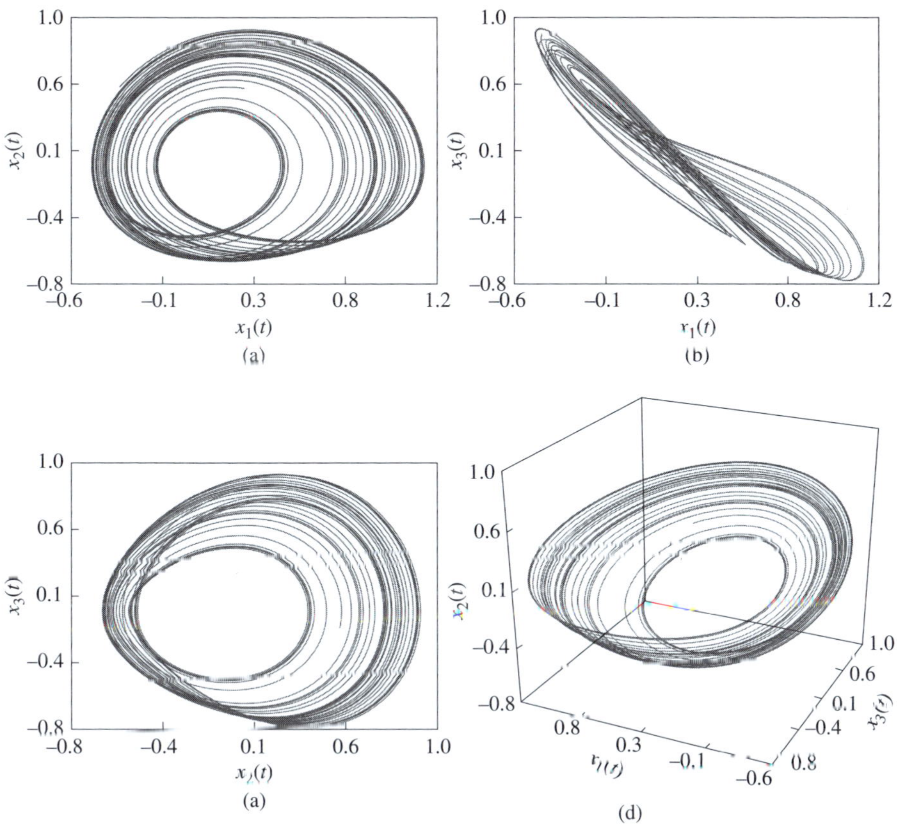

Figure 2.5 Chaotic behaviors of fractional-order Genesio–Tesi system (2.48): (a) $x_1(t)$–$x_2(t)$ plane; (b) $x_1(t)$–$x_3(t)$ plane, (c) $x_2(t)$–$x_3(t)$ plane, (d) $x_3(t)$–$x_1(t)$–$x_2(t)$ space.

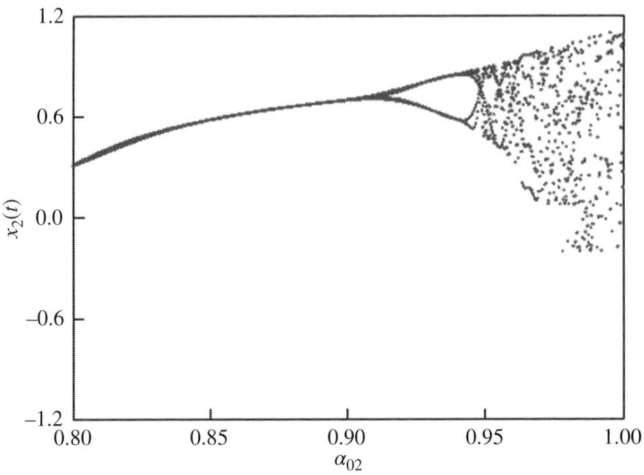

Figure 2.6 Bifurcation diagram of fractional-order Genesio–Tesi system (2.48) for $\alpha_{02} \in [0.8, 1]$.

$$
\begin{aligned}
&D^{\alpha_{01}} x_1(t) = x_2(t),\\
&D^{\alpha_{02}} x_2(t) = x_3(t),\\
&D^{\alpha_{03}} x_3(t) = -\hbar_1 x_1(t) - \hbar_2 x_2(t) - \hbar_3 x_3(t) + x_1^2(t),
\end{aligned}
\tag{2.48}
$$

where α_{01}, α_{02}, and α_{03} are fractional orders, $x_1(t)$, $x_2(t)$, and $x_3(t)$ are system state variables, and $\hbar_1$, $\hbar_2$, and $\hbar_3$ are system parameters. For fractional orders chosen as $\alpha_{01} = \alpha_{02} = 1$ and $\alpha_{03} = 0.95$, system parameters given by $\hbar_1 = 1.1$, $\hbar_2 = 1.1$, and $\hbar_3 = 0.45$, and initial conditions chosen as $x_1(0) = -0.1$, $x_2(0) = 0.5$, and $x_3(0) = 0.2$, the simulation results of the fractional-order Genesio–Tesi system are presented in Figure 2.5. Furthermore, the bifurcation diagram for $\alpha_{02} \in [0.8, 1]$ is presented in Figure 2.6, with the initial conditions $x_1(0) = 0.1$, $x_2(0) = 0.5$, and $x_3(0) = 0.2$. Following the bifurcation diagram (Figure 2.6), the dynamic characteristic of two cycles for $\alpha_{02} = 0.932$ is shown in Figure 2.7.

2.2.4 Fractional-Order Arneodo System

The fractional-order Arneodo system is described as follows [181]:

$$
\begin{aligned}
&D^{\alpha_{01}} x_1(t) = x_2(t),\\
&D^{\alpha_{02}} x_2(t) = x_3(t),\\
&D^{\alpha_{03}} x_3(t) = -\hbar_1 x_1(t) - \hbar_2 x_2(t) - \hbar_3 x_3(t) - x_1^3(t),
\end{aligned}
\tag{2.49}
$$

where α_{01}, α_{02}, and α_{03} are fractional orders, $x_1(t)$, $x_2(t)$, and $x_3(t)$ are system state variables, and $\hbar_1$, $\hbar_2$, and $\hbar_3$ are system parameters. For fractional orders chosen as $\alpha_{01} = 0.97$, $\alpha_{02} = 0.97$, and $\alpha_{03} = 0.96$, system parameters given by $\hbar_1 = -5.5$, $\hbar_2 = 3.5$, and $\hbar_3 = 0.8$, and initial conditions chosen as $x_1(0) = -0.2$, $x_2(0) = 0.5$, and $x_3(0) = 0.2$, the simulation results of the fractional-order Arneodo system are shown in Figure 2.8. Furthermore, the bifurcation diagram for $\alpha_{01} = \alpha_{02} = \alpha_{03} \in [0.8, 1]$ is presented in

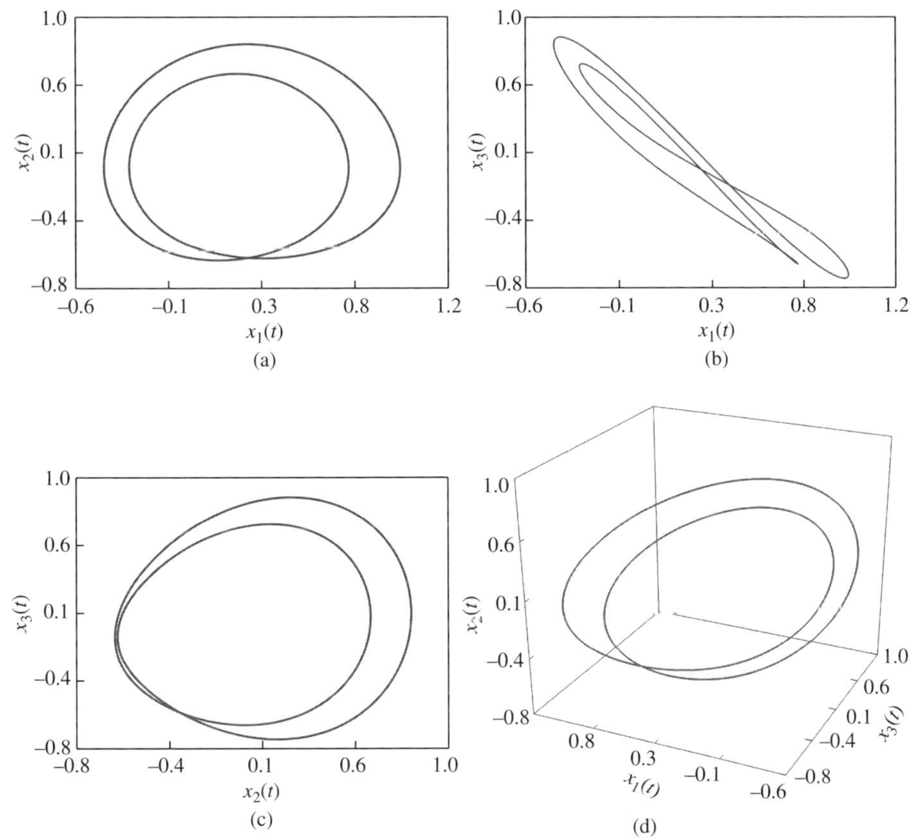

Figure 2.7 Dynamic characteristic of two cycles for fractional-order Genesio–Tesi system (2.48): (a) $x_1(t)$–$x_2(t)$ plane; (b) $x_1(t)$–$x_3(t)$ plane; (c) $x_2(t)$–$x_3(t)$ plane; (d) $x_3(t)$–$x_1(t)$–$x_2(t)$ space.

Figure 2.9. Following the bifurcation diagram (Figure 2.9), the dynamic characteristic of one cycle for $\alpha_{01} = \alpha_{02} = \alpha_{03} = 0.89$ is shown in Figure 2.10.

2.2.5 Fractional-Order Lotka–Volterra System

A two-predator and one-prey generalization of the Lotka–Volterra system was proposed by Samardzija and Greller [182]. Its fractional-order model is given as follows [164]:

$$
\begin{aligned}
D^{\alpha_{01}} x_1(t) &= x_1(t) - x_1(t)x_2(t) + \hbar_1 x_1^2(t) - \hbar_2 x_3(t) x_1^2(t),\\
D^{\alpha_{02}} x_2(t) &= -x_2(t) + x_1(t)x_2(t),\\
D^{\alpha_{03}} x_3(t) &= -\hbar_3 x_3(t) + \hbar_2 x_3(t) x_1^2(t),
\end{aligned}
\tag{2.50}
$$

where α_{01}, α_{02}, and α_{03} are fractional orders, $x_1(t)$, $x_2(t)$, and $x_3(t)$ are system state variables, and $\hbar_1$, $\hbar_2$, and $\hbar_3$ are system parameters. For fractional orders chosen as $\alpha_{01} = \alpha_{02} = \alpha_{03} = 0.95$, system parameters given by $\hbar_1 = 2$, $\hbar_2 = 3$, and $\hbar_3 = 2.7$, and initial conditions chosen as $x_1(0) = 1$, $x_2(0) = 1.4$, and $x_3(0) = 1$, the simulation results of the fractional-order Lotka–Volterra system are presented in Figure 2.11. Furthermore, the bifurcation diagram for $\alpha_{01} = \alpha_{02} = \alpha_{03} \in [0.85, 1.02]$ is presented

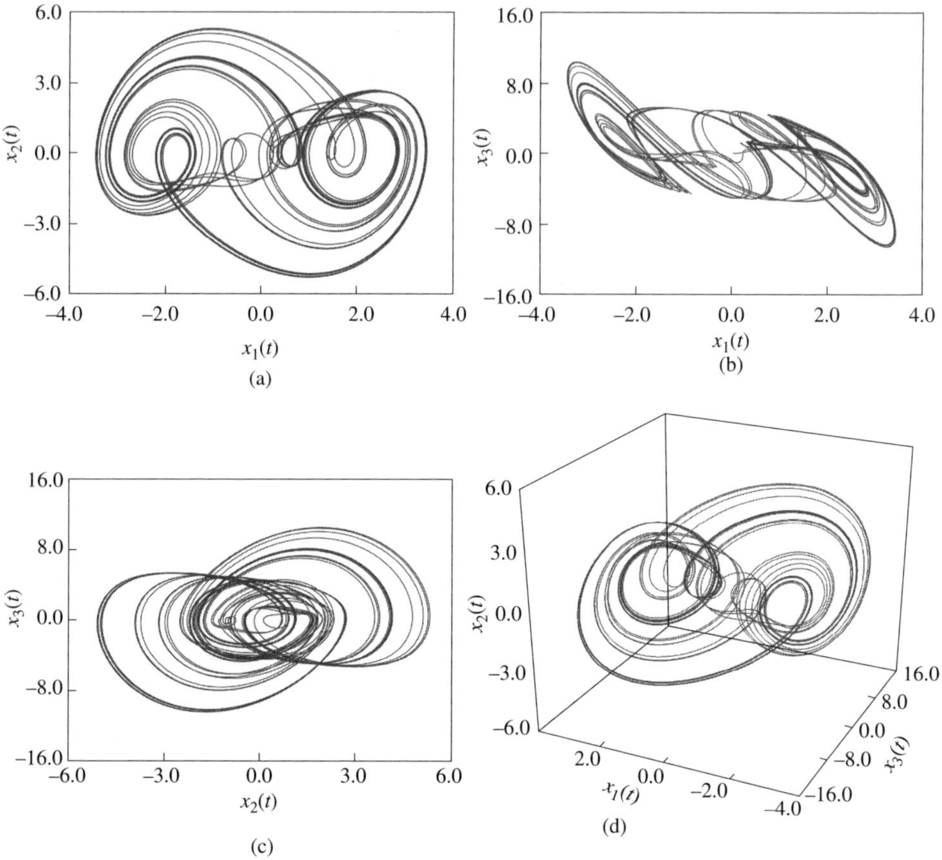

Figure 2.8 Chaotic behaviors of fractional-order Arneodo system (2.49): (a) $x_1(t)$–$x_2(t)$ plane; (b) $x_1(t)$–$x_3(t)$ plane; (c) $x_2(t)$–$x_3(t)$ plane; (d) $x_3(t)$–$x_1(t)$–$x_2(t)$ space.

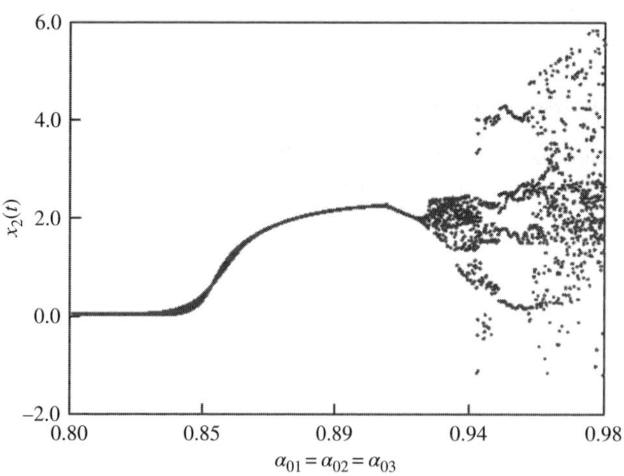

Figure 2.9 Bifurcation diagram of fractional-order Arneodo system (2.49) for $\alpha_{02} \in [0.8, 1]$.

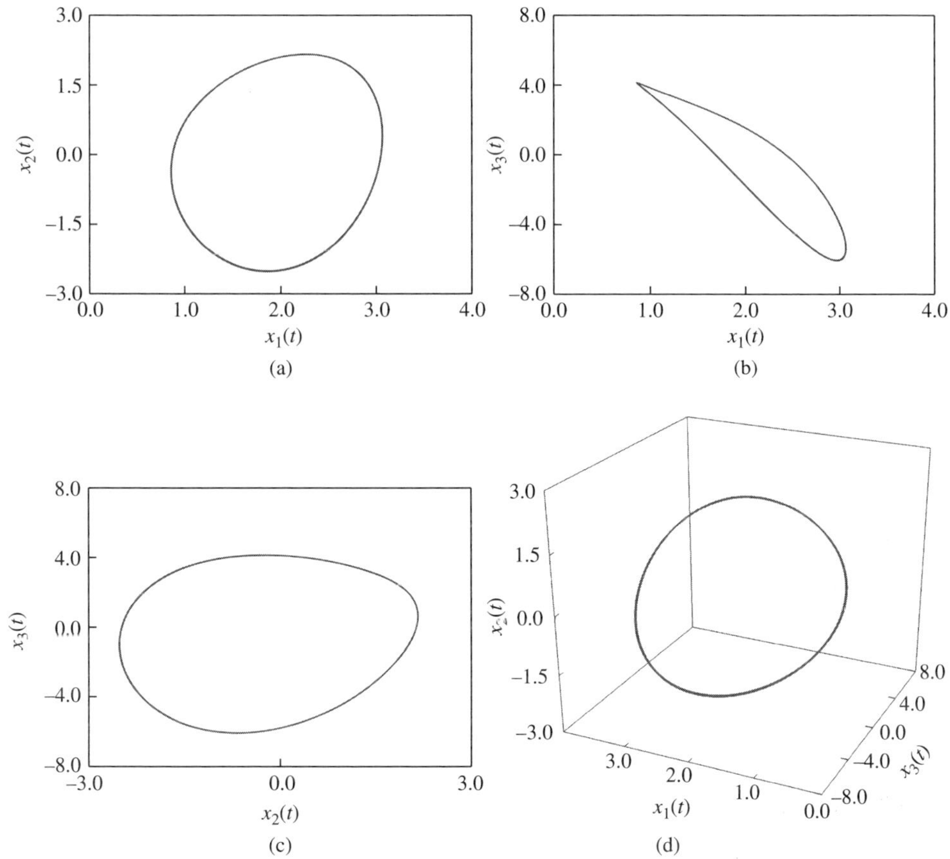

Figure 2.10 Dynamic characteristic of one cycle for fractional-order Arneodo system (2.49): (a) $x_1(t)$–$x_2(t)$ plane; (b) $x_1(t)$–$x_3(t)$ plane; (c) $x_2(t)$–$x_3(t)$ plane; (d) $x_3(t)$–$x_1(t)$–$x_2(t)$ space.

in Figure 2.12. Following the bifurcation diagram (Figure 2.12), the chaotic dynamic characteristic for $\alpha_{01} = \alpha_{02} = \alpha_{03} = 0.92$ is shown in Figure 2.13.

2.2.6 Fractional-Order Financial System

Ma and Chen [183] gave a simplified finance model. According to the integer-order finance model, the fractional-order financial system is described as follows [8]:

$$\begin{aligned}
D^{\alpha_{01}}x_1(t) &= x_3(t) + (x_2(t) - \hbar_1)x_1(t),\\
D^{\alpha_{02}}x_2(t) &= 1 - \hbar_2 x_2(t) - x_1^2(t),\\
D^{\alpha_{03}}x_3(t) &= -x_1(t) - \hbar_3 x_3(t),
\end{aligned} \tag{2.51}$$

where α_{01}, α_{02}, and α_{03} are fractional orders, $x_1(t)$, $x_2(t)$, and $x_3(t)$ are system state variables, and $\hbar_1$, $\hbar_2$, and $\hbar_3$ are system parameters. For fractional orders chosen as $\alpha_{01} = \alpha_{02} = \alpha_{03} = 0.9$, system parameters given by $\hbar_1 = \hbar_3 = 1$ and $\hbar_2 = 0.1$, and initial conditions chosen as $x_1(0) = 2$, $x_2(0) = -1$, and $x_3(0) = 1$, the simulation results of the fractional-order financial system are given in Figure 2.14. Furthermore, the bifurcation

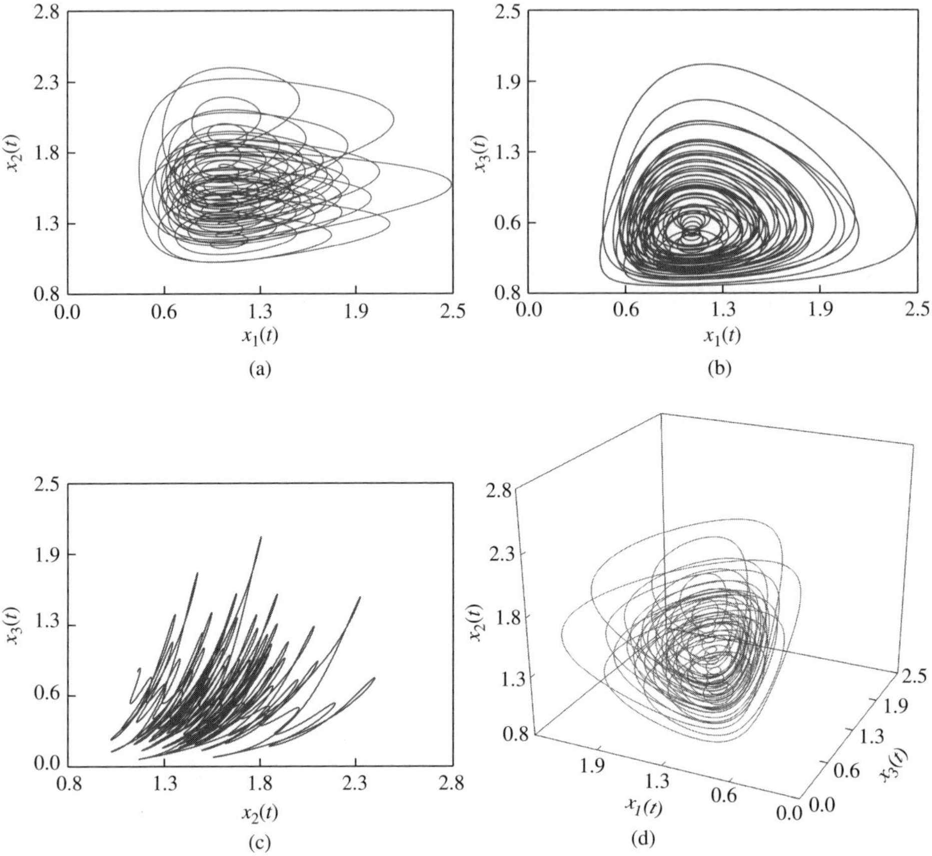

Figure 2.11 Chaotic behaviors of fractional-order Lotka–Volterra system (2.50): (a) $x_1(t)$–$x_2(t)$ plane; (b) $x_1(t)$–$x_3(t)$ plane; (c) $x_2(t)$–$x_3(t)$ plane; (d) $x_3(t)$–$x_1(t)$–$x_2(t)$ space.

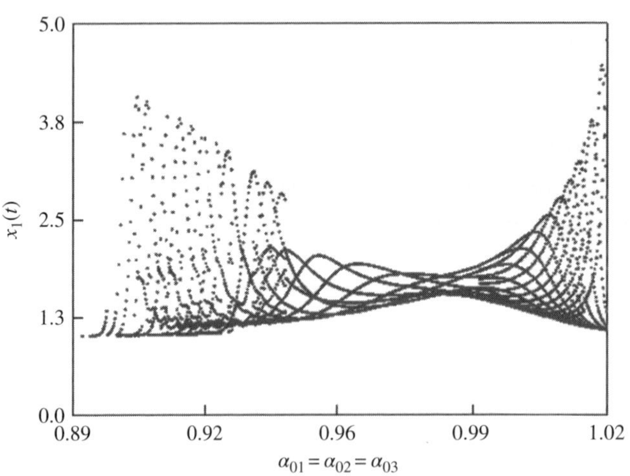

Figure 2.12 Bifurcation diagram of fractional-order Lotka–Volterra system (2.50) for $\alpha_{01} = \alpha_{02} = \alpha_{03} \in [0.85, 1.02]$.

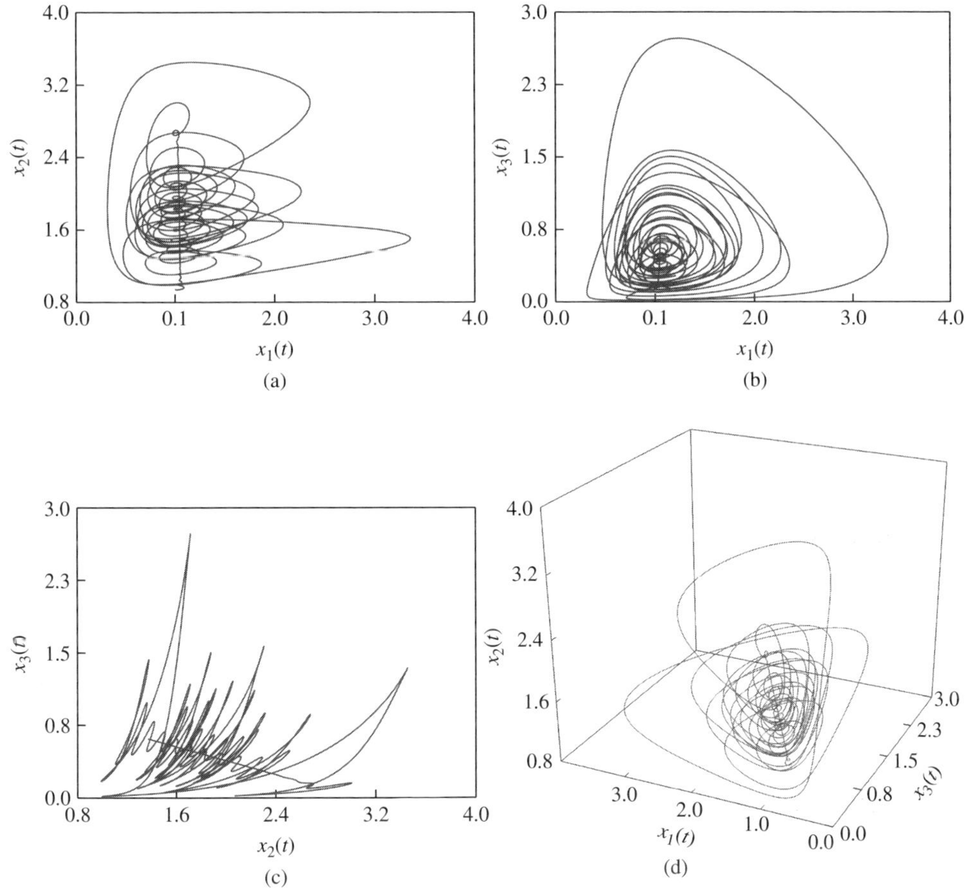

Figure 2.13 Dynamic behaviors of fractional order Lotka–Volterra system (2.50); (a) $x_1(t)$–$x_2(t)$ plane; (b) $x_1(t)$–$x_3(t)$ plane; (c) $x_2(t)$–$x_3(t)$ plane; (d) $x_3(t)$–$x_1(t)$–$x_2(t)$ space.

diagram for $\alpha_{02} \in [0.7, 1.3]$ is presented in Figure 2.15. Following the bifurcation diagram (Figure 2.15), the dynamic characteristic of one cycle for $\alpha_{02} = 1.15$ is shown in Figure 2.16.

2.2.7 Fractional-Order Newton–Leipnik System

The Newton–Leipnik system is described by a nonlinear differential equation in [164]. By considering the fractional calculus, the fractional-order Newton-Leipnik system is given by Sheu *et al.* [184] as follows:

$$
\begin{aligned}
D^{\alpha_{01}} x_1(t) &= -\hbar_1 x_1(t) + x_2(t) + 10 x_2(t) x_3(t), \\
D^{\alpha_{02}} x_2(t) &= -x_1(t) - 0.4 x_2(t) + 5 x_1(t) x_3(t), \\
D^{\alpha_{03}} x_3(t) &= \hbar_2 x_3(t) - 5 x_1(t) x_2(t),
\end{aligned}
\tag{2.52}
$$

where α_{01}, α_{02}, and α_{03} are fractional orders, $x_1(t)$, $x_2(t)$, and $x_3(t)$ are system state variables, and $\hbar_1$ and $\hbar_2$ are system parameters. For fractional orders chosen as

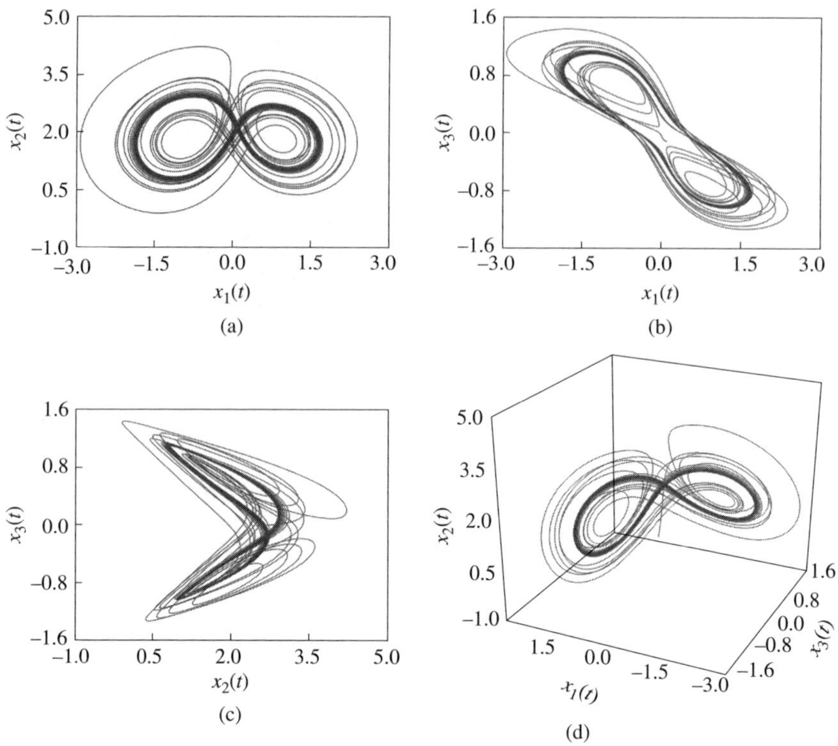

Figure 2.14 Chaotic behaviors of fractional-order financial system (2.51): (a) $x_1(t)$–$x_2(t)$ plane; (b) $x_1(t)$–$x_3(t)$ plane; (c) $x_2(t)$–$x_3(t)$ plane; (d) $x_3(t)$–$x_1(t)$–$x_2(t)$ space.

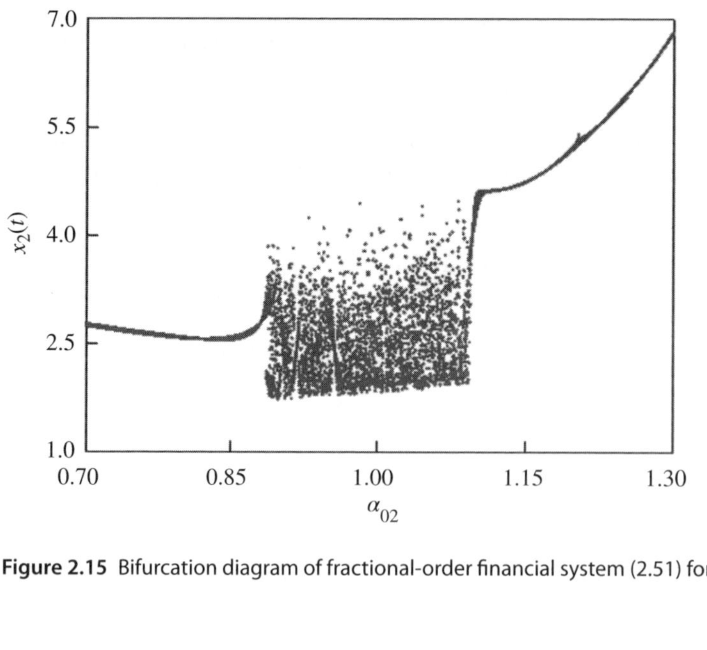

Figure 2.15 Bifurcation diagram of fractional-order financial system (2.51) for $\alpha_{02} \in [0.7, 1.3]$.

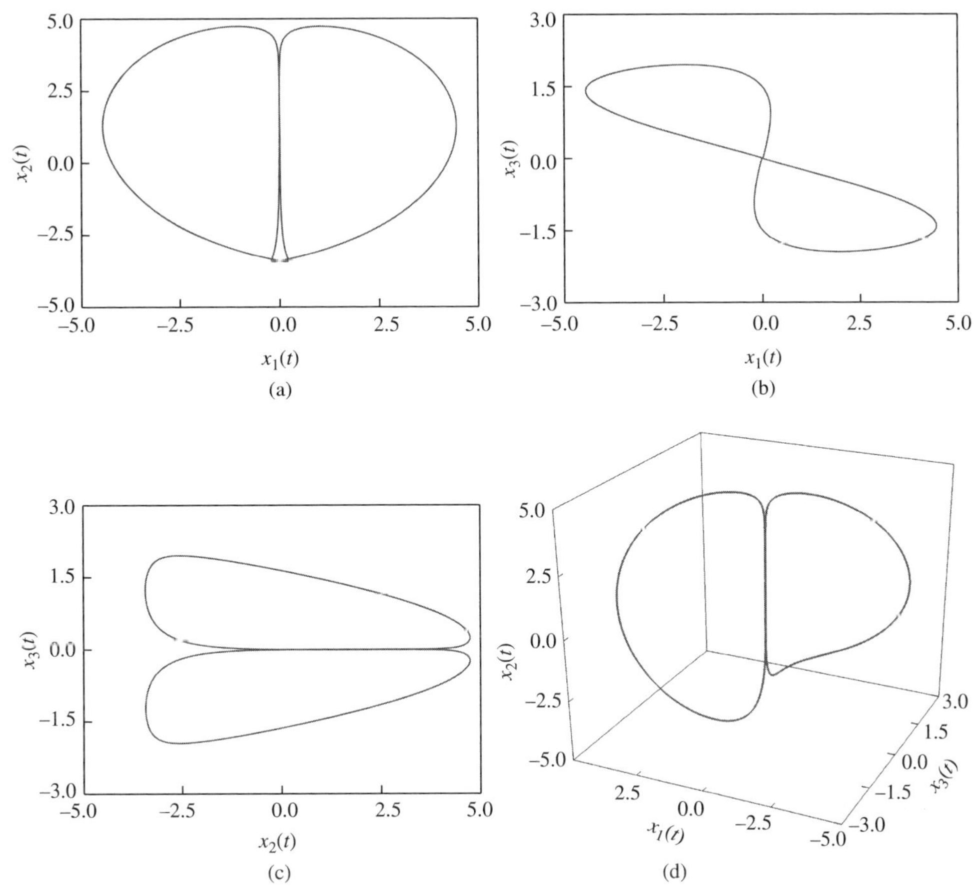

Figure 2.16 Dynamic characteristic of one cycle for fractional-order financial system (2.51). (a) $x_1(t)$–$x_2(t)$ plane; (b) $x_1(t)$–$x_3(t)$ plane; (c) $x_2(t)$–$x_3(t)$ plane; (d) $x_3(t)$–$x_1(t)$–$x_2(t)$ space.

$\alpha_{01} = \alpha_{02} = \alpha_{03} = 0.95$, system parameters given by $\hbar_1 = 0.4$ and $\hbar_2 = 0.175$, and initial conditions chosen as $x_1(0) = 0.19$, $x_2(0) = 0$, and $x_3(0) = -0.18$, the simulation results of the fractional-order Newton–Leipnik system are shown in Figure 2.17. Furthermore, the bifurcation diagram for $\alpha_{03} \in [0.85, 1.2]$ is presented in Figure 2.18. Following the bifurcation diagram (Figure 2.18), the chaotic dynamic characteristic for $\alpha_{03} = 1.11$ is shown in Figure 2.19.

2.2.8 Fractional-Order Duffing System

The fractional-order Duffing system is written as follows [164]:

$$
\begin{aligned}
D^{\alpha_{01}} x_1(t) &= x_2(t), \\
D^{\alpha_{02}} x_2(t) &= x_1(t) - x_1^3(t) - \hbar_1 x_2(t) + \hbar_2 \cos(\hbar_3 t),
\end{aligned}
\tag{2.53}
$$

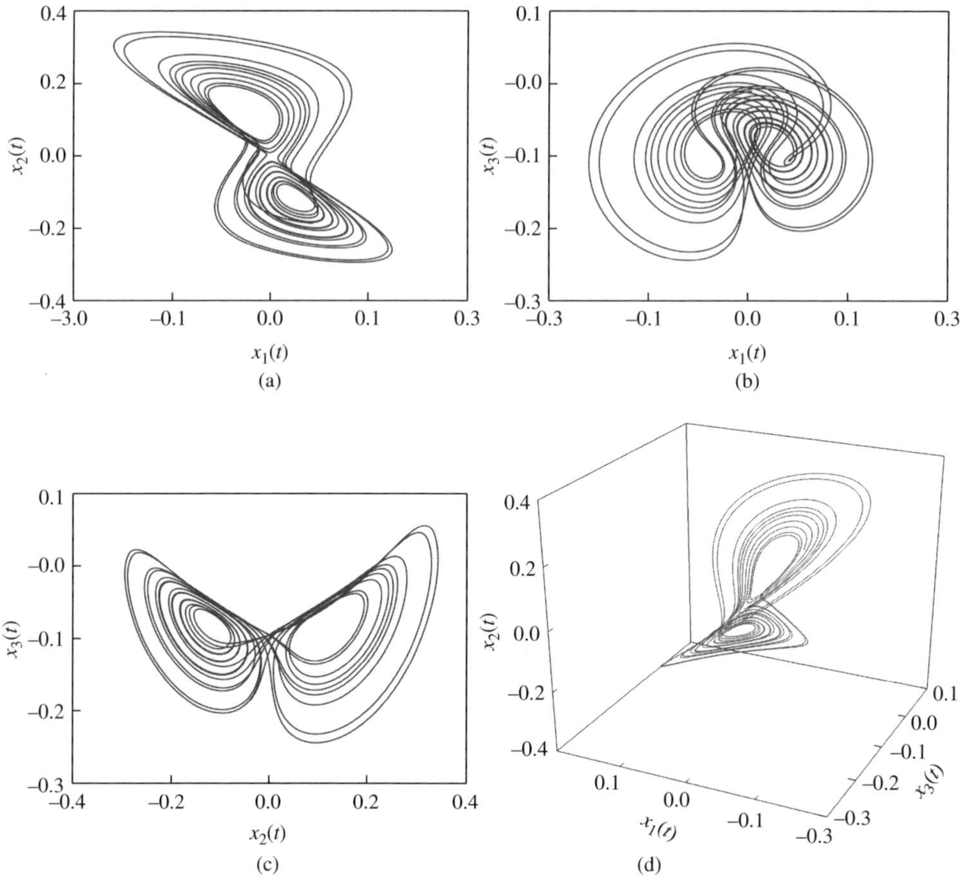

Figure 2.17 Chaotic behaviors of fractional-order Newton–Leipnik system (2.52): (a) $x_1(t)$–$x_2(t)$ plane; (b) $x_1(t)$–$x_3(t)$ plane; (c) $x_2(t)$–$x_3(t)$ plane; (d) $x_3(t)$–$x_1(t)$–$x_2(t)$ space.

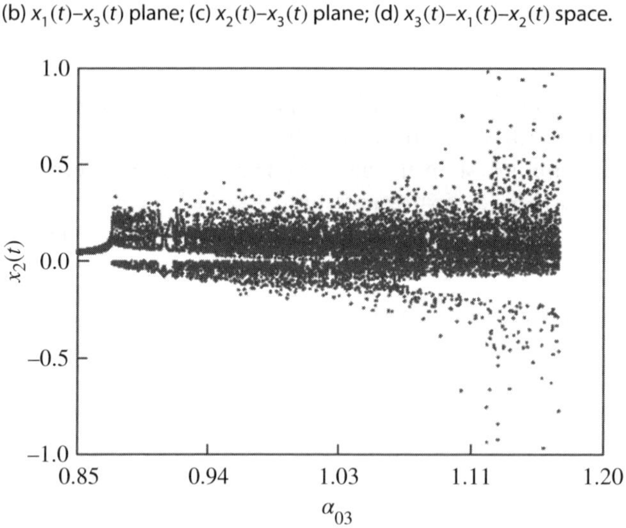

Figure 2.18 Bifurcation diagram of fractional-order Newton–Leipnik system (2.52) for $\alpha_{03} \in [0.85, 1.2]$.

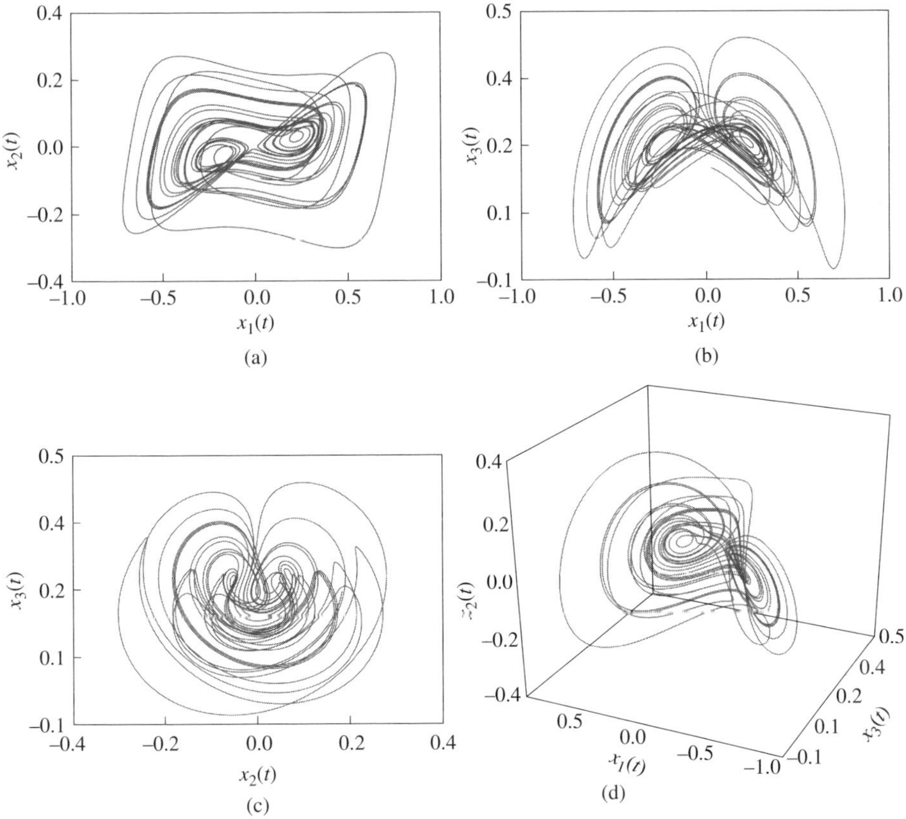

Figure 2.19 Dynamic behaviors of fractional-order Newton–Leipnik system (2.52): (a) $x_1(t)$–$x_2(t)$ plane; (b) $x_1(t)$–$x_3(t)$ plane; (c) $x_2(t)$–$x_3(t)$ plane; (d) $x_3(t)$–$x_1(t)$–$x_2(t)$ space.

where α_{01} and α_{02} are fractional orders, $x_1(t)$ and $x_2(t)$ are system state variables, and $\hbar_1$, $\hbar_2$, and $\hbar_3$ are system parameters. For fractional orders chosen as $\alpha_{01} = 0.9$ and $\alpha_{02} = 1$, system parameters given by $\hbar_1 = 0.15$, $\hbar_2 = 0.3$, and $\hbar_3 = 1$, and initial conditions chosen as $x_1(0) = 0.21$ and $x_2(0) = 0.13$, the simulation results of the fractional-order Duffing system are given in Figure 2.20. Furthermore, the bifurcation diagram for $\alpha_{01} - \alpha_{02} \in [0.8, 1.05]$ is presented in Figure 2.21. Following the bifurcation diagram (Figure 2.21), the dynamic characteristic of one cycle for $\alpha_{01} = \alpha_{02} = 0.938$ is shown in Figure 2.22.

2.2.9 Fractional-Order Lü System

The Lü system [185] is known as a bridge between the Lorenz system [178] and the Chen system [186]. Its fractional-order differential equation is described as follows [187]:

$$
\begin{aligned}
D^{\alpha_{01}} x_1(t) &= \hbar_1 (x_2(t) - x_1(t)),\\
D^{\alpha_{02}} x_2(t) &= -x_1(t)x_3(t) + \hbar_2 x_2(t),\\
D^{\alpha_{03}} x_3(t) &= x_1(t)x_2(t) - \hbar_3 x_3(t),
\end{aligned}
\tag{2.54}
$$

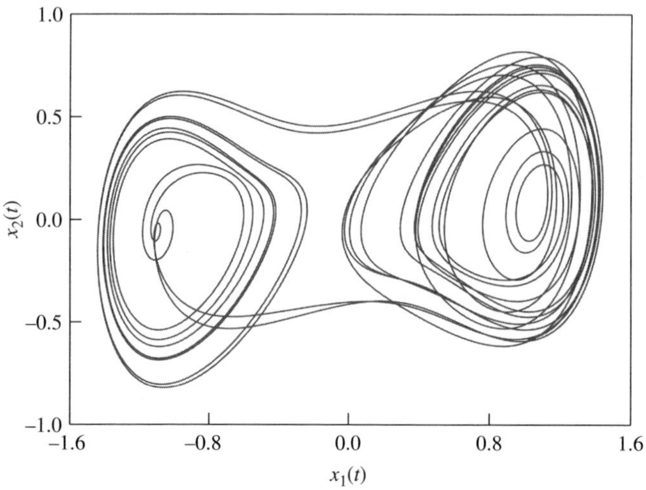

Figure 2.20 Simulation result of fractional-order Duffing system (2.53).

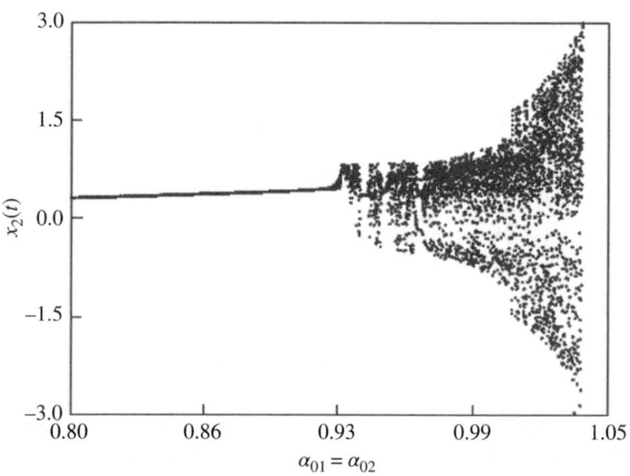

Figure 2.21 Bifurcation diagram of fractional-order Duffing system (2.53) for $\alpha_{01} = \alpha_{02} \in [0.8, 1.05]$.

where α_{01}, α_{02}, and α_{03} are fractional orders, $x_1(t)$, $x_2(t)$, and $x_3(t)$ are system state variables, and $\hbar_1$, $\hbar_2$, and $\hbar_3$ are system parameters. For fractional orders chosen as $\alpha_{01} = \alpha_{02} = \alpha_{03} = 0.95$, system parameters given by $\hbar_1 = 36$, $\hbar_2 = 20$, and $\hbar_3 = 3$, and initial conditions chosen as $x_1(0) = 0.2$, $x_2(0) = 0.5$, and $x_3(0) = 0.3$, the simulation results of the fractional-order Lü system are presented in Figure 2.23. Furthermore, the bifurcation diagram for $\alpha_{02} \in [0.7, 1.7]$ is presented in Figure 2.24. Following the bifurcation diagram (Figure 2.24), the dynamic characteristic of one cycle for $\alpha_{02} = 1.6$ is shown in Figure 2.25.

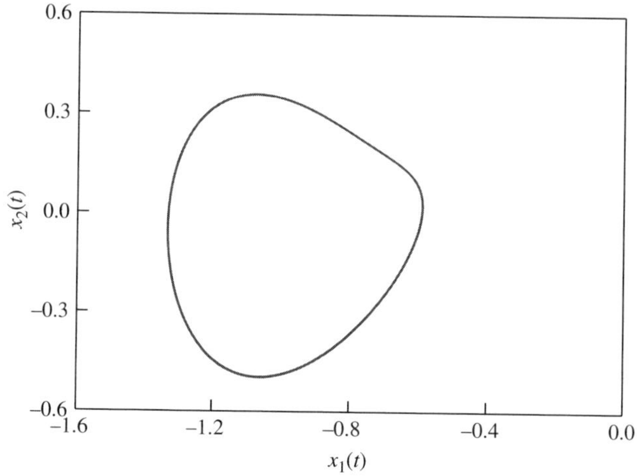

Figure 2.22 Dynamic characteristic of one cycle for fractional-order Duffing system (2.53).

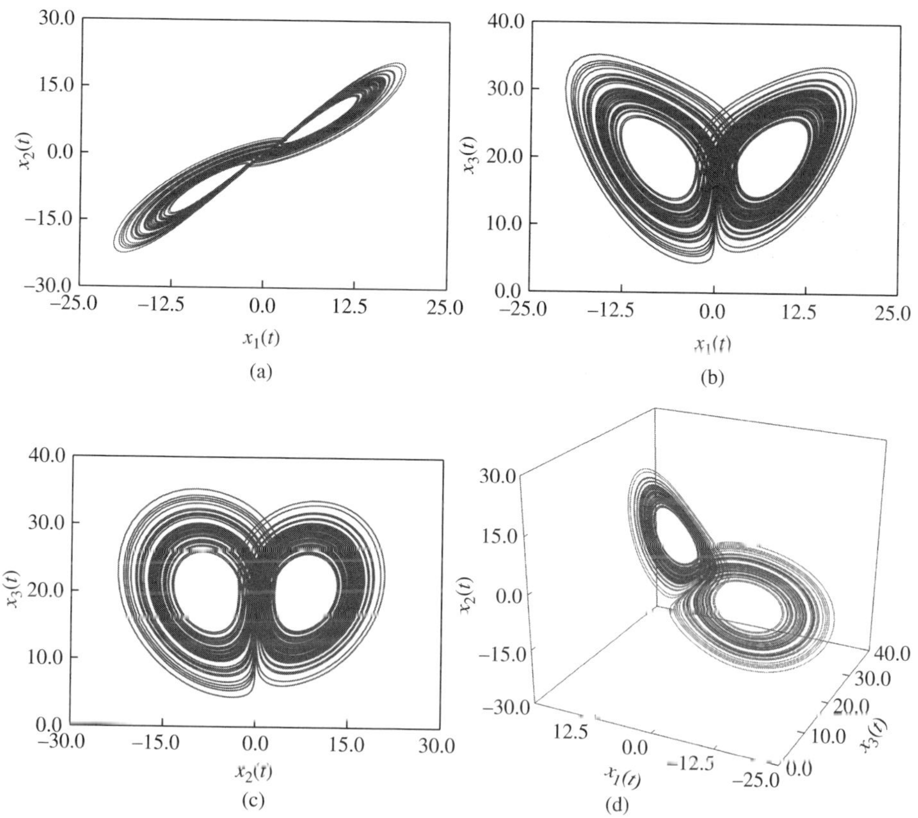

Figure 2.23 Chaotic behaviors of fractional-order Lü system (2.54): (a) $x_1(t)$–$x_2(t)$ plane; (b) $x_1(t)$–$x_3(t)$ plane; (c) $x_2(t)$–$x_3(t)$ plane; (d) $x_3(t)$–$x_1(t)$–$x_2(t)$ space.

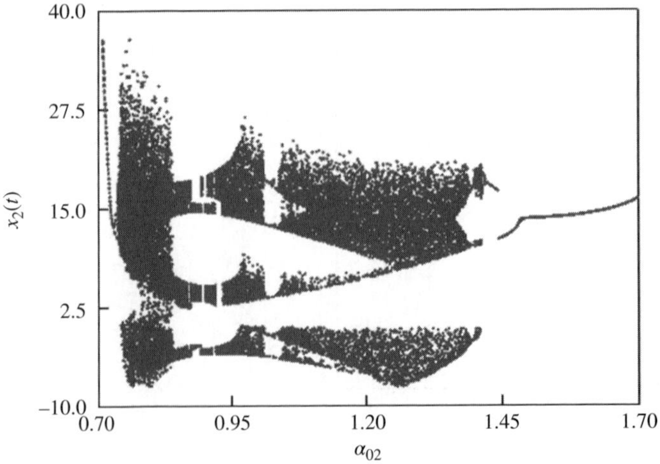

Figure 2.24 Bifurcation diagram of fractional-order Lü system (2.54) for $\alpha_{02} \in [0.7, 1.7]$.

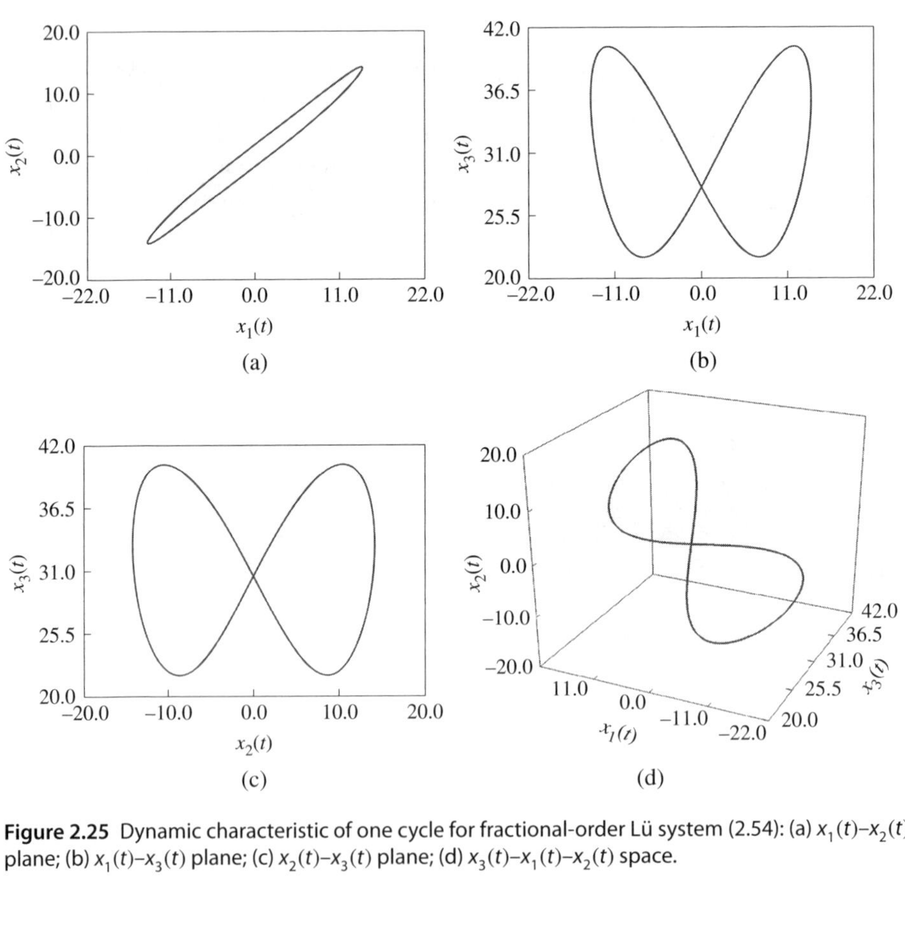

Figure 2.25 Dynamic characteristic of one cycle for fractional-order Lü system (2.54): (a) $x_1(t)$–$x_2(t)$ plane; (b) $x_1(t)$–$x_3(t)$ plane; (c) $x_2(t)$–$x_3(t)$ plane; (d) $x_3(t)$–$x_1(t)$–$x_2(t)$ space.

2.2.10 Fractional-Order Three-Dimensional System

On the basis of the fractional-order Lorenz system [179], a three-dimensional fractional-order system is written as follows [188]:

$$\begin{aligned} D^{\alpha_{01}} x_1(t) &= -\hbar_1 x_1(t) - 2x_2(t)x_3^2(t), \\ D^{\alpha_{02}} x_2(t) &= \hbar_2 x_2(t) + 10x_1(t)x_3(t), \\ D^{\alpha_{03}} x_3(t) &= -\hbar_3 x_3(t) + 5x_2^2(t), \end{aligned} \tag{2.55}$$

where α_{01}, α_{02}, and α_{03} are fractional orders, $x_1(t)$, $x_2(t)$, and $x_3(t)$ are system state variables, and $\hbar_1$, $\hbar_2$, and $\hbar_3$ are system parameters. For fractional orders chosen as $\alpha_{01} = 0.96$, $\alpha_{02} = 0.85$, and $\alpha_{03} = 0.95$, system parameters given by $\hbar_1 = 28$, $\hbar_2 = 12$, and $\hbar_3 = 10$, and initial conditions chosen as $x_1(0) = 0.2$, $x_2(0) = 0.5$, and $x_3(0) = 0.3$, the simulation results of the fractional-order three-dimensional system are shown in Figure 2.26. Furthermore, the bifurcation diagram for $\alpha_{02} \in [0.6, 1.2]$ is presented in Figure 2.27. Following the bifurcation diagram (Figure 2.27), the dynamic characteristic of one cycle for $\alpha_{02} = 0.68$ is shown in Figure 2.28.

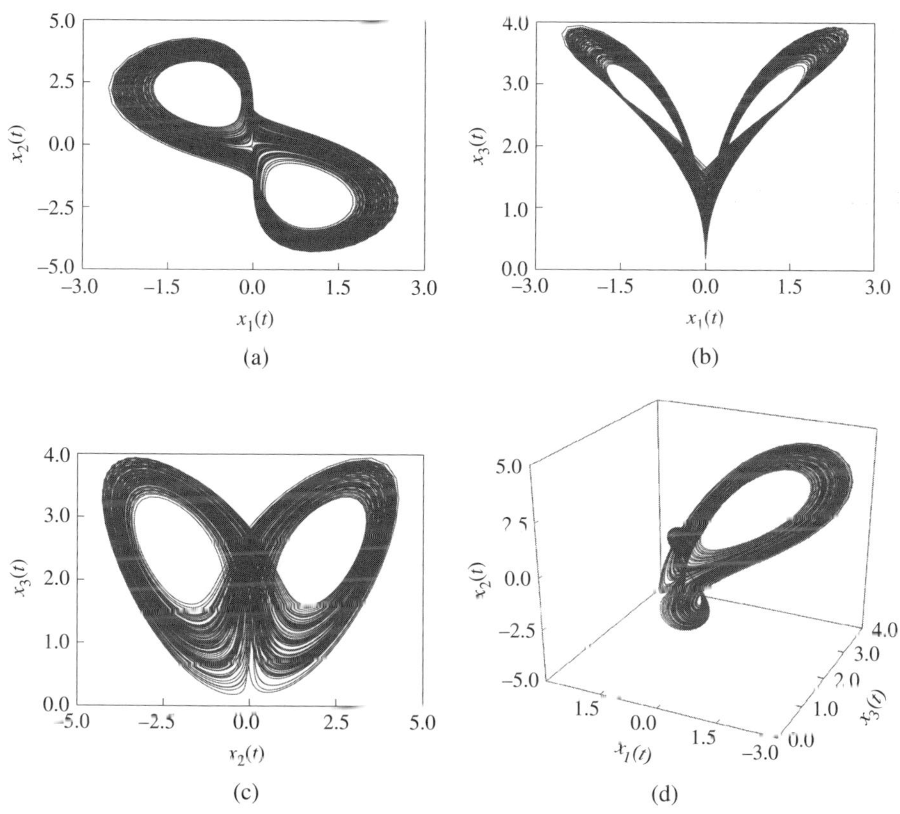

Figure 2.26 Chaotic behaviors of fractional-order three-dimensional system (2.55): (a) $x_1(t)$–$x_2(t)$ plane; (b) $x_1(t)$–$x_3(t)$ plane; (c) $x_2(t)$–$x_3(t)$ plane; (d) $x_3(t)$–$x_1(t)$–$x_2(t)$ space.

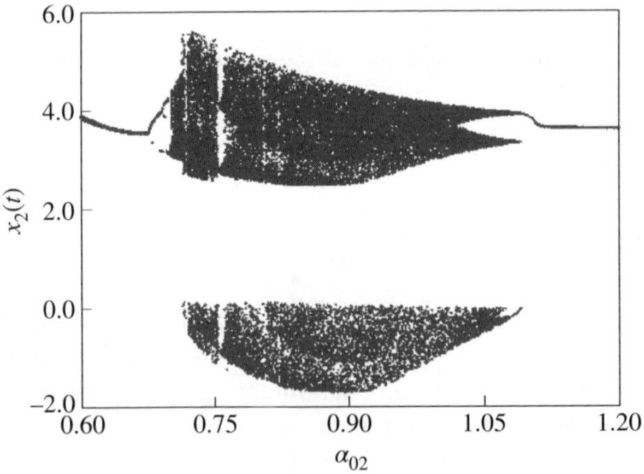

Figure 2.27 Bifurcation diagram of fractional-order three-dimensional system (2.55) for $\alpha_{02} \in [0.6, 1.2]$.

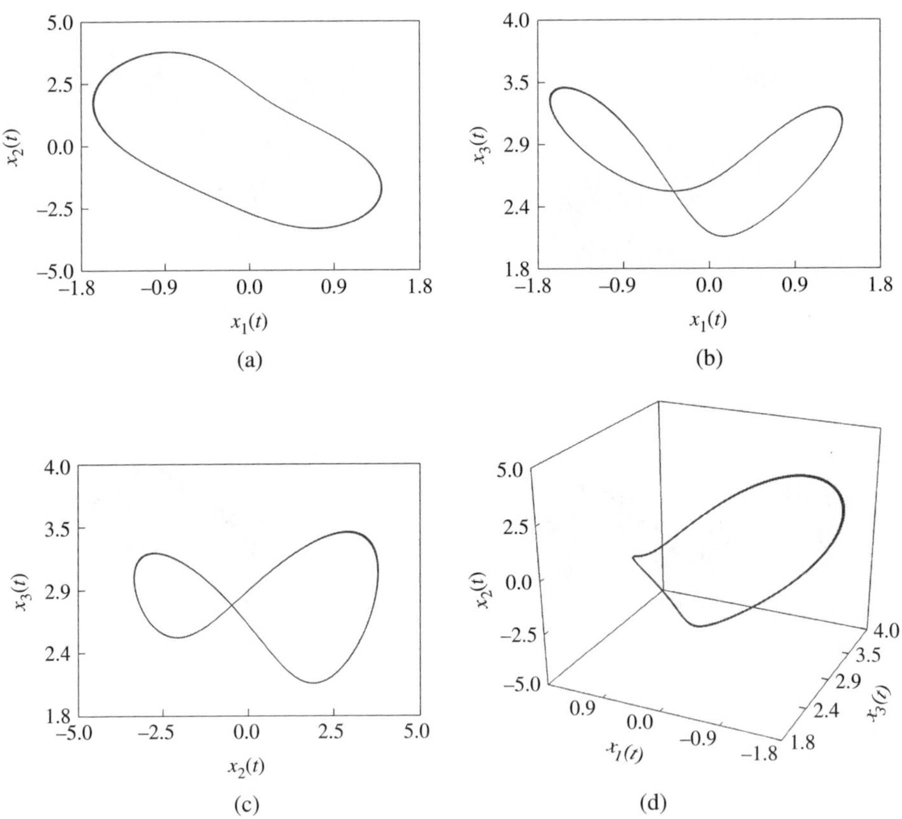

Figure 2.28 Dynamic characteristic of one cycle for fractional-order three-dimensional system (2.55): (a) $x_1(t)$–$x_2(t)$ plane; (b) $x_1(t)$–$x_3(t)$ plane; (c) $x_2(t)$–$x_3(t)$ plane; (d) $x_3(t)$–$x_1(t)$–$x_2(t)$ space.

2.2.11 Fractional-Order Hyperchaotic Oscillator

The fractional-order hyperchaotic oscillator is described by the following form [189]:

$$
\begin{aligned}
D^{\alpha_{01}}x_1(t) &= x_2(t) - G(t),\\
D^{\alpha_{02}}x_2(t) &= -x_1(t) + \hbar_1(x_2(t) - G(t)),\\
D^{\alpha_{03}}x_3(t) &= (1/\hbar_2)(x_3(t) + G(t)),\\
D^{\alpha_{04}}x_4(t) &= (1/\hbar_3)(-x_3(t) + \hbar_4(x_4(t) + G(t))),
\end{aligned} \tag{2.56}
$$

where $G(t) = 0.5\hbar_5(|x_1(t) - x_3(t) + \hbar_6(x_2(t) - x_4(t)) - 1| + x_1(t) - x_3(t) + \hbar_6(x_2(t) - x_4(t)) - 1)$, α_{01}, α_{02}, α_{03}, and α_{04} are fractional orders, $x_1(t)$, $x_2(t)$, $x_3(t)$, and $x_4(t)$ are system state variables, and $\hbar_1$, $\hbar_2$, $\hbar_3$, $\hbar_4$, $\hbar_5$, and $\hbar_6$ are system parameters. For fractional orders chosen as $\alpha_{01} = \alpha_{02} = \alpha_{03} = \alpha_{04} = 0.99$, system parameters given by $\hbar_1 = 0.6$, $\hbar_2 = 0.31$, $\hbar_3 = 0.33$, $\hbar_4 = 0.2$, $\hbar_5 = 1.5$, and $\hbar_6 = 0.1$, and initial conditions chosen as $x_1(0) = 0.2$, $x_2(0) = 0.5$, $x_3(0) = 0.3$, and $x_4(0) = 0.6$, the simulation results of the fractional-order hyperchaotic oscillator are given in Figure 2.29. Furthermore, the bifurcation diagram for $\alpha_{04} \in [0.6, 1.08]$ is presented in Figure 2.30. According to the bifurcation diagram (Figure 2.30), the dynamic characteristic of one cycle for $\alpha_{04} = 0.72$ is shown in Figure 2.31.

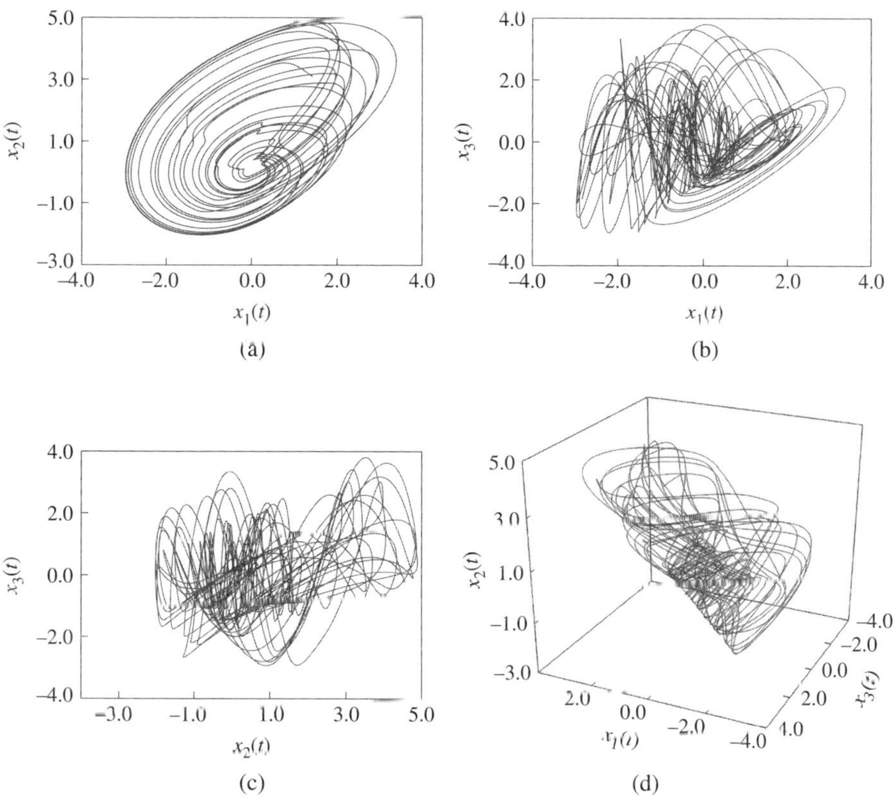

Figure 2.29 Chaotic behaviors of fractional-order hyperchaotic oscillator (2.56): (a) $x_1(t)$–$x_2(t)$ plane; (b) $x_1(t)$–$x_3(t)$ plane; (c) $x_2(t)$–$x_3(t)$ plane; (d) $x_3(t)$–$x_1(t)$–$x_2(t)$ space.

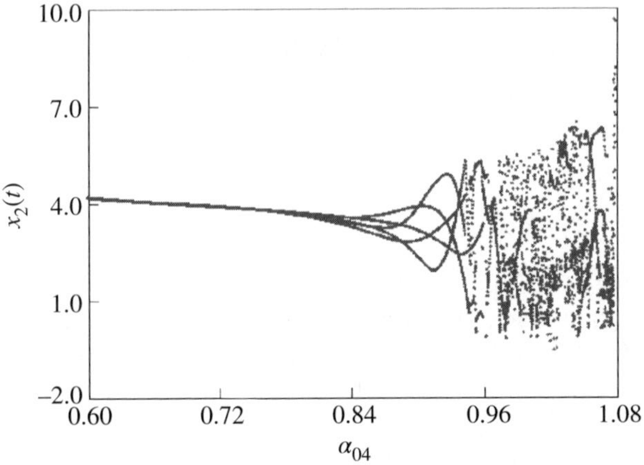

Figure 2.30 Bifurcation diagram of fractional-order hyperchaotic oscillator (2.56) for $\alpha_{04} \in [0.6, 1.08]$.

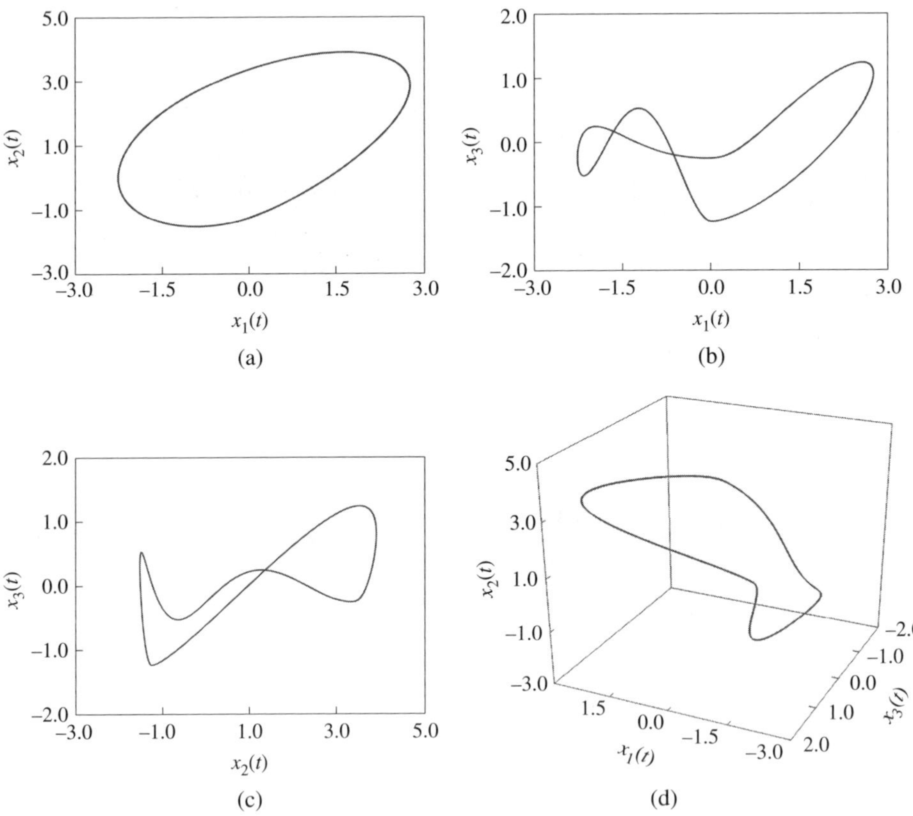

Figure 2.31 Dynamic characteristic of one cycle for fractional-order hyperchaotic oscillator (2.56): (a) $x_1(t)$–$x_2(t)$ plane; (b) $x_1(t)$–$x_3(t)$ plane; (c) $x_2(t)$–$x_3(t)$ plane; (d) $x_3(t)$–$x_1(t)$–$x_2(t)$ space.

2.2.12 Fractional-Order Four-Dimensional Hyperchaotic System

The fractional version of the four-dimensional hyperchaotic system is given by the following [190]:

$$
\begin{aligned}
D^{\alpha_{01}}x_1(t) &= \hbar_1(x_2(t) - x_1(t)),\\
D^{\alpha_{02}}x_2(t) &= \hbar_2 x_1(t) - x_1(t)x_3(t) + \hbar_3 x_2(t) - x_4(t),\\
D^{\alpha_{03}}x_3(t) &= x_1(t)x_2(t) - \hbar_4 x_3(t),\\
D^{\alpha_{04}}x_4(t) &= x_1(t) + \hbar_5,
\end{aligned}
\tag{2.57}
$$

where α_{01}, α_{02}, α_{03}, and α_{04} are fractional orders, $x_1(t)$, $x_2(t)$, $x_3(t)$, and $x_4(t)$ are system state variables, and $\hbar_1$, $\hbar_2$, $\hbar_3$, $\hbar_4$, and $\hbar_5$ are system parameters. For fractional orders chosen as $\alpha_{01} = \alpha_{02} = \alpha_{03} = \alpha_{04} = 0.9$, system parameters given by $\hbar_1 = 36$, $\hbar_2 = -16$, $\hbar_3 = 28$, $\hbar_4 = 3$, and $\hbar_5 = 0.5$, and initial conditions chosen as $x_1(0) = 0.2$, $x_2(0) = 0.5$, $x_3(0) = 0.3$, and $x_4(0) = 0.7$, the simulation results of the fractional-order four-dimensional hyperchaotic system are presented in Figure 2.32. Furthermore, the bifurcation diagram for $\alpha_{01} = \alpha_{02} = \alpha_{03} = \alpha_{04} \in [0.7, 1.02]$ is presented in Figure 2.33. Following the bifurcation diagram (Figure 2.33), the dynamic characteristic of one cycle for $\alpha_{01} = \alpha_{02} = \alpha_{03} = \alpha_{04} = 1$ is shown in Figure 2.34.

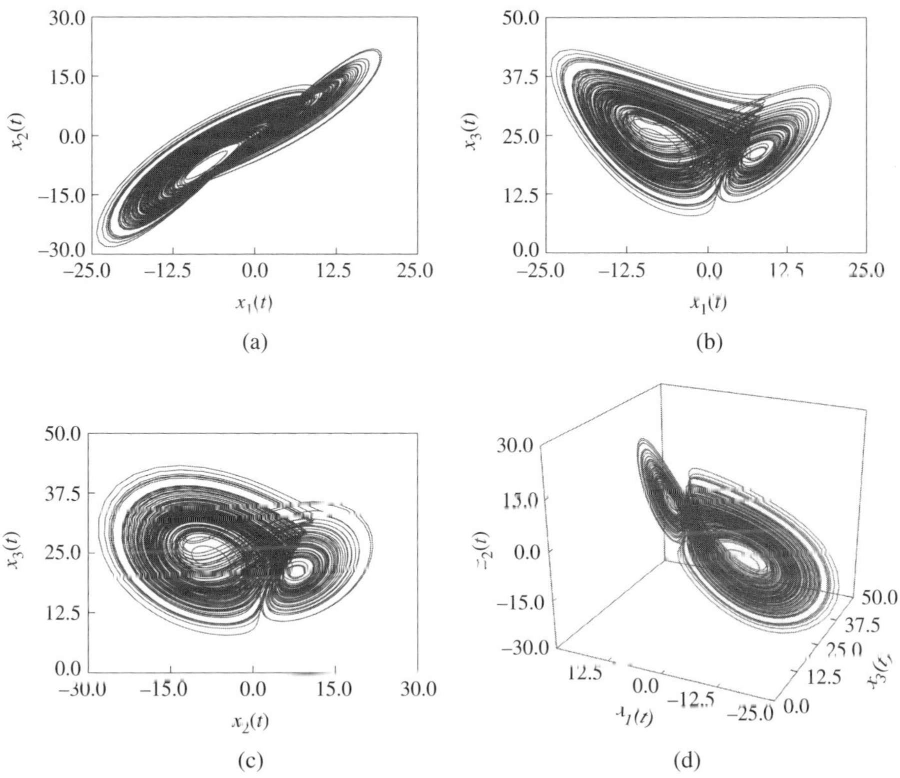

Figure 2.32 Chaotic behaviors of fractional-order four-dimensional hyperchaotic system (2.57): (a) $x_1(t)$–$x_2(t)$ plane; (b) $x_1(t)$–$x_3(t)$ plane; (c) $x_2(t)$–$x_3(t)$ plane; (d) $x_3(t)$–$x_1(t)$–$x_2(t)$ space.

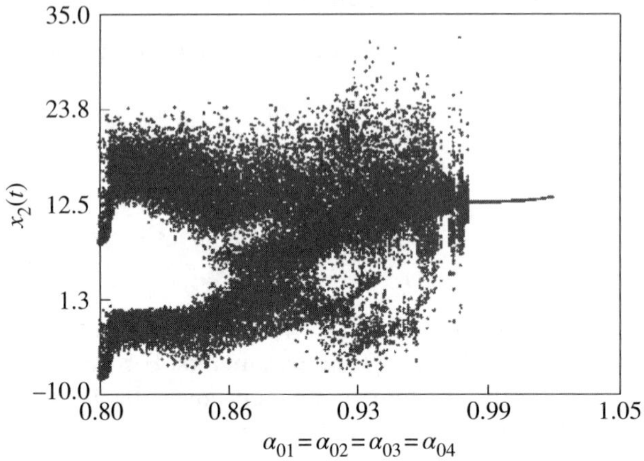

Figure 2.33 Bifurcation diagram of fractional-order four-dimensional hyperchaotic system (2.57) for $\alpha_{01} = \alpha_{02} = \alpha_{03} = \alpha_{04} \in [0.7, 1.02]$.

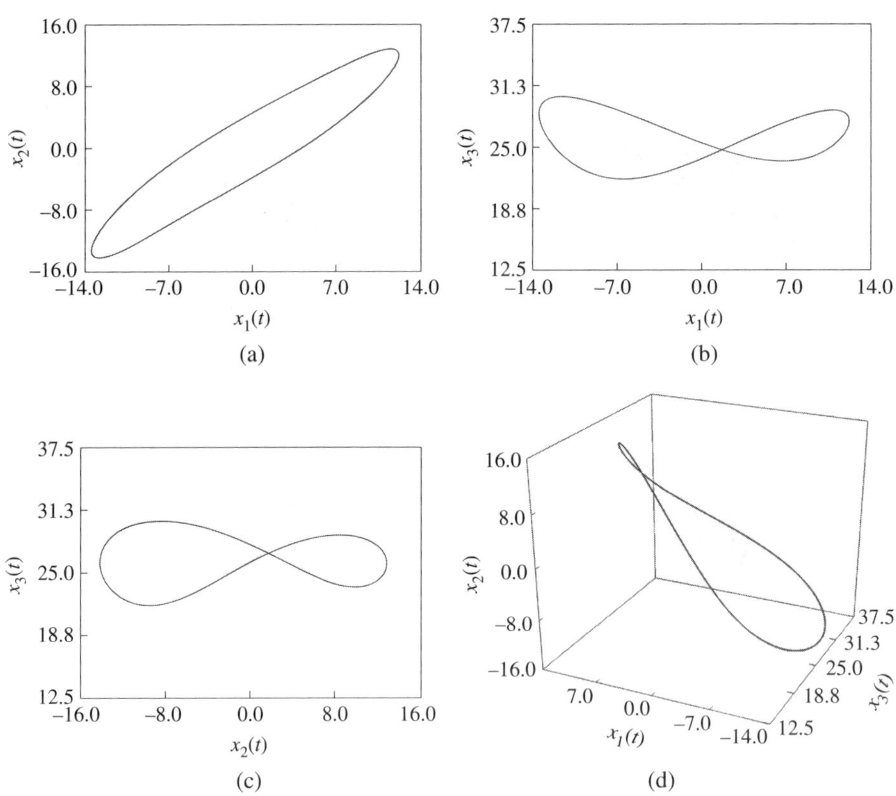

Figure 2.34 Dynamic characteristic of one cycle for fractional-order four-dimensional hyperchaotic system (2.57): (a) $x_1(t)$–$x_2(t)$ plane; (b) $x_1(t)$–$x_3(t)$ plane; (c) $x_2(t)$–$x_3(t)$ plane; (d) $x_3(t)$–$x_1(t)$–$x_2(t)$ space.

2.2.13 Fractional-Order Hyperchaotic Cellular Neural Network

The model of a fractional-order hyperchaotic cellular neural network can be written as follows [191]:

$$\begin{aligned}
D^{\alpha_{01}}x_1(t) &= -x_3(t) - x_4(t),\\
D^{\alpha_{02}}x_2(t) &= \hbar_1 x_2(t) + x_3(t),\\
D^{\alpha_{03}}x_3(t) &= \hbar_2(x_1(t) - x_2(t)),\\
D^{\alpha_{04}}x_4(t) &= \hbar_3(x_1(t) - G(x_4(t))),
\end{aligned} \tag{2.58}$$

where $G(x_4(t)) = x_4(t) - (|x_4(t) - 0.4| - |x_4(t) - 0.8| - |x_4(t) + 0.4| + |x_4(t) + 0.8|)$ α_{01}, α_{02}, α_{03}, and α_{04} are fractional orders, $x_1(t)$, $x_2(t)$, $x_3(t)$, and $x_4(t)$ are system state variables, and $\hbar_1$, $\hbar_2$, and $\hbar_3$ are system parameters. For fractional orders chosen as $\alpha_{01} = \alpha_{02} = \alpha_{03} = \alpha_{04} = 0.97$, system parameters given by $\hbar_1 = 2$, $\hbar_2 = 14$, and $\hbar_3 = 100$, and initial conditions chosen as $x_1(0) = 0.2, x_2(0) = 0.5, x_3(0) = 0.3$, and $x_4(0) = 0.3$, the simulation results of a fractional-order hyperchaotic cellular neural network are shown in Figure 2.35. Furthermore, the bifurcation diagram for $\alpha_{03} \subset [0.7, 1.05]$ is presented in Figure 2.36. Following the bifurcation diagram (Figure 2.36), the chaotic dynamic characteristic for $\alpha_{03} = 1.05$ is shown in Figure 2.37.

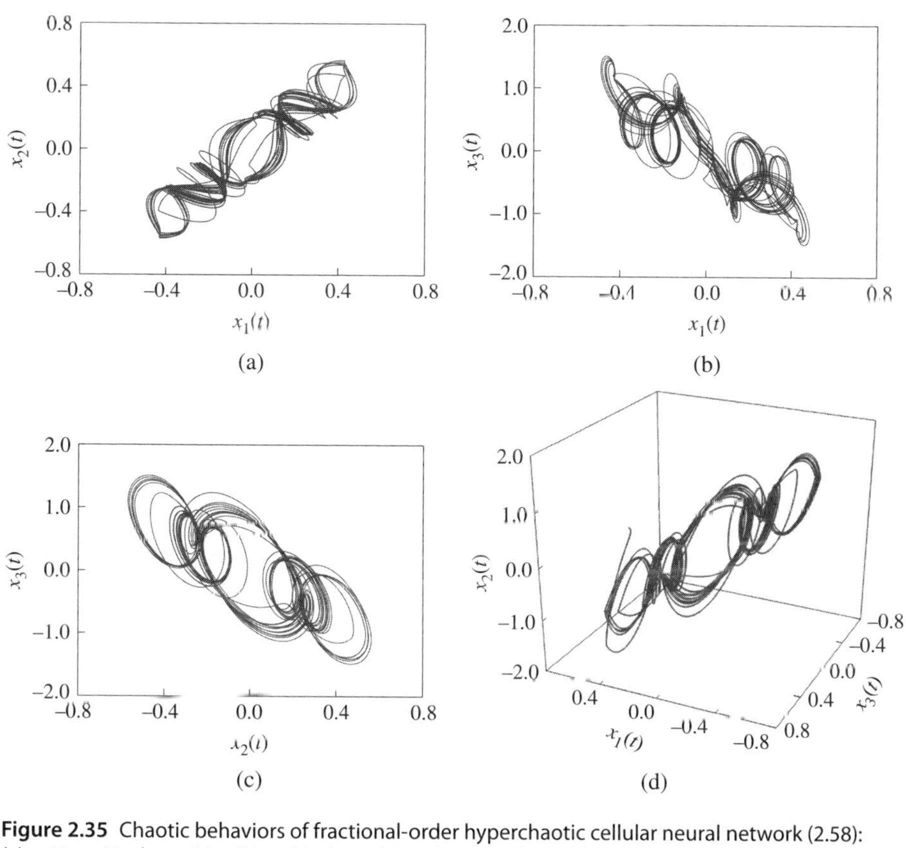

Figure 2.35 Chaotic behaviors of fractional-order hyperchaotic cellular neural network (2.58): (a) $x_1(t)$–$x_2(t)$ plane; (b) $x_1(t)$–$x_3(t)$ plane; (c) $x_2(t)$–$x_3(t)$ plane; (d) $x_3(t)$–$x_1(t)$–$x_2(t)$ space.

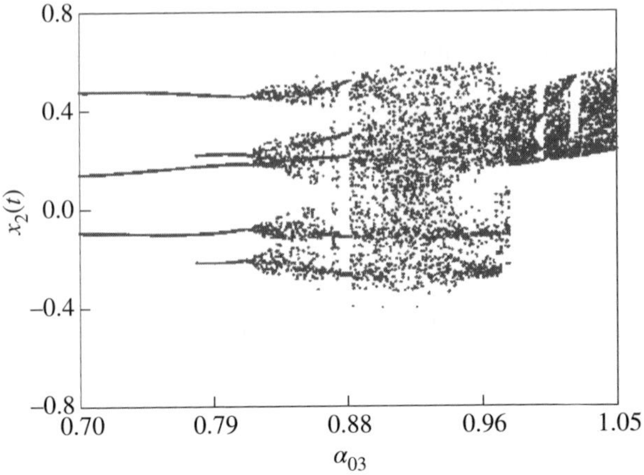

Figure 2.36 Bifurcation diagram of fractional-order hyperchaotic cellular neural network (2.58) for $\alpha_{03} \in [0.7, 1.05]$.

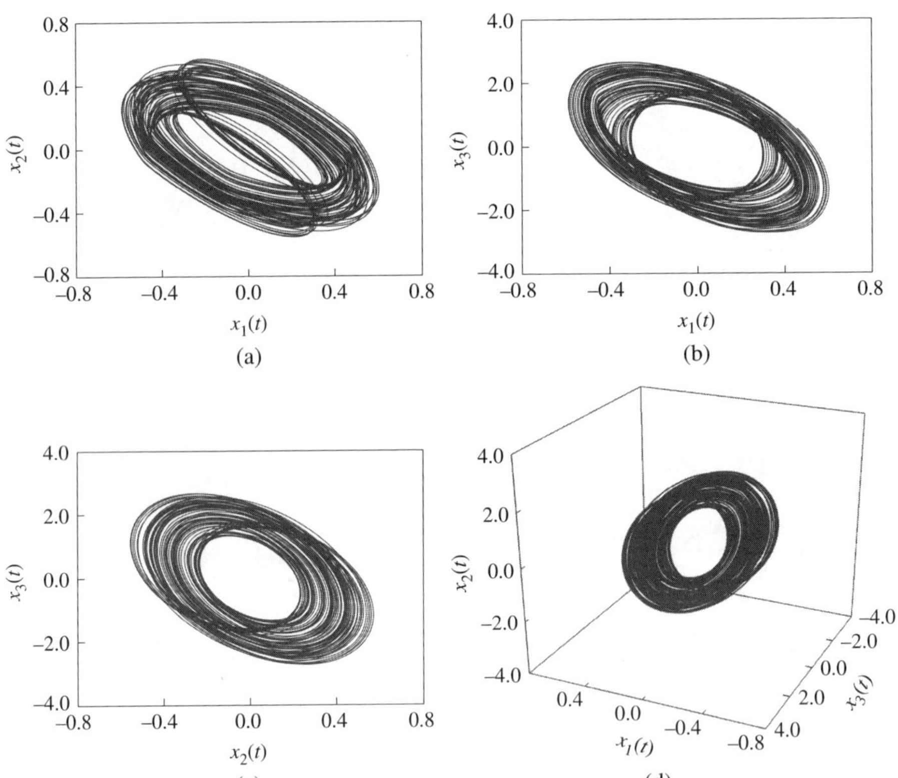

Figure 2.37 Dynamic behaviors of fractional-order hyperchaotic cellular neural network (2.58): (a) $x_1(t)$–$x_2(t)$ plane; (b) $x_1(t)$–$x_3(t)$ plane; (c) $x_2(t)$–$x_3(t)$ plane; (d) $x_3(t)$–$x_1(t)$–$x_2(t)$ space.

2.3 Conclusion

In this chapter, principle definitions of the fractional integral and fractional derivatives have been given. Furthermore, some lemmas have been introduced, along with results on the stability of fractional-order systems. Finally, some fractional-order systems have been listed: fractional-order Lorenz system, fractional-order van der Pol oscillator, fractional-order Genesio–Tesi system, fractional-order Arneodo system, fractional-order Lotka–Volterra system, fractional-order financial system, fractional-order Newton–Leipnik system, fractional-order Duffing system, fractional-order Lü system, fractional-order three-dimensional system, fractional-order hyperchaotic oscillator, fractional-order four-dimensional hyperchaotic system, and fractional-order hyperchaotic cellular neural network.

3

Fractional-Order PID Controller and Fractional-Order Disturbance Observer

3.1 Problem Statement

Over the past decades, some important results on fractional-order controllers have been proposed for various systems [1, 192–194]. Podlubny [1] introduced the concept of a fractional-order $\mathrm{PI}^{\lambda}\mathrm{D}^{\mu}$ controller, including a fractional-order integrator and a fractional-order differentiator. A well-known fractional-order controller was presented by Oustaloup *et al.* [192]. Li *et al.* [193] proposed a novel fractional-order controller, based on Lyapunov stability theory; fractional-order chaotic and hyperchaotic systems were controlled by the proposed fractional-order controller. A novel chaotic system without equilibrium was proposed, and a fractional-order controller was designed to control the chaotic system without equilibrium, by Shao and Chen [194]. Fractional-order controllers are effective in controlled systems, as has been proved in the mentioned works. Furthermore, FODOs have been widely studied for the control of fractional-order systems [169, 195–198]. Shao *et al.* [169] studied FODO-based adaptive sliding-mode synchronization control for a class of fractional-order chaotic systems with unknown bounded disturbances. A robust adaptive neural control was proposed for a fractional-order rotational mechanical system in the presence of system uncertainties and external disturbances by Shao *et al.* [195]. Shao *et al.* [196] investigated a FODO-based sliding-mode control scheme for a class of fractional-order nonlinear systems with external disturbances. The problem of stabilization control was studied for a class of continuous-time fractional systems subject to external constant disturbances based on a disturbance observer by Shao *et al.* [197]. Shao *et al.* [198] proposed a FODO-based adaptive neural tracking control scheme for uncertain fractional-order chaotic systems subject to unknown disturbance and input saturation. According to these discussions, the anti-disturbance ability of control systems can be improved by employing the disturbance observer in the robust control for the uncertain nonlinear system with external disturbance. In this chapter, we mainly introduce a fractional-order $\mathrm{PI}^{\lambda}\mathrm{D}^{\mu}$ controller and a frequency-domain FODO.

Robust Adaptive Control for Fractional-Order Systems with Disturbance and Saturation, First Edition.
Mou Chen, Shuyi Shao, and Peng Shi.

3.2 Fractional-Order PID Controller

In this section, the traditional integer-order PID controller and the fractional-order $PI^{\lambda}D^{\mu}$ controller are introduced. Furthermore, the relationship between the integer-order PID controller and the fractional-order $PI^{\lambda}D^{\mu}$ controller is illustrated.

3.2.1 Integer-Order PID Controller

Since the integer-order PID controller algorithm was proposed by Ziegler-Nichols [199], the integer-order PID controller has been widely applied in the field of industrial control, based on the advantages of simple structure, strong robustness, and so on. The integer-order PID control system is illustrated in Figure 3.1 [200], where k_{p}, k_{i}, and k_{d} are the gain parameters of proportion, integration, and differentiation, respectively, $u(t)$ is the control input signal, $r(t)$ is the desired system output, $e(t) = y(t) - r(t)$ is the system tracking error signal, $y(t)$ is the system output, $C(s)$ is the transfer function of the integer-order PID controller, and $G(s)$ is the transfer function of controlled object.

From Figure 3.1, the transfer function of the integer-order PID controller can be written as [200]

$$C(s) = \frac{U(s)}{E(s)} = k_{\mathrm{p}} + \frac{k_{\mathrm{i}}}{s} + k_{\mathrm{d}}s. \tag{3.1}$$

The control input signal $u(t)$ is then expressed in the time domain as [200]

$$u(t) = k_{\mathrm{p}}e(t) + k_{\mathrm{i}} \int_0^t e(t)\mathrm{d}(t) + k_{\mathrm{d}}\frac{\mathrm{d}e(t)}{\mathrm{d}t}. \tag{3.2}$$

According to different controlled objects and control requirements, the PID controller can be divided up, as PI controller, PD controller, PID controller, incremental PID, integral separation PID, and so on. From the form of the PID controller $u(t)$ (3.2), the orders of the integrator and the differentiator are only one. However, the fractional-order $PI^{\lambda}D^{\mu}$ controller includes an integrator of order λ and a differentiator of order μ, where the orders λ and μ can be adjusted based on the control performance. In the next section, we will introduce the fractional-order $PI^{\lambda}D^{\mu}$ controller.

3.2.2 Fractional-Order $PI^{\lambda}D^{\mu}$ Controller

In the past decades, some fractional-order controllers have been studied, such as PI^{λ}, PD^{μ}, and $PI^{\lambda}D^{\mu}$ [201]. The most common form of a fractional-order PID controller is the

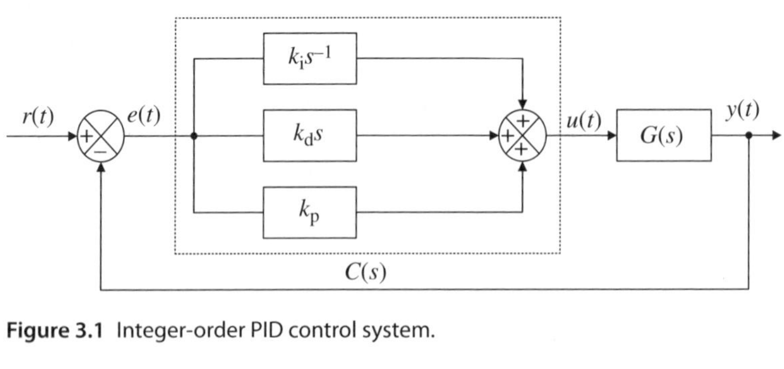

Figure 3.1 Integer-order PID control system.

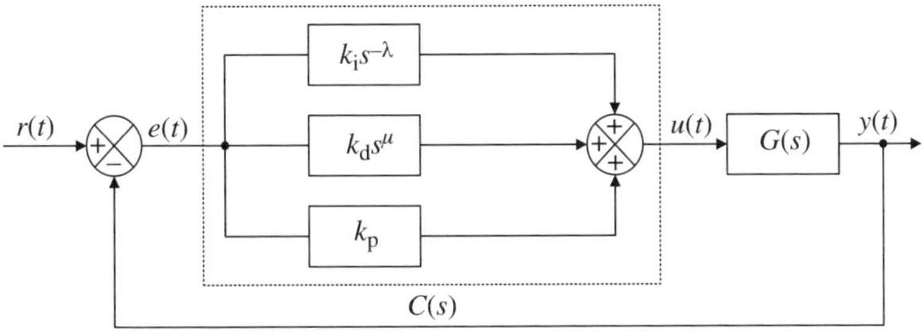

Figure 3.2 Fractional-order $PI^{\lambda}D^{\mu}$ control system.

$PI^{\lambda}D^{\mu}$ controller [1], which includes an integrator of order λ and a differentiator of order μ, where $\lambda \in (0, 2)$ and $\mu \in (0, 2)$. Not only the $PI^{\lambda}D^{\mu}$ controller can be used to control fractional-order systems, but it can also be applied to the control of integer-order systems. A block diagram of the fractional-order $PI^{\lambda}D^{\mu}$ control system is given in Figure 3.2 [200]. From Figure 3.2, the transfer function of the integer-order $PI^{\lambda}D^{\mu}$ controller can be written as [200]

$$C(s) = \frac{U(s)}{E(s)} = k_p + \frac{k_i}{s^{\lambda}} + k_d s^{\mu}. \tag{3.3}$$

The control input signal $u(t)$ is then expressed in the time domain as [200]

$$u(t) = k_p e(t) + k_i D^{-\lambda} e(t) + k_d D^{\mu} e(t), \tag{3.4}$$

where k_p, k_i, and k_d are the gain parameters of proportion, integration, and differentiation, respectively, $u(t)$ is the control input signal, $r(t)$ is the desired system output, $e(t) = y(t) - r(t)$ is the system tracking error signal, $y(t)$ is the system output, $C(s)$ is the transfer function of the fractional-order $PI^{\lambda}D^{\mu}$ controller, and $G(s)$ is the transfer function of the controlled object.

For the integrator of order λ and the differentiator of order μ, if $\lambda > 2$ and $\mu > 0$, the controller will be a high-order controller, which is different from the structure of the classical PID controller. Owing to the adjustable parameters λ and μ, better dynamic characteristics can be obtained through control of the fractional-order $PI^{\lambda}D^{\mu}$ controller than through control of the classical PID controller.

According to Equation (3.3), different types of controller can be realized by the fractional-order $PI^{\lambda}D^{\mu}$ controller when the different values of the orders λ and μ are chosen. As shown in Figure 3.3, by selecting $\lambda = 1$ and $\mu = 1$, a classical integer-order PID controller can be obtained, by choosing $\lambda = 0$ and $\mu = 0$, a P controller can be achieved, by making $\lambda = 1$ and $\mu = 0$, a classical PI controller can be realized, and with $\lambda = 0$ and $\mu = 1$, a classical PD controller can be gained. Because the orders λ and μ can be any real numbers between 0 and 2, all types of integer-order PID controller are special types of the fractional-order $PI^{\lambda}D^{\mu}$ controller.

3.2.3 Control Based on Fractional-Order $PI^{\lambda}D^{\mu}$ Controller

In this section, a fractional-order system is studied using the fractional-order $PI^{\lambda}D^{\mu}$ controller. The transfer function of the fractional-order system is given as follows [1]:

$$G(s) = \frac{1}{0.8s^{2.2} + 0.5s^{0.9} + 1}. \tag{3.5}$$

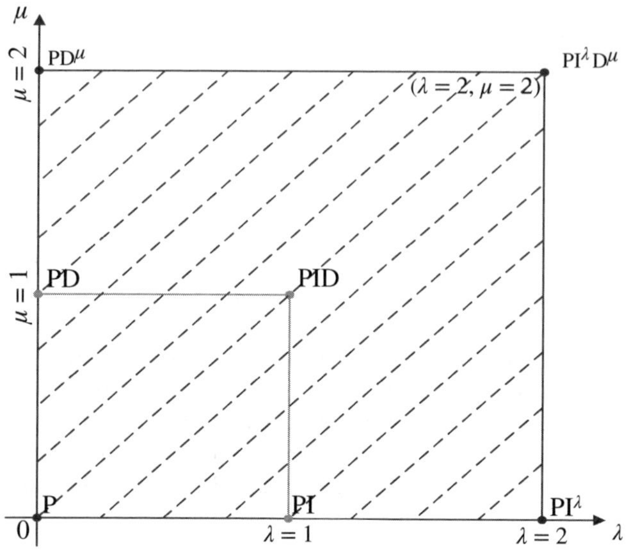

Figure 3.3 Range of values of integer-order PID and fractional-order $PI^{\lambda}D^{\mu}$ control system.

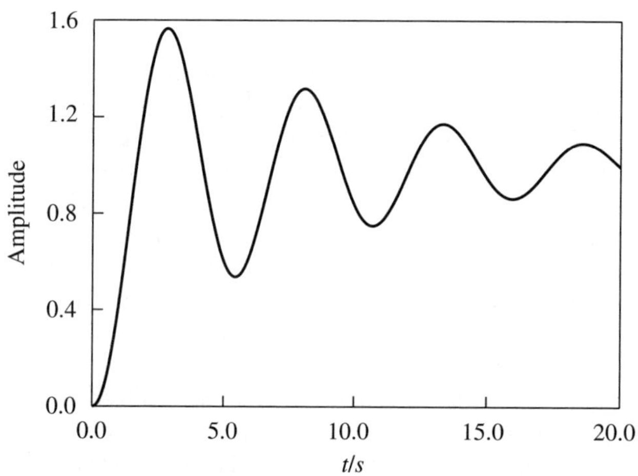

Figure 3.4 Unit step response curve of fractional-order system (3.5).

Following Equation (3.5), the unit step response curve of the fractional-order system is shown in Figure 3.4; this cannot satisfy the control requirement.

To achieve stability of the fractional-order system (3.5), we use the fractional-order $PI^{\lambda}D^{\mu}$ controller. From Xue and Zhao [202], the orders are chosen as $\lambda = 0.2$ and $\mu = 1.1$. Furthermore, the control parameters are selected as $k_{\mathrm{p}} = 138.1817$, $k_{\mathrm{i}} = 2.8914$, and $k_{\mathrm{d}} = 12.382$. Thus, the form of the fractional-order $PI^{\lambda}D^{\mu}$ controller can be written as

$$C(s) = 138.1817 + \frac{2.8914}{s^{0.2}} + 12.382s^{1.1}. \tag{3.6}$$

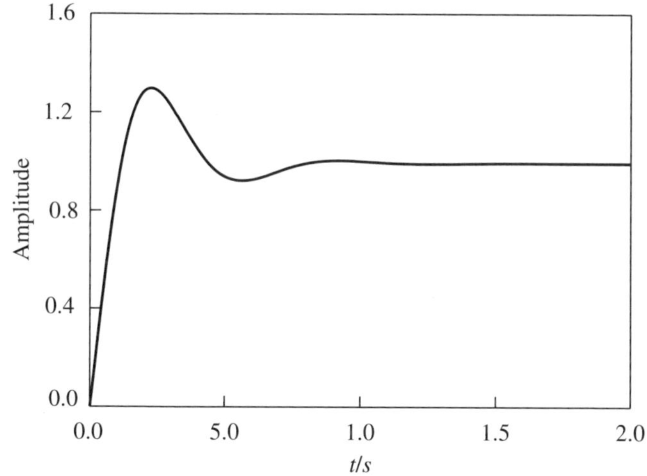

Figure 3.5 Step response of closed-loop fractional-order system (3.5).

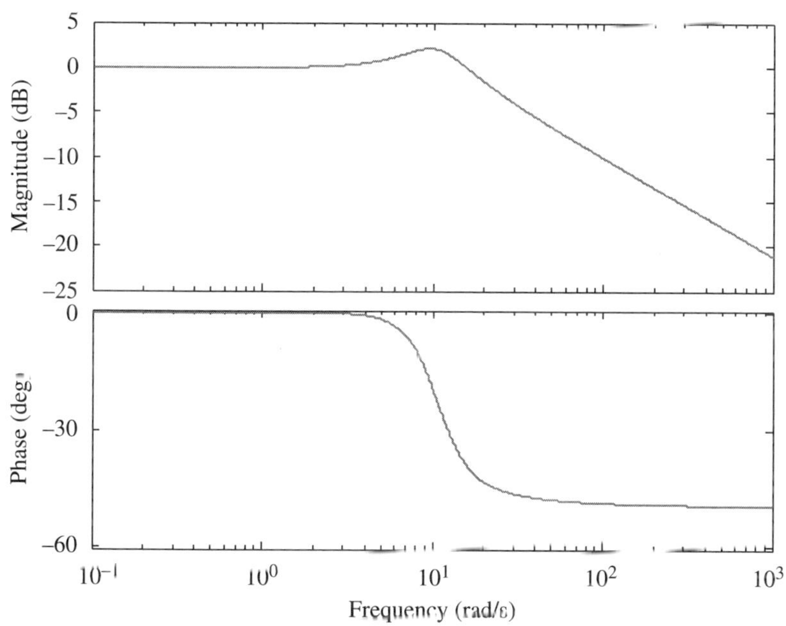

Figure 3.6 Bode diagram of fractional-order system (3.5) with fractional-order $PI^{\lambda}D^{\mu}$ controller (3.6).

Following the fractional-order $PI^{\lambda}D^{\mu}$ controller (3.6), the unit step response of the closed-loop fractional-order system (3.5) is presented in Figure 3.5. Furthermore, the Bode diagram of the fractional-order system (3.5) with the fractional-order $PI^{\lambda}D^{\mu}$ controller (3.6) is shown in Figure 3.6. From these simulation results, we can see that the fractional-order $PI^{\lambda}D^{\mu}$ controller (3.6) can control the fractional-order system (3.5) well.

3.3 Frequency-Domain Fractional-Order Disturbance Observer

3.3.1 Classical Integer-Order Disturbance Observer

As is well known, a practical system is often subjected to external disturbances. These disturbances may lead to oscillations and even increase the instability of a system. To compensate for the effects caused by external disturbances, a disturbance observer can be employed in the control design [203]. The frequency-domain disturbance observer is illustrated in Figure 3.7 [102], in which $C(s)$, $U(s)$, $D(s)$, and $Y(s)$ are the control signal, manipulated variable, external disturbance, and controlled variable, respectively. The signals $\hat{D}(s)$ and $\hat{D}_{\mathrm{f}}(s)$ are the estimates of the lumped disturbances before and after filtration by a Q filter, $G_{\mathrm{P}}(s)$ is the actual plant, $G_{\mathrm{n}}(s)$ is the nominal plant model, and $Q(s)$ is the Q filter. According to Figure 3.7, the transfer function of the output can be represented as follows:

$$Y(s) = \frac{G_{\mathrm{P}}(s)G_{\mathrm{n}}(s)}{G_{\mathrm{n}}(s) + (G_{\mathrm{P}}(s) - G_{\mathrm{n}}(s))Q(s)}C(s) + \frac{G_{\mathrm{P}}(s)G_{\mathrm{n}}(s)(1 - Q(s))}{G_{\mathrm{n}}(s) + (G_{\mathrm{P}}(s) - G_{\mathrm{n}}(s))Q(s)}D(s). \quad (3.7)$$

From Equation (3.7), we can obtain that the design of disturbance observer mainly depends on the design of Q filter, and the time constants of the Q filter are the most important parameters that determine the ability to compensate for disturbances [204]. The following two cases for selecting the Q filter are given [204]:

1) $Q(s) \to 1$. If $Q(s) \to 1$, it can be concluded (3.7) that

$$\frac{G_{\mathrm{P}}(s)G_{\mathrm{n}}(s)}{G_{\mathrm{n}}(s) + (G_{\mathrm{P}}(s) - G_{\mathrm{n}}(s))Q(s)} \to G_{\mathrm{n}}(s)$$

and

$$\frac{G_{\mathrm{P}}(s)G_{\mathrm{n}}(s)(1 - Q(s))}{G_{\mathrm{n}}(s) + (G_{\mathrm{P}}(s) - G_{\mathrm{n}}(s))Q(s)} \to 0.$$

For this case, the dynamic features of the system are similar to that of $G_{\mathrm{P}}(s)$.

2) $Q(s) \to 0$. If $Q(s) \to 0$, we can obtain

$$\frac{G_{\mathrm{P}}(s)G_{\mathrm{n}}(s)}{G_{\mathrm{n}}(s) + (G_{\mathrm{P}}(s) - G_{\mathrm{n}}(s))Q(s)} \to G_{\mathrm{P}}(s)$$

and

$$\frac{G_{\mathrm{P}}(s)G_{\mathrm{n}}(s)(1 - Q(s))}{G_{\mathrm{n}}(s) + (G_{\mathrm{P}}(s) - G_{\mathrm{n}}(s))Q(s)} \to G_{\mathrm{P}}(s).$$

For this case, the system is approximately open-loop.

On the basis of these two cases, the Q filter is designed to be in the form of a low-pass filter ($Q(s) \to 1$) such that the estimate of the external disturbance approximately equals the actual external disturbance in the low-frequency range [204, 205].

3.3.2 Fractional-Order Disturbance Observer

The concept of a FODO was first proposed by Chen *et al.* [206]. The frequency-domain FODO is the same as the classical integer-order disturbance observer, as shown in Figure 3.7. For the FODO design, the Q filter design is very important. The two tuning parameters in the FODO Q filter are the filter bandwidth and the filter order. A conventional integer-order disturbance observer selects the filter order only in the integer domain. However, the FODO can choose the filter order in the real-number domain. The fractional order form of Q filter is as follows:

$$Q(s) = \left(\frac{1}{\sigma s + 1}\right)^{\alpha}, \qquad \alpha \in R, \tag{3.8}$$

where σ is the filter time constant and α is the filter order of the Q filter. To realize the fractional-order Q filter, the broken-line approximation approach is employed to approximate $(1/\sigma s + 1)^{\alpha}$ in the frequency range $[\varpi_{\mathrm{l}}, \varpi_{\mathrm{h}}]$, where $\varpi_{\mathrm{l}} = 1/\sigma$, ϖ_{h} is set as 10^4 to give a sufficiently large frequency range [207]. Let

$$G_{\alpha}(s) = \left(\frac{1 + s/\varpi_{\mathrm{h}}}{1 + s/\varpi_{\mathrm{l}}}\right)^{\alpha}; \tag{3.9}$$

then Equation (3.9) can be approximated by the method of cascading rational functions, and we have

$$G_N(s) = \prod_{i=0}^{N-1} \frac{1 + s/\varpi_i'}{1 + s/\varpi_i}, \qquad G_{\alpha}(s) = \lim_{N\to\infty} G_N(s), \tag{3.10}$$

where

$$\varpi_i = \varpi_{\mathrm{l}} \left(\frac{\varpi_{\mathrm{h}}}{\varpi_{\mathrm{l}}}\right)^{\left(i - \frac{1}{2} - \frac{\alpha}{2}\right)/N}$$

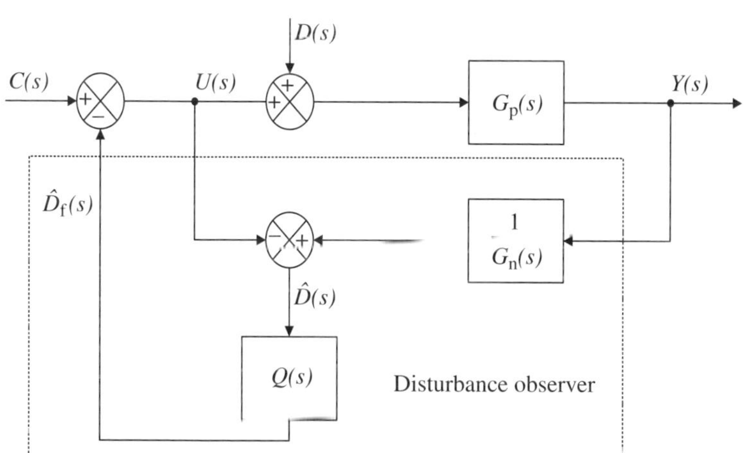

Figure 3.7 Frequency-domain disturbance observer.

and

$$\varpi_i' = \varpi_l \left(\frac{\varpi_h}{\varpi_l} \right)^{\left(i+\frac{1}{2}+\frac{\alpha}{2}\right)/N}.$$

A more accurate approximation will be obtained if a larger approximation order N is selected. Once the approximation order N reaches a certain value, the approximation accuracy would not be proportionally increased [207]. According to Lu *et al.* [207], we choose $\alpha = 0.5$, $\varpi_l = 100$, $\varpi_h = 10^4$ as a case study; Bode plots of the ideal case and

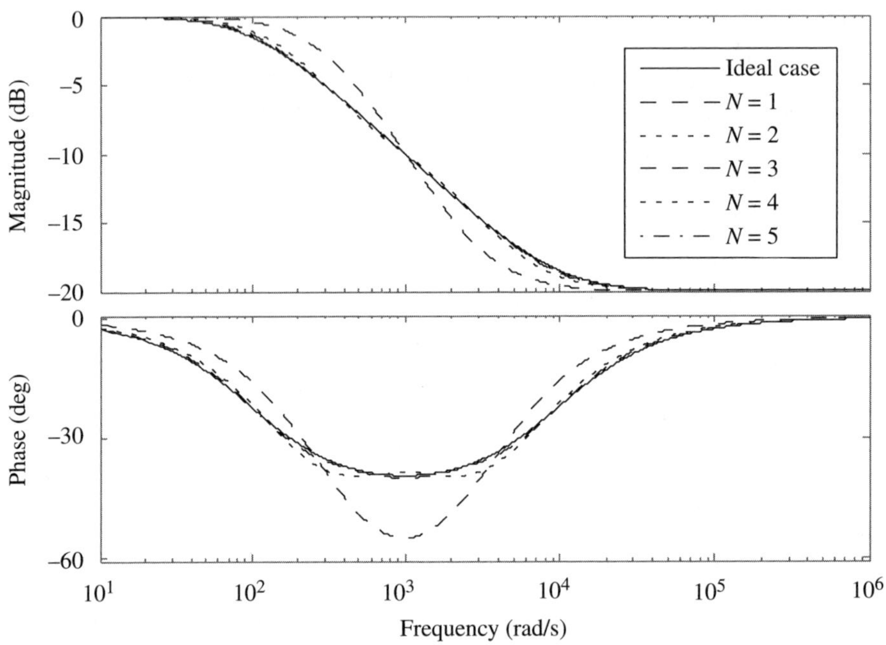

Figure 3.8 Bode plots of ideal case and its approximations with $N = 1, 2, 3, 4, 5$.

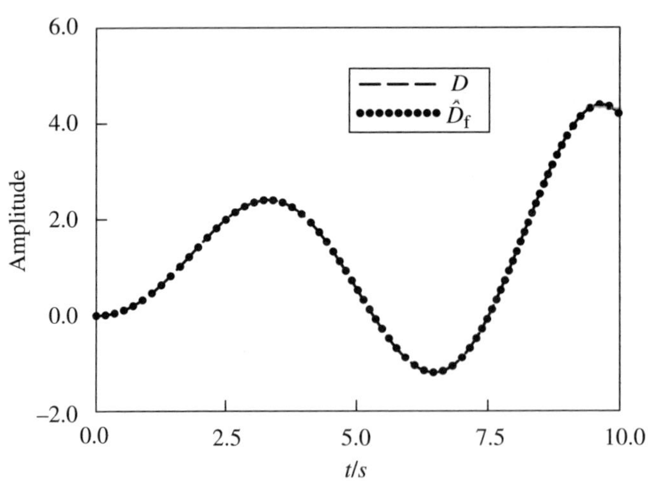

Figure 3.9 Estimation performance of FODO with filter order $\alpha = 0.5$.

its approximations with $N = 1, 2, 3, 4, 5$ are shown in Figure 3.8. From Figure 3.8, it is concluded that a large approximation error only exists in the approximation with $N = 1$. Thus, the fractional-order Q filter with $N = 5$ will be used in the next section as an example.

3.3.3 Estimation Performance of Fractional-Order Disturbance Observer

In this section, the estimation performance of FODO is studied. We assume that the disturbance $D(s) = 1/(s^{0.2} + 1)$. Numerical simulation results are given in Figure 3.9, Figure 3.10, Figure 3.11, and Figure 3.12. Figure 3.9 shows the estimation performance of a FODO with filter order $\alpha = 0.5$; the corresponding estimation error is presented in Figure 3.10. Figure 3.11 shows the estimation performance of a FODO with filter order $\alpha = 1$; the corresponding estimation error is presented in Figure 3.12. From these

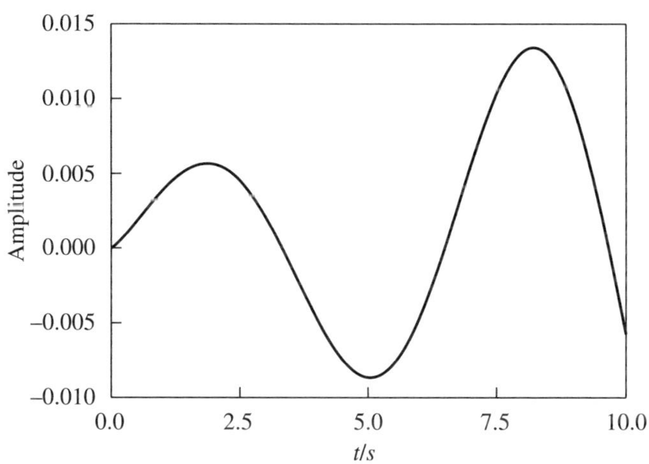

Figure 3.10 Estimation error for output of FODO with filter order $\alpha = 0.5$ and disturbance $D(s) = 1/(s^{0.2} + 1)$.

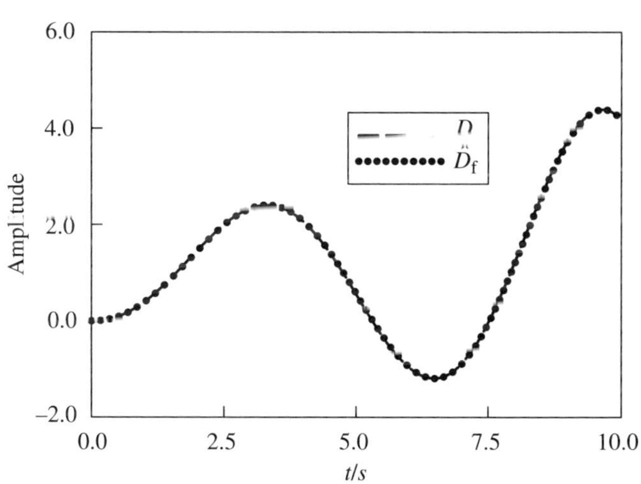

Figure 3.11 Estimation performance of FODO with filter order $\alpha = 1$.

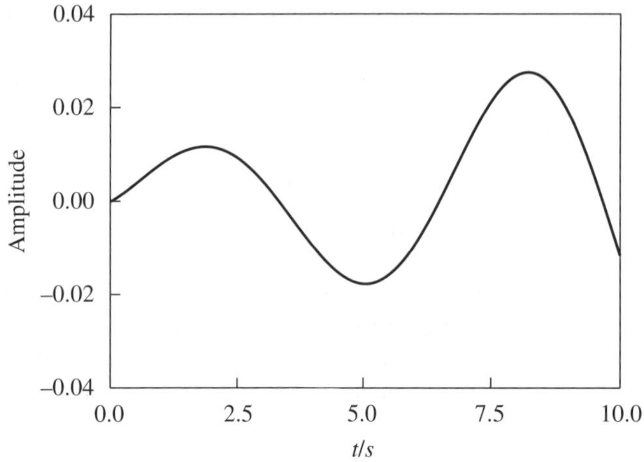

Figure 3.12 Estimation error for output of FODO with filter order $\alpha = 1$ and disturbance $D(s) = 1/(s^{0.2} + 1)$.

figures, it can be seen that a better estimation performance is obtained for a FODO with filter order $\alpha = 0.5$ than for a FODO with filter order $\alpha = 1$.

3.3.4 Control Based on Fractional-Order Disturbance Observer

In this section, the FODO-based control performance is investigated. To study the FODO-based control performance, we consider the disturbance $D(s) = 1/(s^2 + 1)$ and the fractional-order system (3.5). If the FODO is not considered in the control of the fractional-order system (3.5), the numerical simulation result is given in Figure 3.13, based on the designed fractional-order $\mathrm{PI}^{\lambda}\mathrm{D}^{\mu}$ controller (3.6). Figure 3.13 shows that the system output cannot track the desired signal. Then the FODO is employed to control the fractional-order system (3.5), and the control result is shown in Figure 3.14,

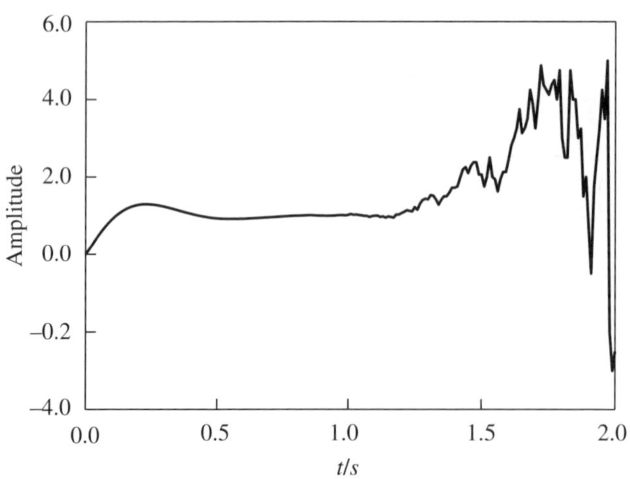

Figure 3.13 Control performance without FODO.

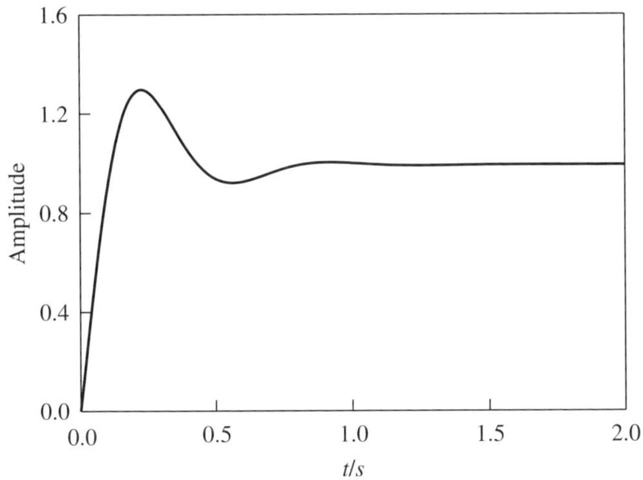

Figure 3.14 Control performance with FODO.

based on the designed fractional-order $PI^{\lambda}D^{\mu}$ controller (3.6) and the designed FODO. From Figure 3.13 and Figure 3.14, it can be seen that better control performance can be obtained if the FODO is considered in the control of the fractional-order system (3.5).

3.4 Conclusion

In this chapter, a fractional-order PID controller and a frequency-domain FODO have been introduced. The relationship between the integer-order PID controller and the fractional-order PID controller has been given. A fractional-order system has been analyzed to illustrate the effectiveness of the fractional-order $PI^{\lambda}D^{\mu}$ controller. Furthermore, the frequency-domain FODO is given. Finally, the estimation performance of the frequency-domain FODO and the FODO-based control performance are investigated with an example.

4

Design of Fractional-Order Controllers for Nonlinear Chaotic Systems and Some Applications

4.1 Fractional-Order Control for a Novel Chaotic System Without Equilibrium

4.1.1 Problem Statement

In this section, a novel chaotic system will be proposed by only considering the straight-line position x_r and the pitch angle θ_p of the two-wheeled self-balancing robot of Googol Technology. A mathematical model related to the two-wheeled self-balancing robot of Googol Technology was established by Son and Anh [208]. The linear mathematical model for x_r and θ_p of the two-wheeled self-balancing robot of Googol Technology is described in the form of

$$\begin{bmatrix} \dot{x}_r \\ \ddot{x}_r \\ \dot{\theta}_p \\ \ddot{\theta}_p \end{bmatrix} = \begin{bmatrix} 0 & 1 & 0 & 0 \\ 0 & 0 & -23.6701 & 0 \\ 0 & 0 & 0 & 1 \\ 0 & 0 & 124.5128 & 0 \end{bmatrix} \begin{bmatrix} x_r \\ \dot{x}_r \\ \theta_p \\ \dot{\theta}_p \end{bmatrix} + \begin{bmatrix} 0 \\ 4.5974 \\ 0 \\ -19.0414 \end{bmatrix} C_\theta, \tag{4.1}$$

where C_θ denotes the pitch torque.

To transform the linear mathematical model into a chaotic system, we consider C_θ as a nonlinear term $\Phi(x_r, \dot{x}_r, \theta_p, \dot{\theta}_p)$, which will be given in next section.

Considering the nonlinear function $\Phi(x_r, \dot{x}_r, \theta_p, \dot{\theta}_p)$ and the control input u, Equation (4.1) can be described as

$$\begin{bmatrix} \dot{x}_r \\ \ddot{x}_r \\ \dot{\theta}_p \\ \ddot{\theta}_p \end{bmatrix} = \begin{bmatrix} 0 & 1 & 0 & 0 \\ 0 & 0 & -23.6701 & 0 \\ 0 & 0 & 0 & 1 \\ 0 & 0 & 124.5128 & 0 \end{bmatrix} \begin{bmatrix} x_r \\ \dot{x}_r \\ \theta_p \\ \dot{\theta}_p \end{bmatrix} + \begin{bmatrix} 0 \\ 4.5974 \\ 0 \\ -19.0414 \end{bmatrix} \Phi(x_r, \dot{x}_r, \dot{\theta}_p, \ddot{\theta}_p) + u, \tag{4.2}$$

where the control input $u = [u_1, u_2, u_3, u_4]^\top$.

In this chapter, we aim to construct a novel chaotic system without equilibrium and to develop a fractional-order control scheme, so that the stabilization of the whole closed-loop system, including the novel chaotic system and the control input, is realized

Robust Adaptive Control for Fractional-Order Systems with Disturbance and Saturation, First Edition.
Mou Chen, Shuyi Shao, and Peng Shi.

based on the designed control strategy. Under the designed fractional-order controller, the state variables of the closed-loop system are asymptotically stable.

In Chapter 2, three types of fractional calculus definitions were introduced. According to the different types of fractional calculus, some important control schemes have been proposed. ML stability theorems have been proposed [209–211] for fractional-order systems. The stability theorem was developed for fractional differential system with the RL derivative [212–214]. In this chapter, a control scheme based on the fractional-order controller will be designed using the Caputo definition (2.17), with the lower limit of the integral as $t_0 = 0$ and fractional order $1 < \alpha < 2$.

4.1.2 Design of Chaotic System and Circuit Implementation

In this chapter, a novel chaotic system without equilibrium has been constructed based on the linear mathematical model (4.1) of the two-wheeled self-balancing robot. For this case, the proposed novel chaotic system can be regarded as an open-loop system of the system (4.2). Furthermore, the chaotic circuit is designed to address the physical realizability of the proposed chaotic system.

4.1.2.1 A Novel Chaotic System

From Equation (4.2), the novel chaotic system can be described as follows:

$$\begin{aligned}
\dot{x}_1 &= x_2, \\
\dot{x}_2 &= -23.6701x_3 + 4.5974\Phi(x), \\
\dot{x}_3 &= x_4, \\
\dot{x}_4 &= 124.5128x_3 - 19.0414\Phi(x),
\end{aligned} \tag{4.3}$$

where $x = [x_1, x_2, x_3, x_4]^{\mathrm{T}}$ is the state vector of the nonlinear system with $x_1 = x_{\mathrm{r}}, x_2 = \dot{x}_{\mathrm{r}}$, $x_3 = \theta_{\mathrm{p}}$ and $x_4 = \dot{\theta}_{\mathrm{p}}$. The nonlinear function $\Phi(x)$ is given by

$$\Phi(x) = \kappa_1(x_2 + x_4 + x_1x_3) + \kappa_2, \tag{4.4}$$

where κ_1 and κ_2 are constants. When $\kappa_1 = 10$ and $\kappa_2 = 0.5$, we obtain the Lyapunov exponents $\lambda_{\mathrm{L}_1} = 0.0177$, $\lambda_{\mathrm{L}_2} = 0$, $\lambda_{\mathrm{L}_3} = -0.0148$, and $\lambda_{\mathrm{L}_4} = -143.8384$ by calculating for the initial conditions $x_{10} = x_{20} = x_{30} = x_{40} = 0.1$. Obviously, the system (4.3) is a chaotic system under such case because $\lambda_{\mathrm{L}_1} > 0$, $\lambda_{\mathrm{L}_2} = 0$, $\lambda_{\mathrm{L}_3} < 0$, and $\lambda_{\mathrm{L}_4} < 0$. On the basis of the system (4.3) and the mentioned parameter values, the simulation results are further presented as shown in Figure 4.1. In addition, to further reflect the properties of chaos, the Poincaré map is shown in Figure 4.2.

To solve the equilibrium of system (4.3), we let $\dot{x}_1 = 0$, $\dot{x}_2 = 0$, $\dot{x}_3 = 0$, and $\dot{x}_4 = 0$; this gives

$$\begin{aligned}
x_2 &= 0, \\
-23.6701x_3 + 4.5974\Phi(x) &= 0, \\
x_4 &= 0, \\
124.5128x_3 - 19.0414\Phi(x) &= 0.
\end{aligned} \tag{4.5}$$

According to system (4.5), we obtain that there is no equilibrium in system (4.3). Furthermore, we ensure that the system (4.3) is dissipative with the following exponential

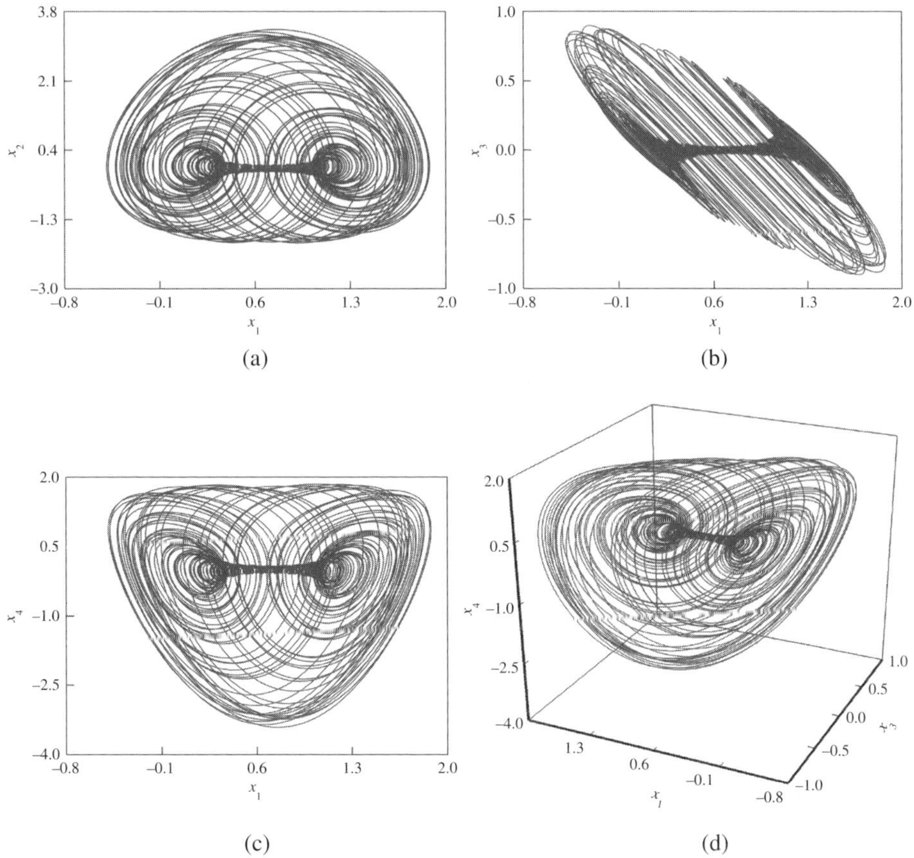

Figure 4.1 Chaotic behaviors of the novel chaotic system: (a) x_1–x_2 plane; (b) x_1–x_3 plane; (c) x_1–x_4 plane; (d) x_3–x_1–x_4 space.

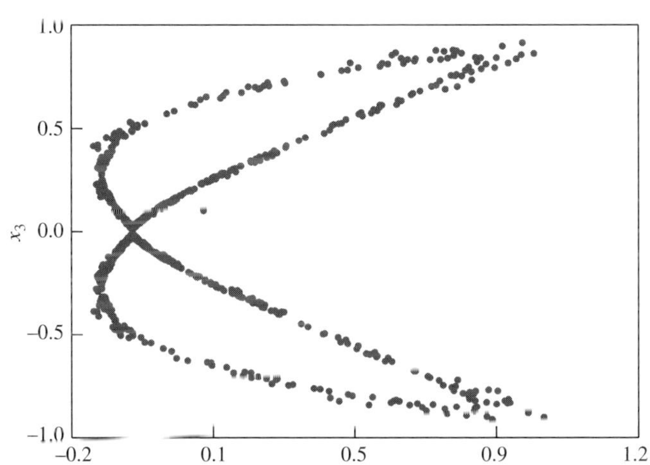

Figure 4.2 Poincaré map in the x_2–x_3 plane.

contraction rate of the exponential function V:

$$\frac{\mathrm{d}V}{\mathrm{d}t} = \mathrm{e}^{-144.44t}, \tag{4.6}$$

with

$$\begin{aligned}\nabla V &= \frac{\partial \dot{x}_1}{\partial x_1} + \frac{\partial \dot{x}_2}{\partial x_2} + \frac{\partial \dot{x}_3}{\partial x_3} + \frac{\partial \dot{x}_4}{\partial x_4} \\ &= -144.44 \\ &< 0. \end{aligned} \tag{4.7}$$

4.1.2.2 Circuit Implementation

To further illustrate the physical realizability of the proposed novel chaotic system (4.3), the system circuit is designed. By using resistors, capacitors, and operational amplifiers (TL082), the designed circuit of the chaotic system is shown in Figure 4.3. According to Figure 4.3, the circuit equation of the chaotic system is described as follows:

$$\dot{x}_1 = \frac{R_{12}}{C_1 R_{11} R_{13}} x_2,$$

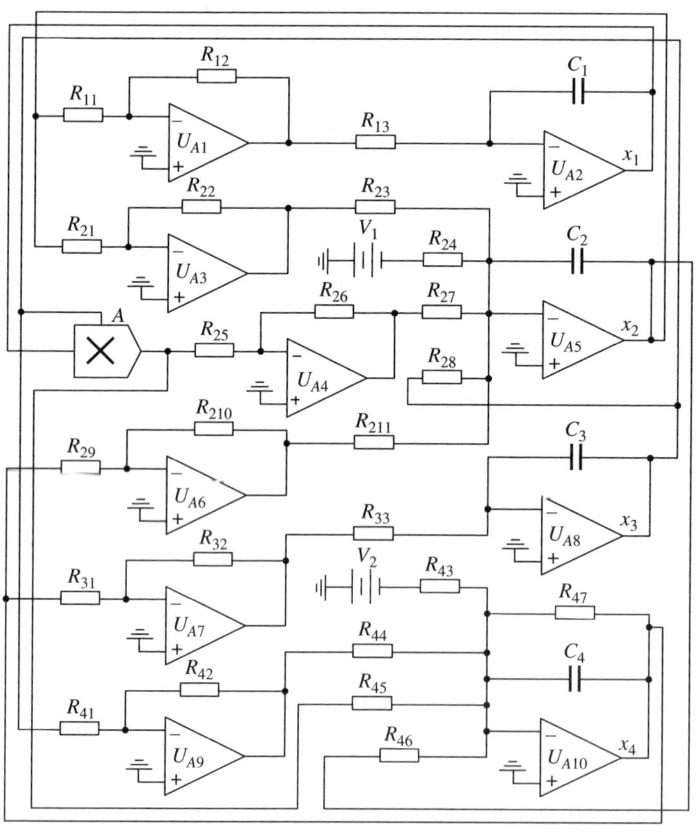

Figure 4.3 Circuit of the novel chaotic system (4.3).

$$\dot{x}_2 = -\frac{1}{C_2R_{28}}x_3 + \frac{R_{22}}{C_2R_{23}R_{21}}x_2 + \frac{R_{26}}{C_2R_{25}R_{27}}x_1x_3 + \frac{R_{210}}{C_2R_{29}R_{211}}x_4 + V_1\frac{1}{C_2R_{24}},$$
$$\dot{x}_3 = \frac{R_{32}}{C_3R_{31}R_{33}}x_4,$$
$$\dot{x}_4 = \frac{R_{42}}{C_4R_{41}R_{44}}x_3 - \frac{1}{C_4R_{46}}x_2 - \frac{1}{C_4R_{45}}x_1x_3 - \frac{1}{C_4R_{47}}x_4 - V_2\frac{1}{C_4R_{43}}. \tag{4.8}$$

By comparing system (4.3) with system (4.8), it can be deduced that all resistance values R_{11}, R_{12}, R_{21}, R_{22}, R_{25}, R_{26}, R_{29}, R_{210}, R_{31}, R_{32}, R_{41}, and R_{42} are 10 kΩ, R_{13} and R_{33} are 1 MΩ, R_{23}, R_{27}, and R_{211} are 21.7514 kΩ, R_{45}, R_{46}, and R_{47} are 5.251715 kΩ, R_{28} is 42.2474 kΩ, R_{44} is 8.031303 kΩ, R_{24} is 435.02849 kΩ, and R_{43} is 105.0343 kΩ. The voltage values are $V_1 = 1$ V and $V_2 = -1$ V. To speed up the circuit response time, we make a time-scale transformation by multiplying by a factor of 100 on the right-hand-side of system (4.3); the capacitance values C_1, C_2, C_3, and C_4 are then 10 nF. In Figure 4.3, $U_{Ai}(i = 1, 2, \dots, 10)$ are operational amplifiers, where A is a unity-gain multiplier.

From the designed circuit of the chaotic system (4.3), experimental circuit phase portraits are presented in Figure 4.4. Comparing Figure 4.1 and Figure 4.4, we observe that there is consistency between numerical simulations and circuit experimental simulations; the circuit simulation results prove the physical realizability of the proposed novel chaotic system (4.3).

4.1.3 Design of Fractional-Order Controller and Stability Analysis

In this section, the control scheme will be proposed for the whole closed-loop system including the constructed chaotic system (4.3) and a fractional-order controller. The goal is to guarantee the stabilization of the closed-loop system under the proposed fractional-order controller.

From Equations (4.3) and (4.4), the chaotic system can be rewritten as

$$\dot{x} = Ax + q(x) + \overline{q}, \tag{4.9}$$

where

$$A = \begin{bmatrix} 0 & 1 & 0 & 0 \\ 0 & 45.974 & -23.6701 & 45.974 \\ 0 & 0 & 0 & 1 \\ 0 & -190.414 & 124.5128 & -190.414 \end{bmatrix},$$

with $x = [x_1, x_2, x_3, x_4]^{\mathrm{T}}$ the state vector, and

$$q(x) = \begin{bmatrix} 0 \\ 45.974x_1x_3 \\ 0 \\ -190.414x_1x_3 \end{bmatrix}, \quad \overline{q} = \begin{bmatrix} 0 \\ 2.2987 \\ 0 \\ -9.5207 \end{bmatrix}.$$

According to the chaotic system (4.9) and considering the control input u, the corresponding system can be written in the following form:

$$\dot{x} = Ax + q(x) + \overline{q} + u, \tag{4.10}$$

where $u = [u_1, u_2, u_3, u_4]^{\mathrm{T}}$ is the designed fractional-order control input.

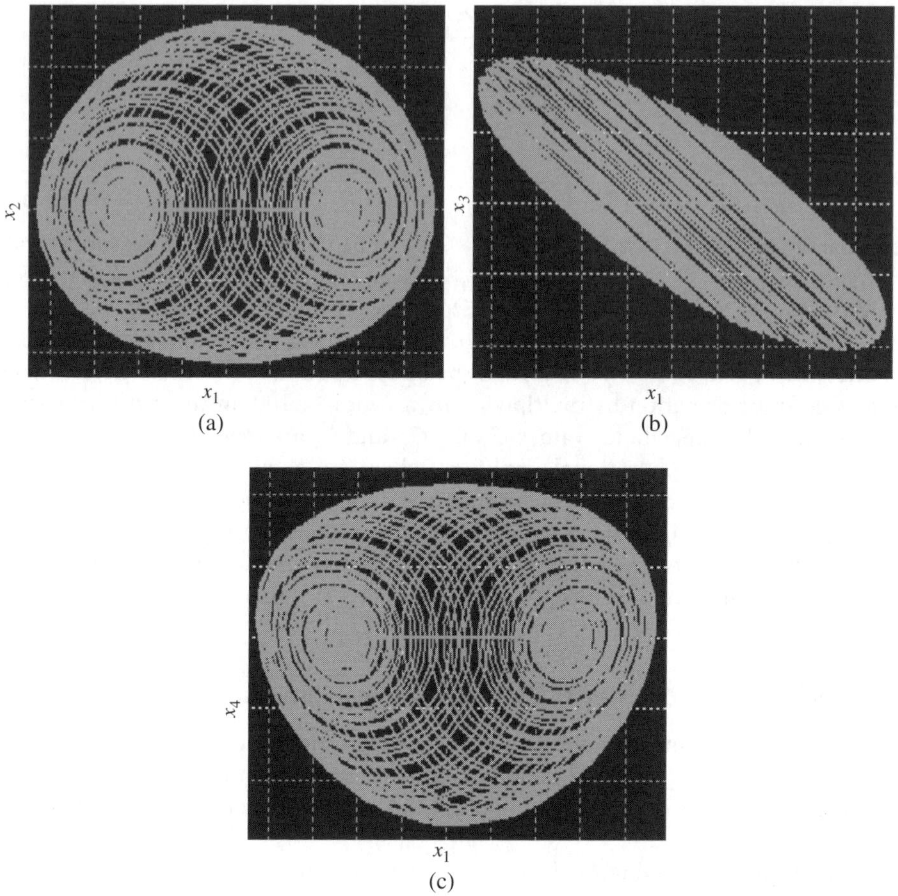

Figure 4.4 Chaotic behaviors of the chaotic circuit: (a) x_1–x_2 plane; (b) x_1–x_3 plane; (c) x_1–x_4 plane.

Based on the state-feedback control method, the controller u is defined as

$$u = -Ax - \overline{q} + KD^{1-\alpha}x, \tag{4.11}$$

where $K = \text{diag}[k_1, k_2, \cdots, k_n]$ is a design control gain matrix and the fractional order satisfies $1 < \alpha < 2$.

According to Equations (4.10) and (4.11), one has

$$\dot{x} = q(x) + KD^{1-\alpha}x. \tag{4.12}$$

To render the stabilization of the closed-loop system (4.12) under the proposed controller (4.11), the following assumption is required.

Assumption 4.1 *The nonlinear function $q(x)$ satisfies that $q(0) = 0$ and $\lim_{\|x\|\to 0} \|q(x)\|/\|x\| = 0$.*

The fractional-order controller based control scheme for the system (4.10) can be summarized in the following theorem.

Theorem 4.1 *For the closed-loop system (4.12), the fractional-order controller is designed based on Equation (4.11). Then the state variables of the closed-loop system (4.12) are asymptotically stabilized to zero when the zero is a stable equilibrium point of the closed-loop system (4.12), if* $\lim_{\|x\|\to 0}\|q(x)\|/\|x\| = 0$, *the fractional order* α: $1 < \alpha < 2$, *and the design matrix* K *satisfies* $\eta = -\max\{\mathrm{Re}\lambda(K)\} > 1$, *where* $\lambda(K)$ *denotes the eigenvalue of* K.

Proof 4.1 *By taking the Laplace transform on system (4.12), we have*

$$sX(s) - x(0) = \mathscr{L}(q(x(t))) + Ks^{1-\alpha}X(s), \tag{4.13}$$

where $X(s)$ *is the Laplace transform of* $x(t)$, $x(0)$ *is the initial condition of Equation (4.6) and* $1 < \alpha < 2$.

Let us multiply both sides of Equation (4.13) by s^α; *this yields*

$$s^{\alpha+1}X(s) - s^\alpha x(0) = s^\alpha \mathscr{L}(q(x(t))) + KsX(s). \tag{4.14}$$

From Equation (4.14), one has

$$X(s) = s^{\alpha-1}(\mathbf{I}s^\alpha - K)^{-1}(x(0) + \mathscr{L}(q(x(t)))), \tag{4.15}$$

where $\mathbf{I}$ *denotes the* 4×4 *identity matrix.*

Taking the Laplace inverse transform of Equation (4.15), one obtains

$$x(t) = E_{\alpha,1}(Kt^\alpha)x(0) + \int_0^t E_{\alpha,1}(K(t-\varphi)^\alpha)q(x(\varphi))\mathrm{d}\varphi. \tag{4.16}$$

On the basis of Lemma 2.11, since K *is a stable matrix,* $-\eta = \max(\mathrm{Re}\lambda(K))(\eta > 0)$, $M \geq 1$, *and* $1 < \alpha < 2$, *Equation (4.16) can be written as*

$$\|x(t)\| \leq M\mathrm{e}^{-\eta t}\|x(0)\| + \int_0^t M\mathrm{e}^{-\eta(t-\varphi)}\|q(x(\varphi))\|\mathrm{d}\varphi. \tag{4.17}$$

Multiplying both sides of Equation (4.17) by $\mathrm{e}^{\eta t}$ *yields*

$$\mathrm{e}^{\eta t}\|x(t)\| \leq M\|x(0)\| + \int_0^t M\mathrm{e}^{\eta\varphi}\|q(x(\varphi))\|\mathrm{d}\varphi. \tag{4.18}$$

According to Assumption 4.1 and the properties of $\lim_{\|x\|\to 0}\|q(x)\|/\|x\| = 0$ *[173, 215], there exists a constant* $\delta > 0$, *such that*

$$\|q(x)\| \leq \frac{1}{M}\|x\| \quad \text{as} \quad \|x\| < \delta. \tag{4.19}$$

Substituting Equation (4.19) into Equation (4.18), one has

$$\mathrm{e}^{\eta t}\|x(t)\| \leq M\|x(0)\| + \int_0^t \mathrm{e}^{\eta\varphi}\|x(\varphi)\|\mathrm{d}\varphi. \tag{4.20}$$

Based on Lemma 2.8, $b(t) = M\|x(0)\|$, $p(\varphi) = 1$, *and* $h(t) = \mathrm{e}^{\eta t}\|x(t)\|$, *we have*

$$\mathrm{e}^{\eta t}\|x(t)\| \leq M\|x(0)\|\mathrm{e}^{t}. \tag{4.21}$$

Inequality (4.21) is equivalent to

$$\|x(t)\| \leq \frac{M\|x(0)\|}{\mathrm{e}^{(\eta-1)t}}. \tag{4.22}$$

When $\eta = -\max\{\mathrm{Re}\lambda(K)\} > 1$, $t \to \infty$, $\|x(t)\|$ *tends asymptotically to zero. This implies that the closed-loop system (4.12) is asymptotically stable. This concludes the proof.* □

4.1.4 Numerical Simulation

In this section, to illustrate and verify the effectiveness of the proposed control scheme, the closed-loop system (4.12) is analyzed. Furthermore, we use the proposed control scheme to stabilize chaotic systems with equilibrium, such as the Chen system [216], Genesio's system [217] and the hyperchaotic Lorenz system [218].

4.1.4.1 Novel Chaotic System

Combining the novel chaotic system (4.3) with the designed controller (4.11), we have the following:

$$\begin{aligned}
\dot{x}_1 &= k_1 D^{1-\alpha} x_1,\\
\dot{x}_2 &= 45.974 x_1 x_3 + k_2 D^{1-\alpha} x_2,\\
\dot{x}_3 &= k_3 D^{1-\alpha} x_3,\\
\dot{x}_4 &= -190.414 x_1 x_3 + k_4 D^{1-\alpha} x_4.
\end{aligned} \tag{4.23}$$

The equilibrium of system (4.23) is obtained by solving $\dot{x}_1 = 0, \dot{x}_2 = 0, \dot{x}_3 = 0$, and $\dot{x}_4 = 0$, which gives the following:

$$\begin{aligned}
k_1 D^{1-\alpha} x_1 &= 0,\\
45.974 x_1 x_3 + k_2 D^{1-\alpha} x_2 &= 0,\\
k_3 D^{1-\alpha} x_3 &= 0,\\
-190.414 x_1 x_3 + k_4 D^{1-\alpha} x_4 &= 0.
\end{aligned} \tag{4.24}$$

According to system (4.24), we obtain that $O = [0, 0, 0, 0]^{\mathrm{T}}$ is the equilibrium of the system (4.23). Furthermore, when the design parameters k_1, k_2, k_3, and k_4 satisfy $k_1 < 0$, $k_2 < 0$, $k_3 < 0$, and $k_4 < 0$, we can guarantee that the equilibrium $O = [0, 0, 0, 0]^{\mathrm{T}}$ is a stable equilibrium based on the stability analysis method of the equilibrium [164]. From Equation (4.23), we have

$$\begin{aligned}
\lim_{\|x\|\to 0} \frac{\|q(x)\|}{\|x\|} &= \lim_{\|x\|\to 0} \frac{\sqrt{38371 x_1^2 x_3^2}}{\sqrt{x_1^2 + x_2^2 + x_3^2 + x_4^2}}\\
&\leq \lim_{\|x\|\to 0} \frac{\sqrt{38371 x_1^2 x_3^2}}{\sqrt{x_3^2}}\\
&= \lim_{\|x\|\to 0} 195.8854 |x_1|\\
&= 0,
\end{aligned} \tag{4.25}$$

which implies that $q(x)$ satisfies Assumption 4.1. On the basis of Theorem 4.1 and the pole placement technique, the feedback control gain matrix K and the order α are chosen as

$$K = \mathrm{diag}[-10, -10, -10, -10], \qquad \alpha = 1.6. \tag{4.26}$$

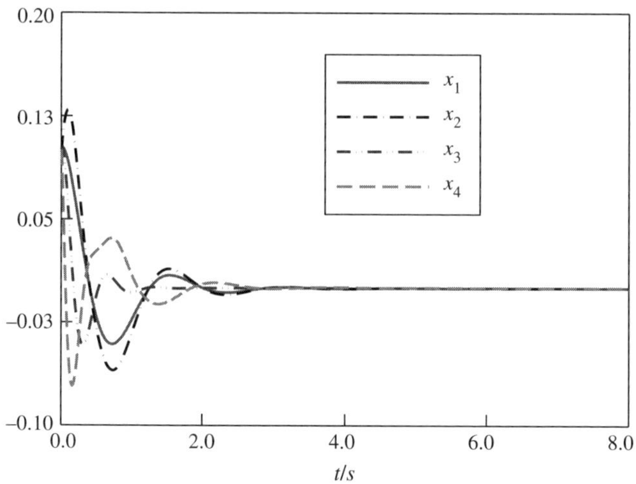

Figure 4.5 Numerical simulation results of the system (4.23).

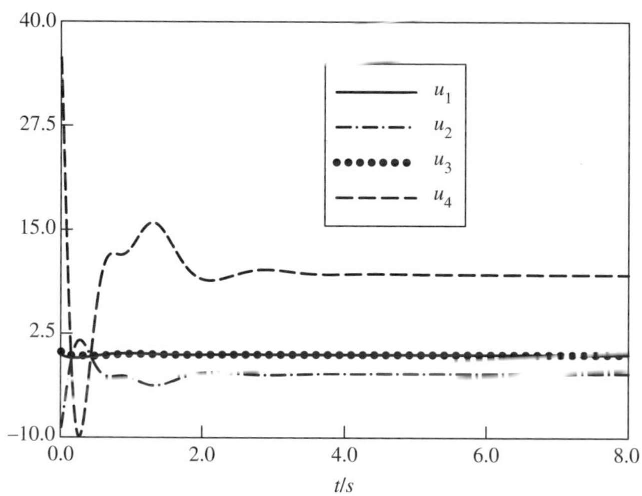

Figure 4.6 Control inputs of the system (4.23).

From this discussion, we have $\|e^{Kt^{\alpha}}\| \leq e^{-10t}$, $M = 1$, and $\eta = -\max\{\mathrm{Re}\lambda(K)\} = 10 > 1$; the conditions of Theorem 4.1 are satisfied. The simulation results are shown in Figure 4.5 and Figure 4.6. According to the numerical simulation results, the closed-loop system (4.23) is asymptotically stable, which implies that the proposed control scheme works effectively.

4.1.4.2 Chaotic Systems with Equilibrium

To further illustrate the effectiveness of the developed control scheme in this chapter, we use the proposed control scheme (4.11) to control the Chen system [216], Genesio's system [217], and the hyperchaotic Lorenz system [218]. We first analyze the following dynamic model of the Chen system [216]:

$$\dot{x}_1 = 35(x_2 - x_1),$$
$$\dot{x}_2 = -7x_1 - x_1x_3 + 28x_2,$$
$$\dot{x}_3 = x_1x_2 - 3x_3. \tag{4.27}$$

From Equation (4.11), the control input u is designed for the Chen system as follows:

$$u_1 = -35(x_2 - x_1) + k_1D^{1-\alpha}x_1,$$
$$u_2 = 7x_1 - 28x_2 + k_2D^{1-\alpha}x_2,$$
$$u_3 = 3x_3 + k_3D^{1-\alpha}x_3. \tag{4.28}$$

Invoking system (4.27), we have

$$\lim_{\|x\|\to 0} \frac{\sqrt{x_1^2x_2^2 + x_1^2x_3^2}}{\sqrt{x_1^2 + x_2^2 + x_3^2}} \le \lim_{\|x\|\to 0} |x_1| = 0, \tag{4.29}$$

where $x = [x_1, x_2, x_3]^\top$. According to Equation (4.29), the nonlinear function in (4.27) can satisfy Assumption 4.1. Therefore, the Chen system (4.27) can be stabilized to zero by choosing appropriate parameters k_1, k_2, and k_3.

Genesio's system is written as follows:

$$\dot{x}_1 = x_2,$$
$$\dot{x}_2 = x_3,$$
$$\dot{x}_3 = -6x_1 - 2.92x_2 - 1.2x_3 + x_1^2. \tag{4.30}$$

To control Genesio's system (4.30), the control input u can be designed based on Equation (4.11), as follows:

$$u_1 = -x_2 + k_1D^{1-\alpha}x_1,$$
$$u_2 = -x_3 + k_2D^{1-\alpha}x_2,$$
$$u_3 = 6x_1 + 2.92x_2 + 1.2x_3 + k_3D^{1-\alpha}x_3. \tag{4.31}$$

From system (4.30), we obtain

$$\lim_{\|x\|\to 0} \frac{\sqrt{x_1^4}}{\sqrt{x_1^2 + x_2^2 + x_3^2}} \le \lim_{\|x\|\to 0} |x_1| = 0, \tag{4.32}$$

where $x = [x_1, x_2, x_3]^\top$.

The nonlinear function in system (4.30) can satisfy Assumption 4.1 based on Equation (4.32). Thus, Genesio's system (4.30) can be stabilized to zero under the appropriate parameters k_1, k_2, and k_3.

The hyperchaotic Lorenz system is given as follows:

$$\begin{aligned}
\dot{x}_1 &= 10(x_2 - x_1),\\
\dot{x}_2 &= 28x_1 - x_1x_3 - x_2,\\
\dot{x}_3 &= x_1x_2 - \frac{8}{3}x_3,\\
\dot{x}_4 &= -x_1x_3 + 1.2x_4.
\end{aligned} \tag{4.33}$$

Combining the hyperchaotic Lorenz system (4.33) and the control law (4.11), the control input u is written as follows:

$$\begin{aligned}
u_1 &= -10(x_2 - x_1) + k_1D^{1-\alpha}x_1,\\
u_2 &= -28x_1 + x_2 + k_2D^{1-\alpha}x_2,\\
u_3 &= \frac{8}{3}x_3 + k_3D^{1-\alpha}x_3,\\
u_4 &= -1.2x_4 + k_4D^{1-\alpha}x_4.
\end{aligned} \tag{4.34}$$

According to system (4.33), we have

$$\lim_{\|x\|\to 0} \frac{\sqrt{2x_1^2x_3^2 + x_1^2x_2^2}}{\sqrt{x_1^2 + x_2^2 + x_3^2}} \le \lim_{\|x\|\to 0} \sqrt{2x_3^2 + x_2^2} = 0, \tag{4.35}$$

where $x = [x_1, x_2, x_3, x_4]^\top$.

On the basis of Equation (4.35), Assumption 4.1 is satisfied for the nonlinear function in system (4.33). By choosing appropriate parameters k_1, k_2, k_3, and k_4, the stabilization of the hyperchaotic Lorenz system (4.33) can be realized.

According to the this discussion and analysis, we obtain that the Chen system (4.27), Genesio's system (4.30), and the hyperchaotic Lorenz system (4.33) are controlled using

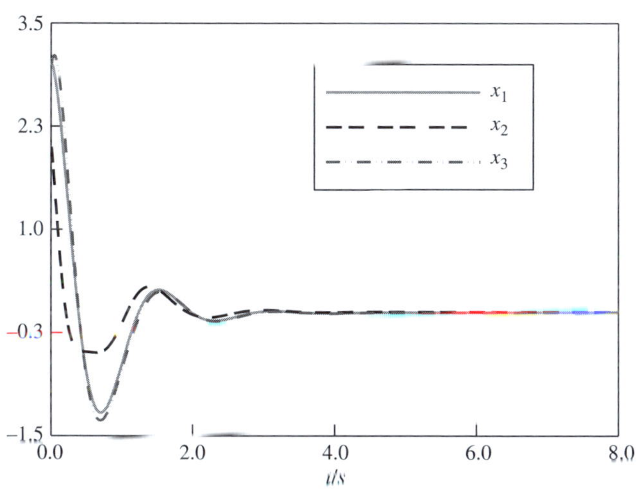

Figure 4.7 Stabilization of Chen system (4.27).

the control scheme described in this chapter. For the numerical simulation of the Chen system (4.27), we choose the control parameters $k_1 = -10$, $k_2 = -10$, and $k_3 = -10$, the initial conditions $x_0 = [3, 2, 3]^T$, and the fractional order $\alpha = 1.6$. For the numerical simulation of Genesio's system (4.30), we set the control parameters $k_1 = -10$, $k_2 = -10$, and $k_3 = -10$, the initial conditions $x_0 = [-1, -1, 0]^T$, and the fractional order $\alpha = 1.6$. In the numerical simulation of the hyperchaotic Lorenz system (4.33), the control parameters are designed as $k_1 = -10$, $k_2 = -10$, $k_3 = -10$, and $k_4 = -10$, the initial conditions are assumed as $x_0 = [0.1, -0.1, 0.3, -0.3]^T$, and the fractional order is chosen as $\alpha = 1.6$.

On the basis of the given simulation conditions, the numerical results are presented in Figure 4.7, Figure 4.8, Figure 4.9, Figure 4.10, Figure 4.11, and Figure 4.12 for the

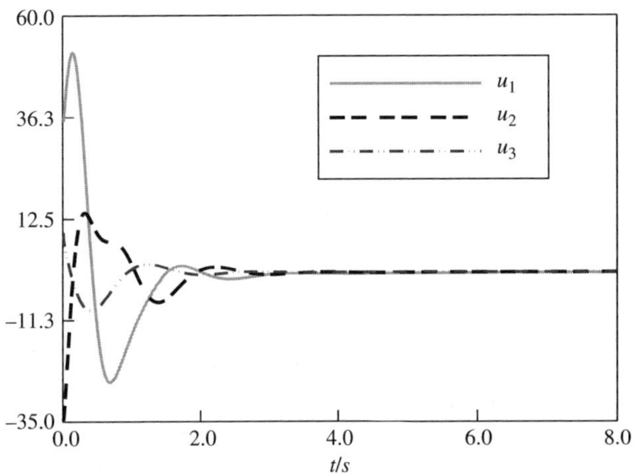

Figure 4.8 Control inputs of Chen system (4.27).

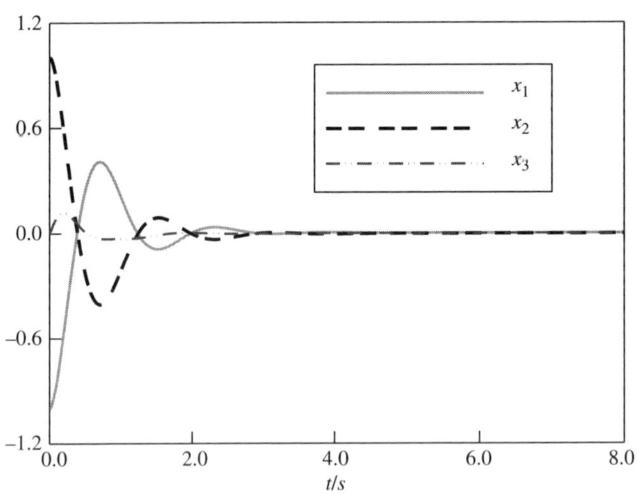

Figure 4.9 Stabilization of Genesio's system (4.30).

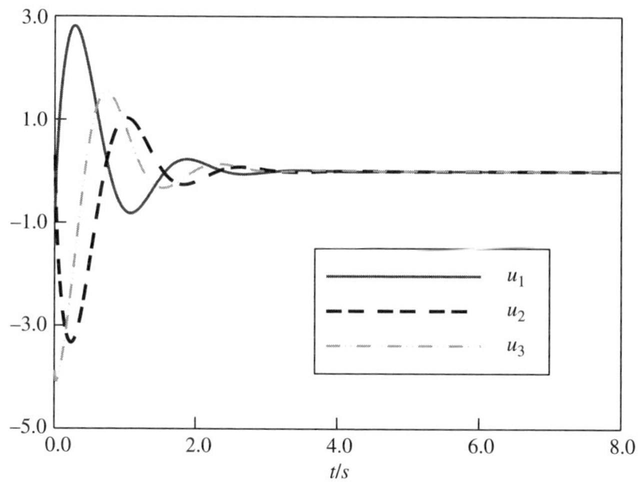

Figure 4.10 Control inputs for Genesio's system (4.30).

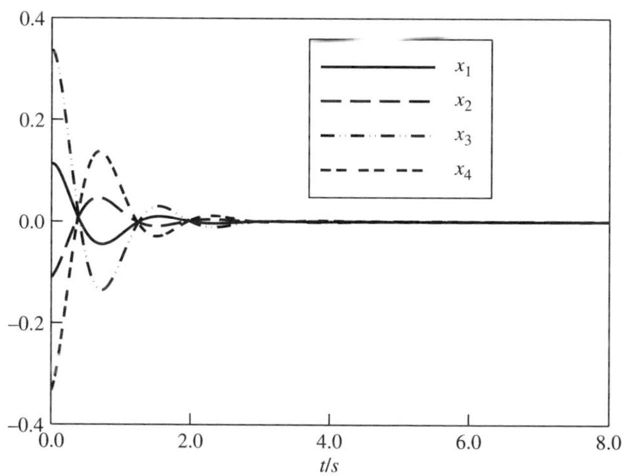

Figure 4.11 Stabilization of hyperchaotic Lorenz system (4.33).

Chen system (4.27), Genesio's system (4.30), and the hyperchaotic Lorenz system (4.33), respectively. The control results of the Chen system (4.27) are shown in Figure 4.7. It is shown that good control performance is obtained under the designed controller (4.28). Figure 4.8 presents the control inputs (4.28). The numerical simulation results of Genesio's system (4.30) are given in Figure 4.9 and Figure 4.10. Figure 4.9 and Figure 4.10 show that the controller (4.31) can stabilize Genesio's system (4.30) well. Finally, Figure 4.11 and Figure 4.12 show that the fractional-order controller (4.34) can control all state variables of the hyperchaotic Lorenz system (4.33) to the origin. Therefore, all the simulation results show that the fractional-order controller can also control the chaotic and hyperchaotic systems with equilibrium.

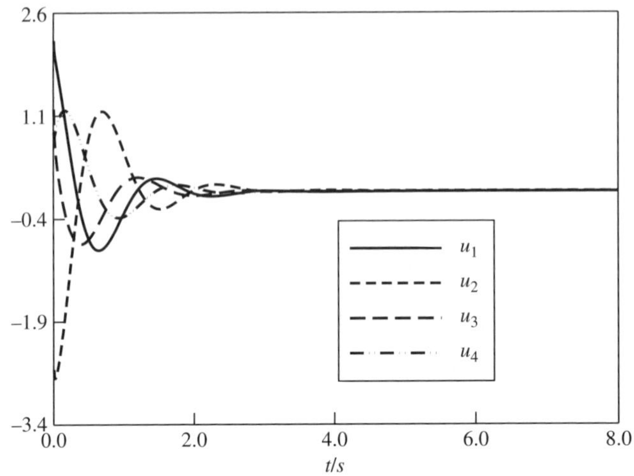

Figure 4.12 Control inputs for hyperchaotic Lorenz system (4.33).

4.2 Application of Chaotic System without Equilibrium in Image Encryption

On the basis of the proposed chaotic system without equilibrium (4.3), the image encryption scheme is developed in this section. Generally, the security will be better if the encrypted image is very fuzzy. The analysis of the effect of encryption for the image encryption scheme [219] will be given by the following four aspects:

1) Histogram analysis. The histogram of the image is an important statistical characteristic of the image, and is the approximation of the density function of the gray image.
2) Correlation of two adjacent pixels. Each pixel of any image has a high correlation with its adjacent pixels, either horizontally, vertically, or diagonally. The correlation of two adjacent pixels can be presented in a diagram. Furthermore, the correlation coefficient of each pair of pixels can also be calculated via the following formulas [219]:

$$E(\overline{x}) = \frac{1}{N_1}\sum_{i=1}^{N_1}\overline{x}_i, \tag{4.36}$$

$$D(\overline{x}) = \frac{1}{N_1}\sum_{i=1}^{N_1}[\overline{x}_i - E(\overline{x})]^2, \tag{4.37}$$

$$\mathrm{cov}(\overline{x},\overline{y}) = \frac{1}{N_1}\sum_{i=1}^{N_1}[\overline{x}_i - E(\overline{x})][\overline{y}_i - E(\overline{y})], \tag{4.38}$$

$$\rho_{\overline{x},\overline{y}} = \frac{\mathrm{cov}(\overline{x},\overline{y})}{\sqrt{D(\overline{x})}\sqrt{D(\overline{y})}}, \tag{4.39}$$

where $\overline{x}_i$ and $\overline{y}_i$ are the gray values of two adjacent pixels in the image, $E(\overline{x})$ denotes the expectation, $D(\overline{x})$ denotes the variance, $\mathrm{cov}(\overline{x},\overline{y})$ denotes the covariance, N_1 denotes the total number of adjacent pairs of pixels, and $\rho_{\bar{x},\bar{y}}$ denotes the correlation coefficient of two adjacent pixels.

3) Anti-attack ability of image encryption scheme. The anti-attack ability is analyzed by the recovery level of the broken image.
4) Sensitivity of keys. A good encryption algorithm should be sensitive to a secret key. Small changes in the decryption key will lead to failed decrypted images.

4.2.1 Image Encryption Scheme

According to the chaotic sequences of the chaotic system without equilibrium (4.3), pixel substitution, and pixel value scrambling methods, the image encryption scheme is given for an image (the size is $M \times N$, where M and N represent the rows and columns of the image). The steps for the encryption algorithm are as follows:

1) Assume a random value $[x_{10}, x_{20}, x_{30}, x_{40}]^\top$ as the initial condition of the chaotic system without equilibrium (4.3). Then four groups of chaotic time sequences with good stochastic properties are obtained.
2) Select a group of chaotic time sequences from the given four groups of chaotic time sequences. Then a subsequence is designed, based on the selected chaotic time sequence, where the size of the subsequence is $M \times N$. Moreover, a new sequence of integers is given according to the subsequence, where the sequence of integers is set as $I_{M\times N}$.
3) The pixel values of the origin image are loaded. Then the time sequence of pixel values ($X_{M\times N}$) is then translated into the binary serialization.
4) The elements in $I_{M\times N}$ and $X_{M\times N}$ are handled by the xor method for each of $I_{M\times N}$ and $X_{M\times N}$. Then a new time sequence $X'_{M\times N}$ is obtained and the binary serialization is translated into decimal format.
5) Select another group of chaotic time sequence from the given four groups of chaotic time sequences. Then a subsequence $L_{M\times N}$ is designed based on the selected chaotic time sequence, where the size of the subsequence is $M \times N$. Furthermore, the subsequence $L_{M\times N}$ is further tackled by $L_{M\times N} = (L_{M\times N} \times 1000)\mathrm{mod}(M \times N)$.
6) A group of new natural sequences T is added. Then the elements in T are initialized, that is, $T = [t_1, t_2, \cdots, t_i, \cdots, t_{M\times N}]^\top = [1, 2, \cdots, M \times N]$. The following processing procedure for the element t_i is given as: (1) select the element l_i in $L_{M\times N}$. Then the ith element in T is exchanged by the l_ith element; (2) the exchanged element is moved back l_i positions.
7) A new time sequence T is obtained based on the given time sequences T and $L_{M\times N}$. Then the new time sequence T is translated into a replacement matrix $T_{M\times N}$.
8) Setting the replacement matrix $T_{M\times N}$ as the mapping address table, and the address of elements for the matrix $X'_{M\times N}$ are rearranged. Then a new pixel matrix $X''_{M\times N}$ is obtained. Therefore, the image encryption scheme is realized.

4.2.2 Histogram Analysis

We assume that the size of the original image is 256×250 and the range of gray levels for each pixel is 0–255. The original image is presented in Figure 4.13a. The initial condition of the chaotic system without equilibrium (4.3) is chosen as $[0.1, 0.1, 0.1, 0.1]^\top$. Then four groups of chaotic time sequences are obtained by numerical simulation. The chaotic time sequence $x_1(t)(20001–85536)$ is selected as the subsequence $I_{M\times N}$ and the original image is encrypted using the pixel substitution method. Furthermore, we select the chaotic time sequence $x_2(t)(20001–85536)$ as the replacement matrix $T_{M\times N}$. Finally, the fully encrypted image is shown in Figure 4.14a.

Figure 4.13b and Figure 4.14b show the corresponding gray distributions of the original image and the image encrypted using the designed algorithm, respectively. The results from Figure 4.13 and Figure 4.14 indicate that the image encryption scheme is valid, that is: (i) the encrypted image in Figure 4.14a is totally different from the original image (Figure 4.13a); (ii) the statistical characteristics of the encrypted image in Figure 4.14b are also different from those of the original image (Figure 4.13b).

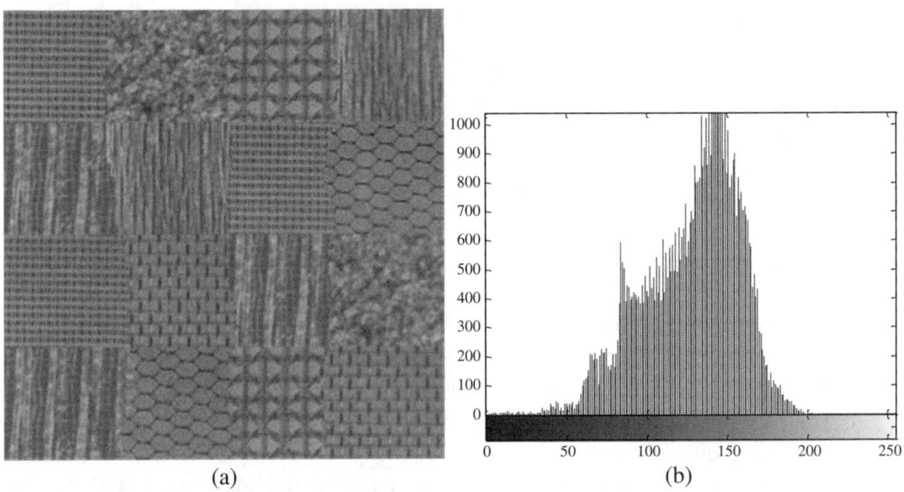

Figure 4.13 (a) Original image; (b) corresponding gray distribution histogram.

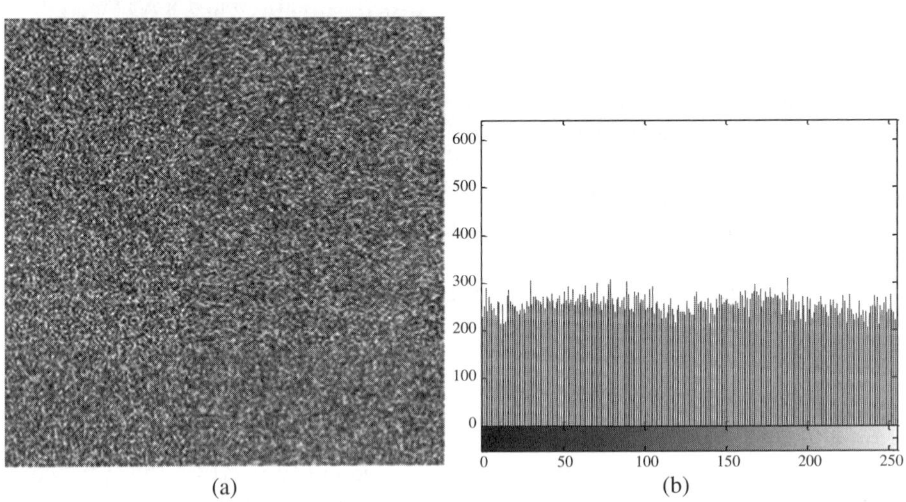

Figure 4.14 (a) Encrypted image; (b) corresponding gray distribution histogram.

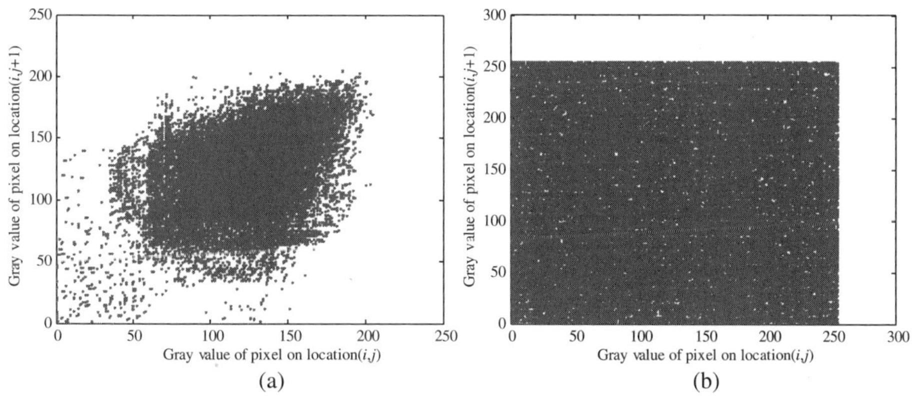

Figure 4.15 Correlations of two horizontal adjacent pixels: (a) in original image; (b) in encrypted image.

4.2.3 Correlation of Two Adjacent Pixels

Figure 4.15a and Figure 4.15b show the correlations of two horizontal adjacent pixels for the original and encrypted images, respectively. Also, Table 4.1 presents the correlation coefficient of the encrypted image and the original image. It is obvious that the correlation coefficient of the encrypted image in any direction is approximately equal to zero. As a result, the correlated relationship is very low.

4.2.4 Anti-Attack Ability of Image Encryption Scheme

Take the cropping attack as the analysis object. We assume that the intermediate region is attacked (the size is 80×80) or that the boundary area is attacked (the size is 256×40), as shown in Figure 4.16. Then the decrypted image, as shown in Figure 4.17, can also present the shape of the original image. Furthermore, Figure 4.17 is further handled by a median filter algorithm. The restored image is shown in Figure 4.18, which shows that a better restored effect is presented. It is obvious from Figure 4.16, Figure 4.17, and Figure 4.18 that the attacked image can also be restored using the designed image encryption scheme.

4.2.5 Sensitivity Analysis of Key

Key sensitivity is an essential property for any good cryptosystem, and ensures the security of the cryptosystem against brute-force attacks. The encrypted image produced by the cryptosystem should be sensitive to the secret key. That is to say, if the attacker uses two slightly different keys to decrypt the same plain image, the two encrypted

Table 4.1 Correlation coefficients in original and encrypted images.

	Original image	Encrypted Image
Horizontal	0.7467	−0.005
Vertical	0.8319	−0.0107
Diagonal	0.3468	−0.0027

Figure 4.16 Attacked images.

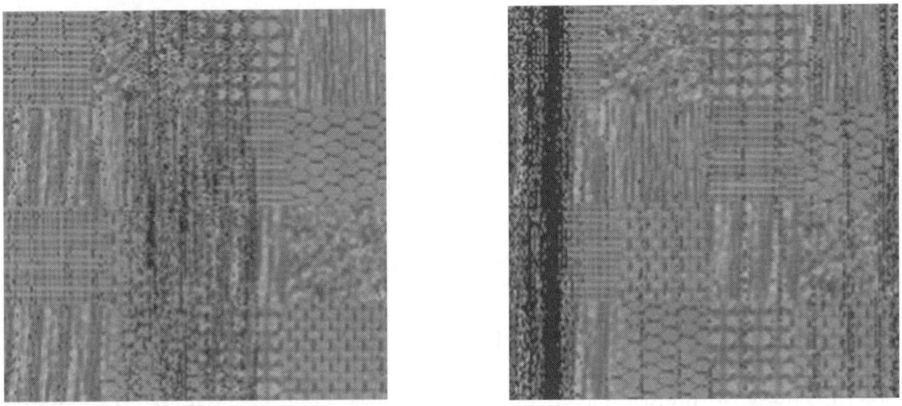

Figure 4.17 Decrypted images.

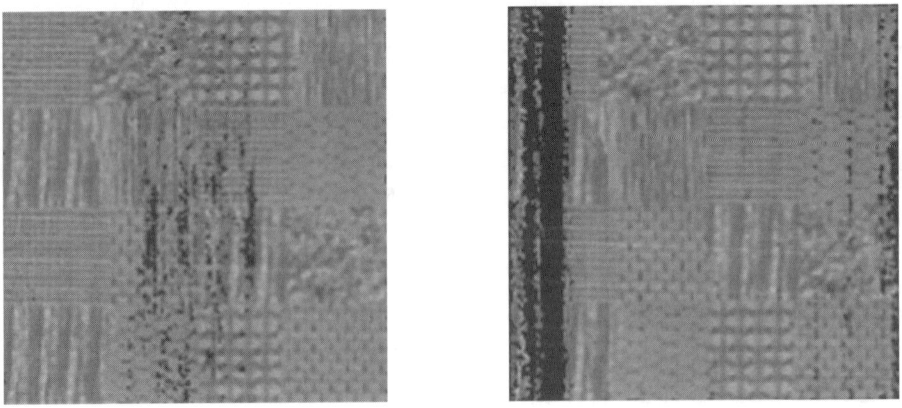

Figure 4.18 Restored images.

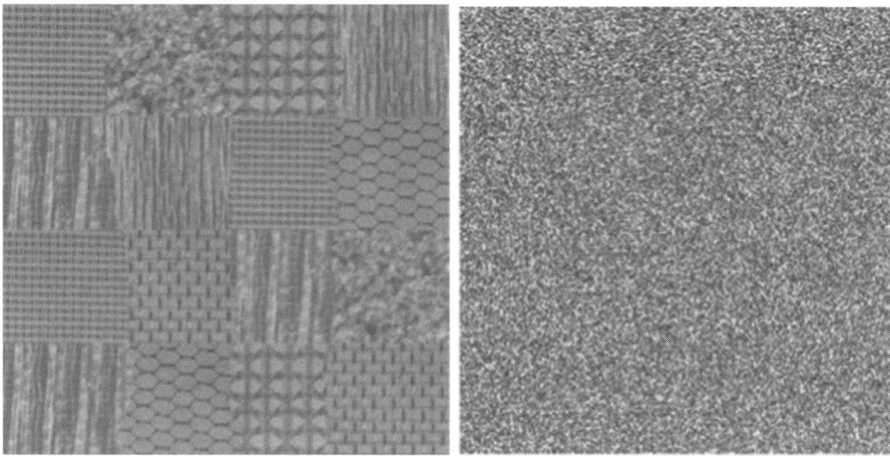

(a) (b)

Figure 4.19 Decryption attempt results using: (a) correct key; (b) incorrect key.

images should be completely independent of each other [220]. The encrypted image of Figure 4.14a is decrypted using the secret keys $[0.1, 0.1, 0.1, 0.1]^\top$ and the decrypted image is shown in Figure 4.19a. Now, we attempt to decrypt the encrypted image using other keys (the key is assumed as $[0.1 + 10^{-3}, 0.1 + 10^{-3}, 0.1 + 10^{-3}, 0.1 + 10^{-3}]^\top$). The decrypted image is shown in Figure 4.19b. Obviously, the decrypted image is different from the original image (Figure 4.19a). We can conclude that the decrypted image cannot be restored via incorrect keys.

4.3 Synchronization Control for Fractional-Order Nonlinear Chaotic Systems

4.3.1 Problem Description

According to the Caputo fractional derivative (2.17), the fractional-order system can be written as [221]

$$D^\alpha x(t) = Bx(t) + g(x(t)), \tag{4.40}$$

where $x(t) = [x_1(t), x_2(t), \cdots, x_n(t)]^\top \in R^n$ is the state vector, $B \in R^{n\times n}$ is a constant matrix, and the fractional order is $0 < \alpha < 1$. If $y(t) \in R^n$ denotes another state vector, the nonlinear functions $g(x(t)) \in R^n$ and $g(y(t)) \in R^n$ satisfy the following condition:

$$g(y(t)) - g(x(t)) = g_l(x(t))(y(t) - x(t)) + g_{nl}(y(t) - x(t)), \tag{4.41}$$

where $g_l(x(t))(y(t) - x(t))$ and $g_{nl}(y(t) - x(t))$ denote the linear and nonlinear parts, respectively.

This section aims to develop a synchronization control scheme based on the fractional-order controller. On the basis of the designed controller, the response system can synchronize the drive system well under the proper conditions.

4.3.2 Design of Synchronization Controller

Since synchronization can be applied to secure communications and signal processing, synchronization control is an important problem for investigations of fractional-order chaotic systems. Synchronization of fractional-order chaotic systems is realized by designing an appropriate fractional-order controller.

The fractional-order chaotic system (4.40) is taken as the master system, and the slave system is defined as

$$D^{\alpha}y(t) = By(t) + g(y(t)) + u(t), \tag{4.42}$$

where $u(t) = [u_1(t), u_1(t), \cdots, u_n(t)]^{\top} \in R^n$ denotes the control input vector.

Define the synchronization error as

$$e(t) = y(t) - x(t), \tag{4.43}$$

where $e(t) = [e_1(t), e_2(t), \cdots, e_n(t)]^{\top} \in R^n$.

According to Equations (4.40), (4.42), and (4.43), the synchronization error system can be described as

$$D^{\alpha}e(t) = Be(t) + g(y(t)) - g(x(t)) + u(t). \tag{4.44}$$

Furthermore, the control input is designed by

$$u(t) = -g_l(x(t))e(t) + MD^{\alpha}e(t), \tag{4.45}$$

where $M = \text{diag}[m_1, m_2, \cdots, m_n]$ denotes the design matrix, $\mathbf{I} - M$ is an invertible matrix, diag(·) denotes a diagonal matrix, and $\mathbf{I} \in R^{n\times n}$ denotes an identity matrix.

This synchronization control scheme for fractional-order nonlinear chaotic systems can be summarized in the following theorem.

Theorem 4.2 *If the design control matrix M can render $P = (\mathbf{I} - M)^{-1}B$ and $U = (\mathbf{I} - M)^{-1}$ to satisfy $\alpha > \|U\|/\|P\|$ and $|\arg(spec(P))| > \alpha\pi/2$, and the nonlinear function $g_{\text{nl}}(e(t))$ satisfies*

$$\lim_{e(t)\to 0} \|g_{\text{nl}}(e(t))\|/\|e(t)\| = 0,$$

the synchronization error system (4.44) can be stabilized under the designed controller (4.45). Thus, the synchronization is realized between the master system (4.40) and the slave system (4.42).

Proof 4.2 *According to Equations (4.41), (4.44), and (4.45), we obtain*

$$(\mathbf{I} - M)D^{\alpha}e(t) = Be(t) + g_{\text{nl}}(e(t)), \tag{4.46}$$

*where **I** is an identity matrix and $\mathbf{I} - M$ is an invertible matrix.*

By taking the Laplace transform on system (4.46), we have

$$s^{\alpha}\text{E}(s) = s^{\alpha-1}e(0) + P\text{E}(s) + U\mathscr{L}(g_{\text{nl}}), \tag{4.47}$$

where $\text{E}(s) = \mathscr{L}(e(t))$ and $g_{\text{nl}} = g_{\text{nl}}(e(t))$.

Furthermore, Equation (4.47) can be written as

$$e(t) = E_{\alpha,1}(Pt^{\alpha})e(0) + \int_0^t (t-\tau)^{\alpha-1}E_{\alpha,\alpha}(P(t-\tau)^{\alpha})Ug'\text{d}\tau, \tag{4.48}$$

where $g' = g_{\text{nl}}(e(\tau))$.

From Lemma 2.7 and Equation (4.48), we have

$$\|e(t)\| \leq \frac{N}{1+\|P\|t^{\alpha}}\|e(0)\| + \int_0^t \frac{(t-\tau)^{\alpha-1}N_0}{1+\|P\|(t-\tau)^{\alpha}}\|U\|\,\|g'\|\,\mathrm{d}\tau, \tag{4.49}$$

where $N > 0$ and $N_0 > 0$ are constants.

If the nonlinear function $g_{\mathrm{nl}}(e(t))$ satisfies

$$\lim_{e(t)\to 0}\|g_{\mathrm{nl}}(e(t))\|/\|e(t)\| = 0,$$

there exists a constant $\delta > 0$ based on the results of Chen et al. [222], such that

$$\|g_{\mathrm{nl}}(e(t))\| \leq \frac{1}{N_0}\|e(t)\|, \qquad \|e(t)\| < \delta. \tag{4.50}$$

Substituting Equation (4.50) into Equation (4.49), we have

$$\|e(t)\| \leq \frac{N}{1+\|P\|t^{\alpha}}\|e(0)\| + \int_0^t \frac{(t-\tau)^{\alpha-1}}{1+\|P\|(t-\tau)^{\alpha}}\|U\|\|e(\tau)\|\mathrm{d}\tau. \tag{4.51}$$

On the basis of Lemma 2.8 and Equation (4.51), we obtain that $\rho(t) = \|e(t)\|$, $\sigma(t) = N\|e(0)\|/(1+\|P\|t^{\alpha})$, and $\varsigma(\tau) = \|U\|(t-\tau)^{\alpha-1}/(1+\|P\|(t-\tau)^{\alpha})$. Thus, Equation (4.51) can be written as

$$\begin{aligned}\|e(t)\| &\leq \frac{N}{1+\|P\|t^{\alpha}}\|e(0)\| \\ &\quad + \int_0^t \frac{N\|U\|\|e(0)\|(t-\tau)^{\alpha-1}}{(1+\|P\|\tau^{\alpha})(1+\|P\|(t-\tau)^{\alpha})} \times \exp\left(\int_\tau^t \frac{\|U\|(t-r)^{\alpha-1}}{1+\|P\|(t-r)^{\alpha}}\mathrm{d}r\right)\mathrm{d}\tau \\ &= \frac{N}{1+\|P\|t^{\alpha}}\|e(0)\| + \int_0^t \frac{N\|U\|\|e(0)\|(t-\tau)^{\alpha-1}}{(1+\|P\|\tau^{\alpha})(1+\|P\|(t-\tau)^{\alpha})^{1-(\|U\|/(\alpha\|P\|))}}\mathrm{d}\tau.\end{aligned} \tag{4.52}$$

According to the stability analysis method [222] for fractional-order nonlinear systems, $\|e(t)\|$ satisfies the following form:

$$\begin{aligned}\|e(t)\| &\leq \frac{N}{1+\|P\|t^{\alpha}}\|e(0)\| \\ &\quad + N\|U\|\|e(0)\|\|P\|^{(\|U\|/(\alpha\|P\|))-2} \times \frac{\Gamma(\|U\|/\|P\|)\Gamma(1-\alpha)}{\Gamma(1+(\|U\|/\|P\|)-\alpha)}t^{(\|U\|/\|P\|)-\alpha}.\end{aligned} \tag{4.53}$$

Combining $\alpha > \|U\|/\|P\|$ with Equation (4.53), we have

$$\lim_{t\to\infty}\|e(t)\| = 0. \tag{4.54}$$

Therefore, $\|e(t)\|$ asymptotically tends to zero as $t \to \infty$. This implies that the synchronization is realized between the master system (4.40) and the slave system (4.42). This concludes the proof. □

4.3.3 Simulation Examples

To illustrate the effectiveness of the designed synchronization controller, the fractional-order Chen system [223] and the fractional-order Lorenz system [179] are investigated.

4.3.3.1 Fractional-Order Chen System

The fractional-order Chen system is given as follows [223]:

$$
\begin{aligned}
D^{\alpha}x_1(t) &= c_1(x_2(t) - x_1(t)),\\
D^{\alpha}x_2(t) &= (c_2 - c_1)x_1(t) - x_1(t)x_3(t) + c_2x_2(t),\\
D^{\alpha}x_3(t) &= x_1(t)x_2(t) - c_3x_3(t),
\end{aligned} \tag{4.55}
$$

where α is the fractional order, c_1, c_2, and c_3 are system parameters, and $x_1(t)$, $x_2(t)$, and $x_3(t)$ are system state variables. The parameters are chosen as $c_1 = 35$, $c_2 = 28$ and $c_3 = 3$, the fractional order is given by $\alpha = 0.9$, and the initial conditions are chosen as $[-9, -5, 14]^{\mathrm{T}}$. Then the chaotic behaviors of the fractional-order Chen system (4.55) are shown in Figure 4.20.

Furthermore, the fractional-order Chen system (4.55) can be written as

$$
\begin{pmatrix} D^{\alpha}x_1(t)\\ D^{\alpha}x_2(t)\\ D^{\alpha}x_3(t)\end{pmatrix} = \begin{pmatrix} -35 & 35 & 0\\ -7 & 28 & 0\\ 0 & 0 & -3\end{pmatrix}\begin{pmatrix} x_1(t)\\ x_2(t)\\ x_3(t)\end{pmatrix} + \begin{pmatrix} 0\\ -x_1(t)x_3(t)\\ x_1(t)x_2(t)\end{pmatrix}. \tag{4.56}
$$

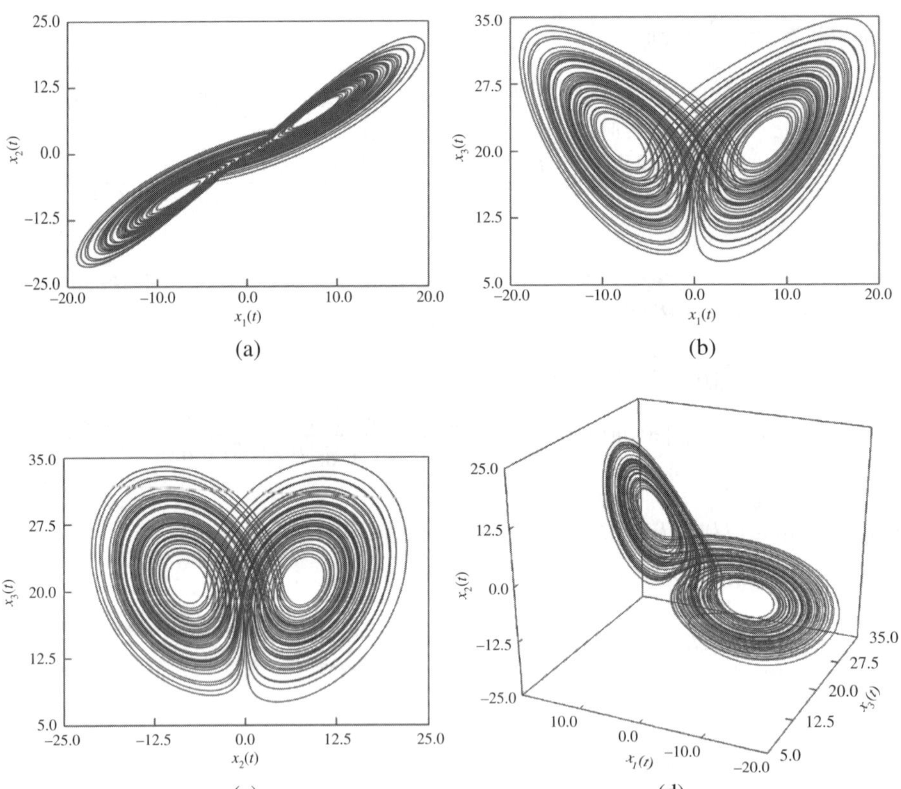

Figure 4.20 Chaotic behaviors of fractional-order Chen system: (a) x_1–x_2 plane; (b) x_1–x_3 plane; (c) x_2–x_3 plane; (d) x_3–x_1–x_2 space.

The fractional-order Chen system (4.55) is taken as the master system, and the slave system is described by

$$\begin{pmatrix} D^{\alpha}y_1(t) \\ D^{\alpha}y_2(t) \\ D^{\alpha}y_3(t) \end{pmatrix} = \begin{pmatrix} -35 & 35 & 0 \\ -7 & 28 & 0 \\ 0 & 0 & -3 \end{pmatrix} \begin{pmatrix} y_1(t) \\ y_2(t) \\ y_3(t) \end{pmatrix} + \begin{pmatrix} 0 \\ -y_1(t)y_3(t) \\ y_1(t)y_2(t) \end{pmatrix} + \begin{pmatrix} u_1(t) \\ u_2(t) \\ u_3(t) \end{pmatrix}, \tag{4.57}$$

where $y_1(t)$, $y_2(t)$, and $y_3(t)$ are system state variables and $u_1(t)$, $u_2(t)$, and $u_3(t)$ are control inputs.

According to Equations (4.56) and (4.57), we obtain

$$\begin{aligned} g(y(t)) - g(x(t)) &= \begin{pmatrix} 0 \\ -y_1(t)y_3(t) \\ y_1(t)y_2(t) \end{pmatrix} - \begin{pmatrix} 0 \\ -x_1(t)x_3(t) \\ x_1(t)x_2(t) \end{pmatrix} \\ &= \begin{pmatrix} 0 & 0 & 0 \\ -x_3(t) & 0 & -x_1(t) \\ x_2(t) & x_1(t) & 0 \end{pmatrix} \begin{pmatrix} e_1(t) \\ e_2(t) \\ e_3(t) \end{pmatrix} + \begin{pmatrix} 0 \\ -e_1(t)e_3(t) \\ e_1(t)e_2(t) \end{pmatrix}. \end{aligned} \tag{4.58}$$

On the basis of Equation (4.58), we have

$$\begin{aligned} g_1(x(t)) &- \begin{pmatrix} 0 & 0 & 0 \\ -x_3(t) & 0 & -x_1(t) \\ x_2(t) & x_1(t) & 0 \end{pmatrix}, \\ g_{\text{nl}}(e(t)) &= \begin{pmatrix} 0 \\ -e_1(t)e_3(t) \\ e_1(t)e_2(t) \end{pmatrix}. \end{aligned} \tag{4.59}$$

From Equation (4.59), we obtain

$$\begin{aligned} \lim_{e(t)\to 0} \|g_{\text{nl}}(e(t))\| / \|e(t)\| &= \lim_{e(t)\to 0} \sqrt{\frac{(e_1(t)e_3(t))^2 + (e_1(t)e_2(t))^2}{(e_1(t))^2 + (e_2(t))^2 + (e_3(t))^2}} \\ &\leq \lim_{e(t)\to 0} \sqrt{\frac{(e_1(t)e_3(t))^2 + (e_1(t)e_2(t))^2}{(e_1(t))^2}} \\ &= \lim_{e(t)\to 0} \sqrt{(e_3(t))^2 + (e_2(t))^2} \\ &= 0. \end{aligned} \tag{4.60}$$

Referring to the designed controller of Equations (4.45), (4.56), (4.57), and (4.59), on the basis of Equation (4.58), we have

$$D^{\alpha}e(t) = \begin{pmatrix} -35 & 35 & 0 \\ -7 & 28 & 0 \\ 0 & 0 & -3 \end{pmatrix} \begin{pmatrix} e_1(t) \\ e_2(t) \\ e_3(t) \end{pmatrix} + \begin{pmatrix} 0 \\ -e_1(t)e_3(t) \\ e_1(t)e_2(t) \end{pmatrix} + MD^{\alpha}e(t), \tag{4.61}$$

where $M = \text{diag}[m_1, m_2, m_3]$ is a design control matrix. If the conditions in Theorem 4.2 are satisfied, the error system (4.61) will tend to zero. Then synchronization is realized between the master system (4.56) and the slave system (4.57).

For the numerical simulation, we choose the design parameters as $m_1 = -4$, $m_2 = 6$, and $m_3 = -1$ and the initial conditions as $x(0) = [-9, -5, 14]^{\mathrm{T}}$ and $y(0) = [-10, -7, 15]^{\mathrm{T}}$.

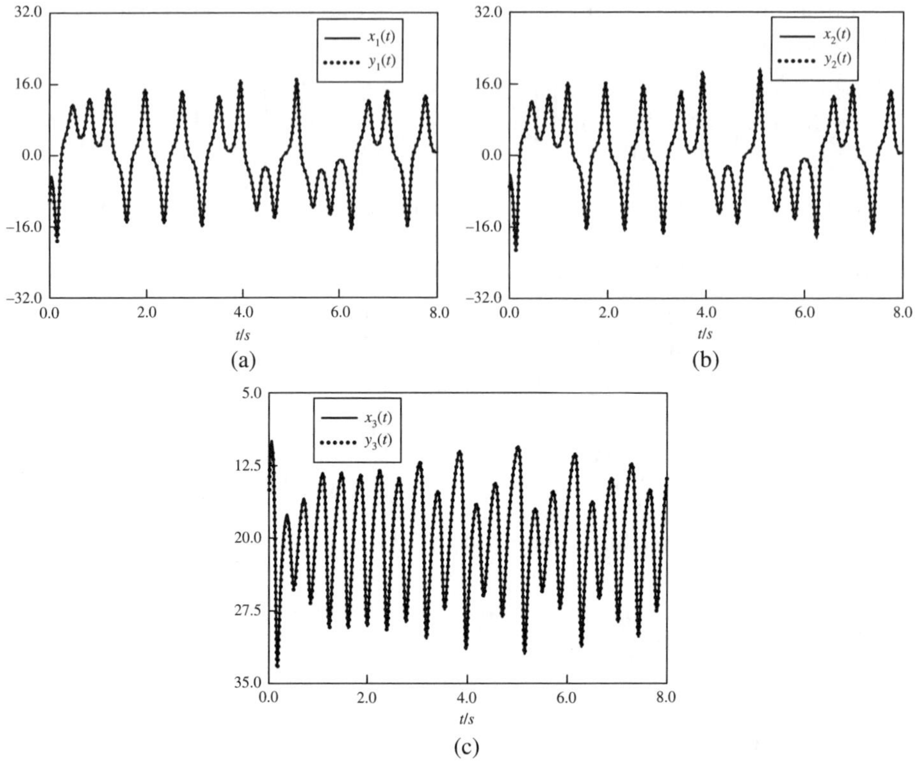

Figure 4.21 Synchronization results of state variables of two fractional-order Chen systems: (a) $x_1(t)$ and $y_1(t)$; (b) $x_2(t)$ and $y_2(t)$; (c) $x_3(t)$ and $y_3(t)$.

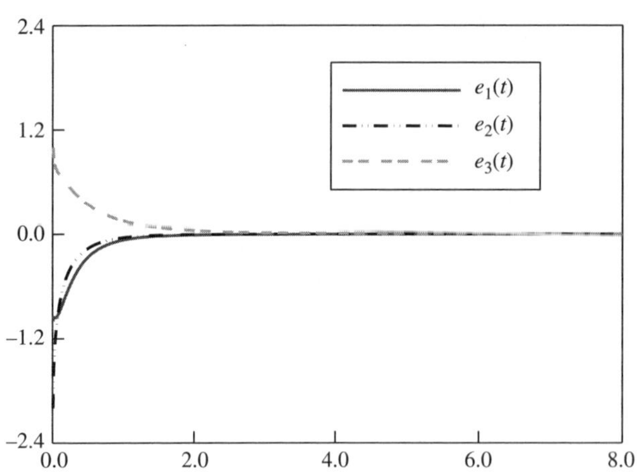

Figure 4.22 Synchronization errors $e_1(t)$, $e_2(t)$, and $e_3(t)$ of two fractional-order Chen systems.

Then we have

$$P = \begin{bmatrix} -7 & 7 & 0 \\ 1.4 & -5.6 & 0 \\ 0 & 0 & -1.5 \end{bmatrix}, \qquad U = \begin{bmatrix} 0.2 & 0 & 0 \\ 0 & -0.2 & 0 \\ 0 & 0 & 0.5 \end{bmatrix}. \tag{4.62}$$

According to Equation (4.57), we obtain that $|\arg(\mathrm{spec}(P))| > \alpha\pi/2 = 1.4137$ and $0.9 = \alpha > \|U\|/\|P\| = 0.0448$, thus the conditions in Theorem 4.2 are satisfied. The numerical results are shown in Figure 4.21 and Figure 4.22 under the proposed synchronization control scheme. The state synchronization results of the master system (4.56) and the slave system (4.57) are given in Figure 4.21a–c. It is shown that the good synchronization performance is achieved. Figure 4.22 shows that the synchronization errors $e_1(t)$, $e_2(t)$, and $e_3(t)$ are convergent. According to the simulation results, the master system (4.56) and the slave system (4.57) are synchronous under the designed fractional-order controller (4.45). Therefore, the proposed synchronization control scheme is valid for fractional-order chaotic systems.

4.3.3.2 Fractional-Order Lorenz System

According to Equation (2.46), the fractional-order Lorenz system can be written as follows:

$$\begin{aligned} D^\alpha x_1(t) &= \hbar_1(x_2(t) - x_1(t)), \\ D^\alpha x_2(t) &= x_1(t)(\hbar_2 - x_3(t)) - x_2(t), \\ D^\alpha x_3(t) &= x_1(t)x_2(t) - \hbar_3 x_3(t), \end{aligned} \tag{4.63}$$

where α is the fractional order, $x_1(t)$, $x_2(t)$, and $x_3(t)$ are system state variables, and $\hbar_1$, $\hbar_2$, and $\hbar_3$ are system parameters. For the fractional order given by $\alpha = 0.995$, system parameters chosen as $\hbar_1 = 10$, $\hbar_2 = 28$, and $\hbar_3 = 8/3$, and initial conditions chosen as $x_1(0) = x_2(0) = x_3(0) = 0.1$, the simulation results of the fractional-order Lorenz system are shown in Figure 2.1.

From Equation (4.63), we have

$$\begin{pmatrix} D^\alpha x_1(t) \\ D^\alpha x_2(t) \\ D^\alpha x_3(t) \end{pmatrix} = \begin{pmatrix} -10 & 10 & 0 \\ 28 & -1 & 0 \\ 0 & 0 & -\frac{8}{3} \end{pmatrix} \begin{pmatrix} x_1(t) \\ x_2(t) \\ x_3(t) \end{pmatrix} + \begin{pmatrix} 0 \\ -x_1(t)x_3(t) \\ x_1(t)x_2(t) \end{pmatrix}. \tag{4.64}$$

The fractional-order Lorenz system (4.64) is taken as the master system, and the slave system is given as

$$\begin{pmatrix} D^\alpha y_1(t) \\ D^\alpha y_2(t) \\ D^\alpha y_3(t) \end{pmatrix} = \begin{pmatrix} -10 & 10 & 0 \\ 28 & -1 & 0 \\ 0 & 0 & -\frac{8}{3} \end{pmatrix} \begin{pmatrix} y_1(t) \\ y_2(t) \\ y_3(t) \end{pmatrix} + \begin{pmatrix} 0 \\ -y_1(t)y_3(t) \\ y_1(t)y_2(t) \end{pmatrix} + \begin{pmatrix} u_1(t) \\ u_2(t) \\ u_3(t) \end{pmatrix}, \tag{4.65}$$

where $y_1(t)$, $y_2(t)$, and $y_3(t)$ are system state variables and $u_1(t)$, $u_2(t)$, and $u_3(t)$ are control inputs.

From Equations (4.64) and (4.65), we obtain

$$g(y(t)) - g(x(t)) = \begin{pmatrix} 0 \\ -y_1(t)y_3(t) \\ y_1(t)y_2(t) \end{pmatrix} - \begin{pmatrix} 0 \\ -x_1(t)x_3(t) \\ x_1(t)x_2(t) \end{pmatrix}$$

$$= \begin{pmatrix} 0 & 0 & 0 \\ -x_3(t) & 0 & -x_1(t) \\ x_2(t) & x_1(t) & 0 \end{pmatrix} \begin{pmatrix} e_1(t) \\ e_2(t) \\ e_3(t) \end{pmatrix} + \begin{pmatrix} 0 \\ -e_1(t)e_3(t) \\ e_1(t)e_2(t) \end{pmatrix}. \tag{4.66}$$

According to Equation (4.66), we have

$$g_l(x(t)) = \begin{pmatrix} 0 & 0 & 0 \\ -x_3(t) & 0 & -x_1(t) \\ x_2(t) & x_1(t) & 0 \end{pmatrix},$$
$$g_{nl}(e(t)) = \begin{pmatrix} 0 \\ -e_1(t)e_3(t) \\ e_1(t)e_2(t) \end{pmatrix}. \tag{4.67}$$

On the basis of Equation (4.67), we obtain

$$\begin{aligned} \lim_{e(t)\to 0} \|g_{nl}(e(t))\| / \|e(t)\| &= \lim_{e(t)\to 0} \sqrt{\frac{(e_1(t)e_3(t))^2 + (e_1(t)e_2(t))^2}{(e_1(t))^2 + (e_2(t))^2 + (e_3(t))^2}} \\ &\le \lim_{e(t)\to 0} \sqrt{\frac{(e_1(t)e_3(t))^2 + (e_1(t)e_2(t))^2}{(e_1(t))^2}} \\ &= \lim_{e(t)\to 0} \sqrt{(e_3(t))^2 + (e_2(t))^2} \\ &= 0. \end{aligned} \tag{4.68}$$

Invoking Equations (4.45), (4.64)–(4.66), and (4.67), we have

$$D^\alpha e(t) = \begin{pmatrix} -10 & 10 & 0 \\ 28 & -1 & 0 \\ 0 & 0 & -\frac{8}{3} \end{pmatrix} \begin{pmatrix} e_1(t) \\ e_2(t) \\ e_3(t) \end{pmatrix} + \begin{pmatrix} 0 \\ -e_1(t)e_3(t) \\ e_1(t)e_2(t) \end{pmatrix} + MD^\alpha e(t), \tag{4.69}$$

where $M = \text{diag}[m_1, m_2, m_3]$ is a design matrix. The error system (4.69) will tend to zero when the conditions in Theorem 4.2 are satisfied. Then synchronization is realized between the master system (4.64) and the slave system (4.65).

For the numerical simulation, the design parameters are designed as $m_1 = -3$, $m_2 = 3$, and $m_3 = -1$, the initial conditions are chosen as $x(0) = [-5, -7, 10]^\top$ and $y(0) = [-7, -8, 11]^\top$. Then we have

$$P = \begin{bmatrix} -2.5 & 2.5 & 0 \\ -14 & 0.5 & 0 \\ 0 & 0 & -1.333 \end{bmatrix}, \qquad U = \begin{bmatrix} 0.25 & 0 & 0 \\ 0 & -0.5 & 0 \\ 0 & 0 & 0.5 \end{bmatrix}. \tag{4.70}$$

According to Equation (4.70), we obtain that $|\arg(\text{spec}(P))| > \alpha\pi/2 = 1.4137$ and $0.995 = \alpha > \|U\|/\|P\| = 0.0351$, thus the conditions in Theorem 4.2 are satisfied. According to the proposed synchronization control scheme, the numerical results are given in Figure 4.23 and Figure 4.24. The state synchronization results are presented in Figure 4.23a–c. From Figure 4.23, the synchronization performance is good. Figure 4.24 shows that the synchronization errors $e_1(t)$, $e_2(t)$, and $e_3(t)$ are convergent. On the basis of the simulation results, the master system (4.64) and the slave system (4.65) are synchronous, based on the designed fractional-order controller (4.45). Therefore, the proposed synchronization control scheme is effective for fractional-order chaotic systems.

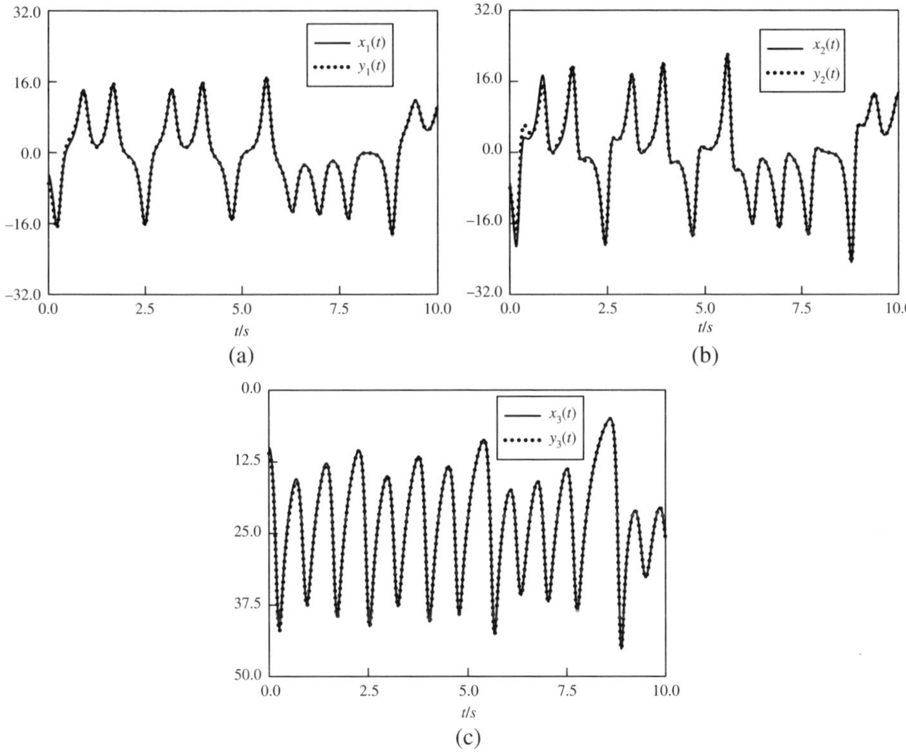

Figure 4.23 Synchronization results of state variables of two fractional-order Lorenz systems: (a) $x_1(t)$ and $y_1(t)$; (b) $x_2(t)$ and $y_2(t)$; (c) $x_3(t)$ and $y_3(t)$.

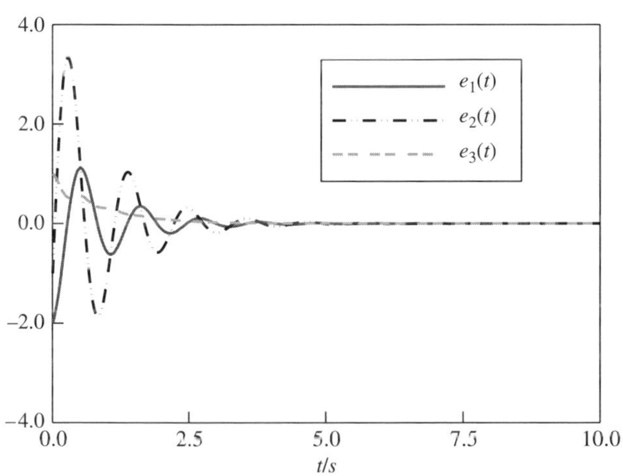

Figure 4.24 Synchronization errors $e_1(t)$, $e_2(t)$, and $e_3(t)$ of two fractional-order Lorenz systems.

4.3.4 Application of Synchronization Control Scheme in Secure Communication

In this section, the synchronization control scheme is applied in secure communication through the chaotic masking technology. The information signal can be concealed and recovered.

To illustrate the effectiveness of the synchronization control scheme in secure communication, synchronization of the fractional-order Chen system is applied in secure communication based on the chaotic masking technology [224] illustrated in

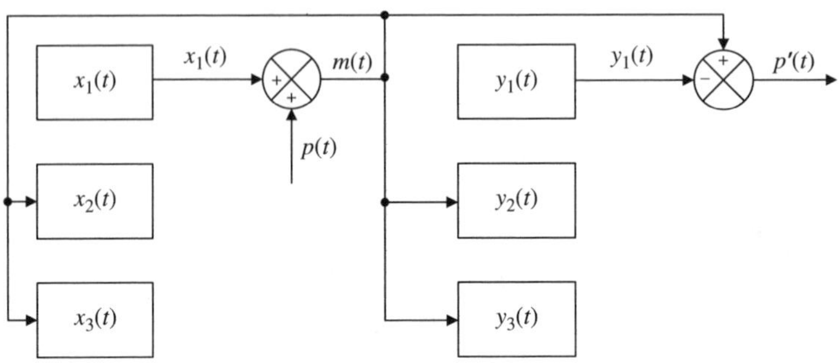

Figure 4.25 Chaotic masking technology.

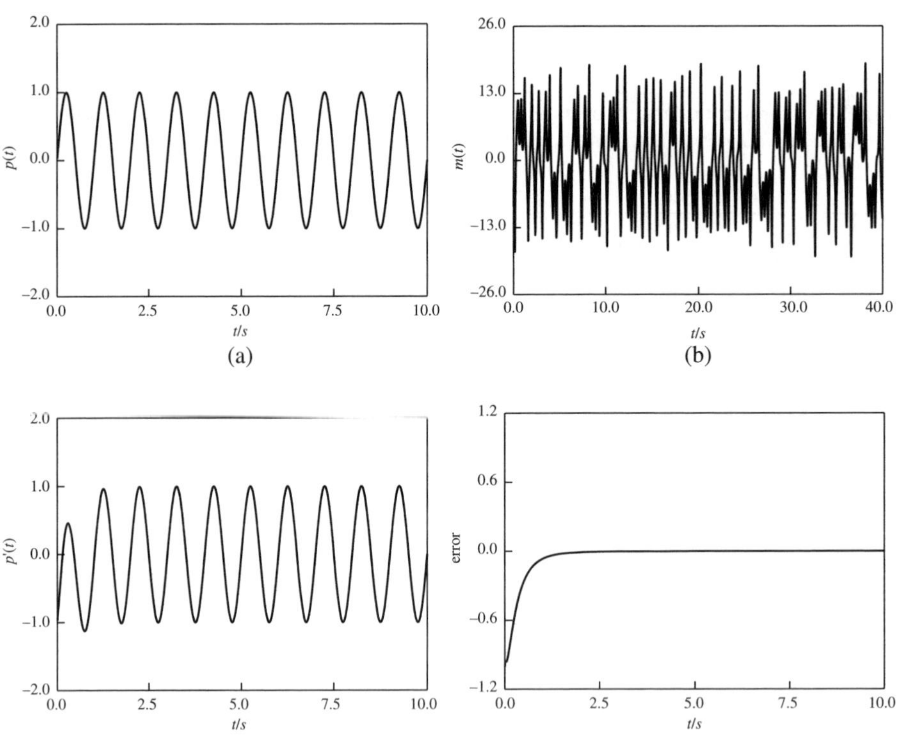

Figure 4.26 Simulation results: (a) information signal $p(t)$; (b) mixture signal $m(t)$; (c) recovered signal $p'(t)$; (d) error signal $p(t) - p'(t)$.

Figure 4.25. For secure communication, we assume that the master system (4.55) is regarded as the transmitter of the secure communication system and the slave system (4.57) is regarded as the acceptor of the secure communication system. Then we use the chaotic synchronization signals $x_1(t)$ and $y_1(t)$ to realize encryption transmission of the information signal $p(t)$. If synchronization is achieved for the chaotic signals $x_1(t)$ and $y_1(t)$, the synchronization error $e_1(t)$ will tend to zero and the recovered signal $p'(t)$ can be described as

$$p'(t) = x_1(t) + p(t) - y_1(t) \rightarrow p(t). \tag{4.71}$$

According to Equation (4.71), we can obtain that the encrypted signal can be recovered by the acceptor of the secure communication system when synchronization is realized. For the numerical simulation, we assume that the information signal $p(t) = \sin(2\pi t)$. The mixture signal of $x_1(t)$ and $p(t)$ can be written as $m(t) = x_1(t) + p(t)$. Then the secure communication system is realized based on synchronization of the fractional-order Chen system. The simulation results are presented in Figure 4.26. From Figure 4.26, we can see that the information signal $p(t)$ can be masked by the chaotic signal $x(t)$. Furthermore, the acceptor can recover the encrypted signal well. Thus, the effectiveness of the synchronization control scheme is illustrated in secure communication.

4.4 Conclusion

In this chapter, a novel chaotic system without equilibrium has been proposed. The Lyapunov exponents and the Poincaré map of the proposed chaotic system have been given. Meanwhile, the dissipativeness of the new chaotic system has been illustrated. The chaotic circuit has been designed to demonstrate the physical realizability of the novel chaotic system. In addition, on the basis of the Gronwall inequality, the Laplace transform, the ML function, and the state-feedback method, a stability theorem for the chaotic system without equilibrium has been given. The designed controller has been developed to realize the stabilization of the closed-loop systems. The proposed control scheme has been developed to control the chaotic and hyperchaotic systems with equilibrium, i.e. the Chen system, Genesio's system, and the hyperchaotic Lorenz system. An image encryption scheme has been developed by using the proposed chaotic system without equilibrium. Furthermore, on the basis of the Gronwall inequality, the Laplace transform, the ML function, and the state-feedback method, a synchronization control scheme has been given for a class of fractional-order nonlinear systems. The synchronization control scheme has been used to realize the synchronization of two fractional-order Chen systems and two fractional-order Lorenz systems, respectively. Numerical simulation results further illustrate the effectiveness of the developed synchronization control scheme. Finally, the proposed synchronization control scheme has been applied in secure communication through chaotic masking technology and the effectiveness of synchronization control scheme in secure communication is illustrated.

5

Sliding-Mode Control for Fractional-Order Nonlinear Systems Based on Disturbance Observer

5.1 Problem Statement

According to the Caputo definition of the fractional derivative (2.17), a class of fractional-order nonlinear systems subjected to external disturbances and control inputs is considered, which can be described as follows:

$$\begin{aligned} D^{\alpha}x &= F(x) + u + d, \\ y &= x, \end{aligned} \tag{5.1}$$

where α is the fractional order with $0 < \alpha < 1$, $x = [x_1, x_2, \cdots, x_n]^{\top} \in R^n$ is the state vector of the fractional-order system (5.1), $F(x) = [f_1(x), f_2(x), \cdots, f_n(x)]^{\top} \in R^n$ is the known function vector, $u(t) = [u_1, u_2, \cdots, u_n]^{\top} \in R^n$ is the control input, $d(t) = [d_1, d_2, \cdots, d_n]^{\top} \in R^n$ is the external disturbance vector, and $y \in R^n$ is the system output vector.

In this chapter, we design a nonlinear sliding-mode tracking control scheme to track the desired output of the fractional-order nonlinear system (5.1) based on a designed fractional-order SMDO. Under the proposed control scheme, the given desired trajectory x_{d} can be followed by x under the effect of the unknown external disturbances. The proposed control scheme will be rigorously shown to guarantee all the signals convergent in the closed-loop system; meanwhile tracking errors converge to zeros.

To proceed with the design of the sliding-mode control for the fractional-order nonlinear system (5.1) with external disturbances, the following assumptions are required.

Assumption 5.1 *For the given desired trajectory x_{d}, the Caputo fractional derivative of x_{d} exists.*

Assumption 5.2 *The external disturbance $d_i (i = 1, 2, \ldots, n)$ is upper bounded and $|d_i| \le \eta_i$, where η_i is a positive constant.*

Robust Adaptive Control for Fractional-Order Systems with Disturbance and Saturation, First Edition.
Mou Chen, Shuyi Shao, and Peng Shi.

5.2 Adaptive Control Design Based on Fractional-Order Sliding-Mode Disturbance Observer

In this chapter, a sliding-mode control scheme is developed based on the fractional-order SMDO for the fractional-order nonlinear system (5.1) in the presence of unknown bounded external disturbances.

5.2.1 Design of Fractional-Order Sliding-Mode Disturbance Observer

According to the fractional-order nonlinear system (5.1), we have

$$D^{\alpha}x_i = f_i(x) + u_i + d_i, \tag{5.2}$$

where x_i is the ith element of x, $f_i(x)$ is the ith element of $F(x)$, u_i is the ith element of u, d_i is the ith element of d, and $i = 1, 2, \dots, n$.

Since the disturbance d is unknown in Equation (5.1), the designed tracking controller cannot contain the external disturbance d. To overcome this problem, a fractional-order SMDO is designed to estimate the disturbance.

For the system (5.2), the fractional-order SMDO is designed as follows:

$$\begin{aligned} s_i &= x_i - r_i, \\ D^{\alpha}r_i &= u_i - \chi_i, \\ \chi_i &= -k_{0i}\text{sign}(s_i) - \hat{\eta}_i\text{sign}(s_i) - |f_i(x)|\text{sign}(s_i), \\ \hat{d}_i &= -\chi_i - f_i(x), \end{aligned} \tag{5.3}$$

where $k_{0i} > 0$ is a designed positive constant, s_i is the auxiliary sliding-mode surface, r_i and χ_i are the state variables of the fractional-order SMDO (5.3), $\hat{d}_i$ is the output of the disturbance observer, $\hat{\eta}_i$ is the estimate of η_i, and sign($\cdot$) is the sign function.

According to Equations (5.2) and (5.3), we have

$$D^{\alpha}s_i = D^{\alpha}x_i - D^{\alpha}r_i = f_i(x) + d_i + \chi_i. \tag{5.4}$$

Consider the following Lyapunov candidate function:

$$V_1 = \frac{1}{2}s_i^2 + \frac{1}{2k_{1i}}\tilde{\eta}_i^2, \tag{5.5}$$

where $k_{1i} > 0$ is a design constant, $\tilde{\eta}_i$ is the estimation error, and $\tilde{\eta}_i = \eta_i - \hat{\eta}_i$.

From Lemma 2.1 and Equation (5.5), we have

$$D^{\alpha}V_1 \le s_iD^{\alpha}s_i + \frac{1}{k_{1i}}(\eta_i - \hat{\eta}_i)D^{\alpha}(\eta_i - \hat{\eta}_i). \tag{5.6}$$

In particular, if the Caputo derivative of a constant function is 0, we have

$$D^{\alpha}(\eta_i - \hat{\eta}_i) = -D^{\alpha}\hat{\eta}_i. \tag{5.7}$$

Substituting Equation (5.7) into Equation (5.6), we obtain

$$D^{\alpha}V_1 \le s_iD^{\alpha}s_i - \frac{1}{k_{1i}}(\eta_i - \hat{\eta}_i)D^{\alpha}\hat{\eta}_i. \tag{5.8}$$

Invoking Equation (5.4), Equation (5.8) can be rewritten as

$$D^\alpha V_1 \le s_i(f_i(x) + d_i + \chi_i) - \frac{1}{k_{1i}}(\eta_i - \hat{\eta}_i)D^\alpha \hat{\eta}_i. \tag{5.9}$$

On the basis of Assumption 5.2, we have

$$D^\alpha V_1 \le |s_i|\eta_i + s_i(f_i(x) + \chi_i) - \frac{1}{k_{1i}}(\eta_i - \hat{\eta}_i)D^\alpha \hat{\eta}_i. \tag{5.10}$$

From Equation (5.10), we obtain

$$D^\alpha V_1 \le |s_i|\eta_i + |s_i||f_i(x)| + s_i\chi_i - \frac{1}{k_{1i}}(\eta_i - \hat{\eta}_i)D^\alpha \hat{\eta}_i. \tag{5.11}$$

Furthermore, the adaptive law for $\hat{\eta}_i$ is chosen as

$$D^\alpha \hat{\eta}_i = k_{1i}|s_i|. \tag{5.12}$$

According to Equations (5.3), (5.11), and (5.12), we obtain

$$\begin{aligned} D^\alpha V_1 &\le |s_i|\eta_i + |s_i||f_i(x)| - s_i(k_{0i}\text{sign}(s_i) \\ &\quad + \hat{\eta}_i\text{sign}(s_i) + |f_i(x)|\text{sign}(s_i)) - \eta_i|s_i| + \hat{\eta}_i|s_i| \\ &= -k_{0i}|s_i| < 0, \qquad \forall s_i \ne 0. \end{aligned} \tag{5.13}$$

From Equation (5.13), we can see that the fractional derivative of the Lyapunov function V_1 is negative definite. Thus, it can be concluded from Lemma 2.2 that the origin of the sliding surface s_i is asymptotically stable. When sliding surface s_i is stable, we obtain that $s_i = 0$ and $D^\alpha s_i = 0$. According to Equations (5.3) and (5.4), $\hat{d}_i = -\chi_i - f_i(x)$ can approximate the disturbance d_i.

This design procedure of fractional-order SMDO can be summarized in the following theorem.

Theorem 5.1 *Consider the fractional-order nonlinear system (5.2) in the presence of the unknown time-varying external disturbance d_i, which is estimated via the fractional-order SMDO (5.3). The adaptive law for $\hat{\eta}_i$ is chosen as Equation (5.12). By appropriately adjusting the control parameters k_{0i} and k_{1i}, the disturbance d_i can be approximated well based on the fractional-order SMDO (5.3).*

On the basis of these analyses, Theorem 5.1 can be easily proven.

5.2.2 Controller Design and Stability Analysis

In this subsection, the fractional-order SMDO-based sliding-mode tracking control scheme will be studied for the fractional-order nonlinear system (5.1). To design the sliding-mode tracking control scheme for the studied fractional-order nonlinear system (5.1), we define an auxiliary variable as

$$\zeta_i = x_{di} - r_i, \tag{5.14}$$

where x_{di} is the ith element of x_d.

In accordance with Equations (5.2) and (5.3), the Caputo fractional derivative of ξ_i is

$$\begin{aligned} D^\alpha \xi_i &= D^\alpha x_{\mathrm{d}i} - D^\alpha x_i + D^\alpha s_i \\ &= -f_i(x) - u_i - \mathrm{d}_i + D^\alpha x_{\mathrm{d}i} + D^\alpha s_i. \end{aligned} \tag{5.15}$$

Using the designed SMDO, the fractional-order SMDO-based sliding-mode tracking control scheme is proposed as

$$u_i = -f_i(x) - \hat{d}_i + D^\alpha x_{\mathrm{d}i} + k_{2i}\mathrm{sign}(\xi_i), \tag{5.16}$$

where $k_{2i} > 0$ is a design positive constant.

This development of the sliding-mode control using the fractional-order SMDO for the fractional-order nonlinear system (5.1) is summarized as follows.

Theorem 5.2 *Considering a fractional-order nonlinear system (5.1) subjected to the disturbance, the fractional-order SMDO is designed as Equation (5.3). Using the developed disturbance observer-based sliding-mode tracking control scheme (5.16), the convergence of all closed-loop signals is guaranteed.*

Proof 5.1 *Considering Equations (5.15) and (5.16), we have*

$$D^\alpha \xi_i = -k_{2i}\mathrm{sign}(\xi_i) - \tilde{d}_i + D^\alpha s_i. \tag{5.17}$$

From Equation (5.3), we obtain

$$D^\alpha s_i = \tilde{d}_i. \tag{5.18}$$

Substituting Equation (5.18) into Equation (5.17), we have

$$D^\alpha \xi_i = -k_{2i}\mathrm{sign}(\xi_i). \tag{5.19}$$

Consider the Lyapunov function candidate

$$V_2 = \frac{1}{2}\xi_i^2. \tag{5.20}$$

Invoking Equations (5.19) and (5.20), the Caputo fractional derivative of V_2 is

$$\begin{aligned} D^\alpha V_2 &\le \xi_i D^\alpha \xi_i \\ &= -k_{2i}|\xi_i| < 0, \qquad \forall \xi_i \neq 0. \end{aligned} \tag{5.21}$$

According to Lemma 2.2 and Equation (5.21), the fractional derivative of the Lyapunov function V_2 is negative definite, which means that the origin of the auxiliary variable ξ_i is asymptotically stable. When ξ_i is stable, we obtain that $\xi_i = 0$. From the definition of r_i, we know $x_{\mathrm{d}i} - x_i + s_i = 0$. According to the analysis of s_i, we obtain that $x_{\mathrm{d}i} - x_i$ is stable with $x_{\mathrm{d}i} = x_i$ while ξ_i and s_i are stable.

For the whole closed-loop system, the Lyapunov function candidate is chosen as

$$V = \sum_{i=1}^{n} \frac{1}{2}s_i^2 + \sum_{i=1}^{n} \frac{1}{2k_{1i}}\tilde{\eta}_i^2 + \sum_{i=1}^{n} \frac{1}{2}\xi_i^2. \tag{5.22}$$

Considering Equations (5.13) and (5.21), we have

$$\begin{aligned} D^\alpha V &\le -\sum_{i=1}^{n} k_{0i}|s_i| - \sum_{i=1}^{n} k_{2i}|\xi_i| \\ &< 0, \qquad \forall s_i \neq 0, \xi_i \neq 0. \end{aligned} \tag{5.23}$$

From Equation (5.23) and Lemma 2.2, we know that the convergence of all closed-loop system signals can be guaranteed. This concludes the proof. □

5.3 Simulation Examples

In this section, simulation results of two fractional-order nonlinear systems [167] are presented to illustrate the effectiveness of the proposed fractional-order SMDO-based sliding-mode tracking control scheme for the fractional-order nonlinear system with external disturbances.

5.3.1 Example 1

The fractional-order nonlinear system in the presence of external disturbances and control inputs can be described as follows:

$$\begin{aligned} D^{\alpha}x_1 &= -x_1 + x_2^3 + u_1 + d_1, \\ D^{\alpha}x_2 &= -x_1 - x_2 + u_2 + d_2, \end{aligned} \tag{5.24}$$

where x_1 and x_2 are system state variables, u_1 and u_2 are control inputs, and d_1 and d_2 are external disturbances.

For the numerical simulation of the fractional-order system (5.24), to reduce sliding-mode chattering, the saturation function will be used instead of the sign function; this has the following form:

$$\operatorname{sat}\left(\frac{s_i}{\vartheta_{0i}}\right) = \begin{cases} 1 & s_i > \vartheta_{0i}, \\ \frac{s_i}{\vartheta_{0i}} & |s_i| \le \vartheta_{0i}, \\ -1 & s_i < -\vartheta_{0i}, \end{cases} \tag{5.25}$$

$$\operatorname{sat}\left(\frac{\xi_i}{\vartheta_{1i}}\right) = \begin{cases} 1 & \xi_i > \vartheta_{1i}, \\ \frac{\xi_i}{\vartheta_{1i}} & |\xi_i| \le \vartheta_{1i}, \\ -1 & \xi_i < -\vartheta_{1i}, \end{cases} \tag{5.26}$$

where $\vartheta_{0i} > 0$ and $\vartheta_{1i} > 0$ are design parameters.

In the simulation, the fractional order is chosen as $\alpha = 0.98$ and the initial condition of the system (5.24) is given by $x(0) = [-6, -4]^{\mathrm{T}}$. The control parameters are chosen as $k_{01} = k_{02} = 55$, $k_{11} = k_{12} = 0.03$, $k_{21} = k_{22} = 1$, $\vartheta_{01} = \vartheta_{02} = 0.3$, and $\vartheta_{11} = \vartheta_{12} = 0.2$. The desired trajectories are chosen as $x_{1d} = 30\sin(t)$ and $x_{2d} = 4\sin(t)$. The external disturbances are assumed as $d_1(t) = 50\sin(t)$ and $d_2(t) = 25\sin(t)$. According to the result of Ishteva [225], we have $\lambda_1 D^{\alpha}\sin(\lambda_2 t) = -\lambda_1 \frac{1}{2}\mathrm{i}(\mathrm{i}\lambda_2)^m t^{m-\alpha}(E_{1,m-\alpha+1}(\mathrm{i}\lambda t) - (-1)^n E_{1,m-\alpha+1}(-\mathrm{i}\lambda t))$, where i denotes the square root of minus one and λ_1 and λ_2 are arbitrary numbers. In this simulation, the parameter $m = 1$ and the fractional order $\alpha = 0.98$. Thus, Assumption 5.1 and Assumption 5.2 are satisfied.

The tracking results of the fractional-order nonlinear system (5.24) with external disturbances are shown in Figure 5.1 under the proposed sliding-control scheme. From the simulation results, the tracking performance of fractional-order nonlinear system (5.24) is satisfactory, as shown in Figure 5.2. Furthermore, the observation performance of the proposed fractional-order SMDO (5.3) is presented in Figure 5.3 and Figure 5.4. It is

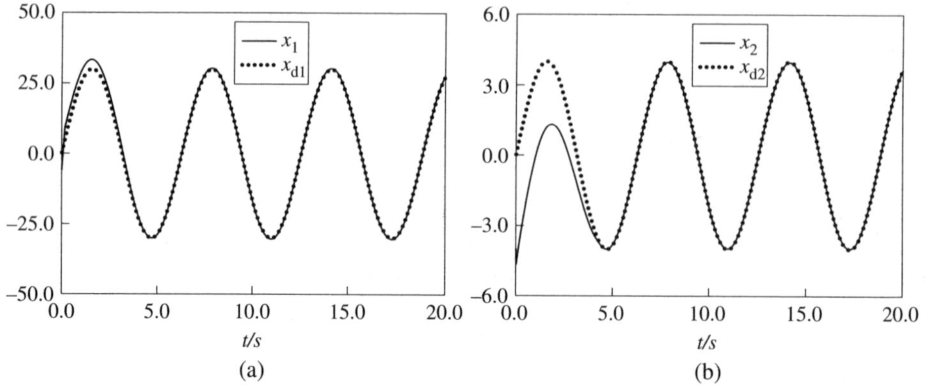

Figure 5.1 Output y of fractional-order system (5.24) follows desired trajectory x_d: (a) x_1 and x_{d1}; (b) x_2 and x_{d2}.

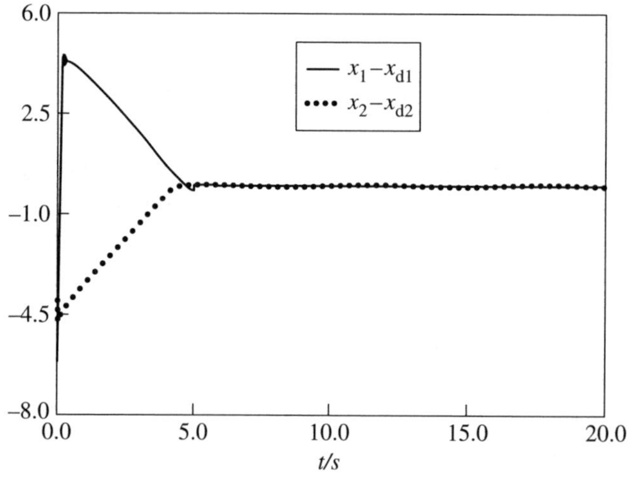

Figure 5.2 Tracking errors of fractional-order system (5.24): $x_1 - x_{d1}$ and $x_2 - x_{d2}$.

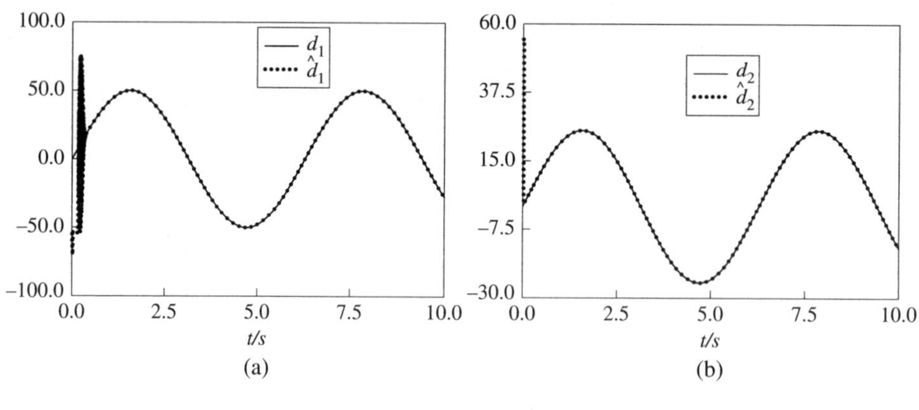

Figure 5.3 Disturbance $d(t)$ and approximate output of $\hat{d}(t)$ for the simulation of fractional-order system (5.24): (a) d_1 and $\hat{d}_1$; (b) d_2 and $\hat{d}_2$.

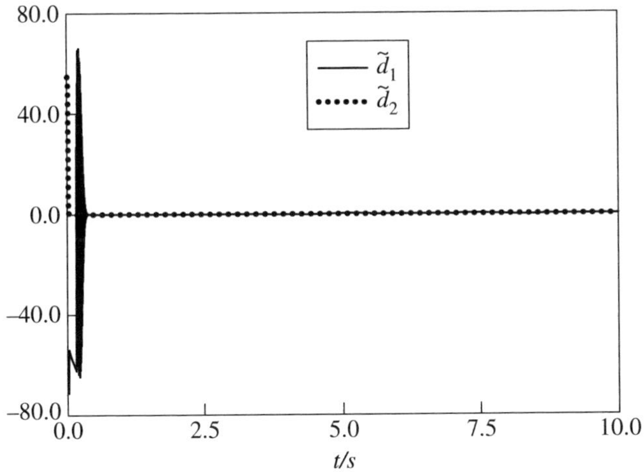

Figure 5.4 Estimation errors $\tilde{d}_1$ and $\tilde{d}_2$ for disturbances d_1 and d_2 in fractional-order system (5.24).

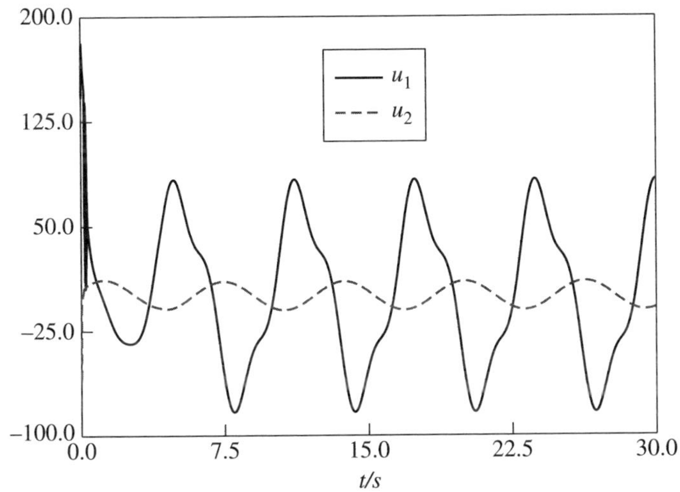

Figure 5.5 Control inputs u_1 and u_2 of fractional-order system (5.24).

evident from Figure 5.3 and Figure 5.4 that the disturbance observer is effective and feasible. The control input signals are shown in Figure 5.5. It is concluded from these simulation results that the proposed sliding-mode control technique for the fractional-order nonlinear system using fractional-order SMDO is effective.

5.3.2 Example 2

Consider the following fractional order nonlinear time-varying system with the following external disturbances and control inputs:

$$\begin{aligned} D^{\alpha}x_1 &= -\sin^2(t)x_1 - \sin(t)\cos(t)x_2 + u_1 + d_1, \\ D^{\alpha}x_2 &= -\sin(t)\cos(t)x_1^2 - \cos^2(t)x_2 + u_2 + d_2, \end{aligned} \tag{5.27}$$

where x_1 and x_2 are system state variables, u_1 and u_2 are control inputs, and d_1 and d_2 are external disturbances.

To reduce sliding-mode chattering, the saturation function will be used instead of the sign function; this has the following form:

$$\text{sat}\left(\frac{s_i}{\vartheta_{2i}}\right) = \begin{cases} 1 & s_i > \vartheta_{2i}, \\ \frac{s_i}{\vartheta_{2i}} & |s_i| \le \vartheta_{2i}, \\ -1 & s_i < -\vartheta_{2i}, \end{cases} \tag{5.28}$$

$$\text{sat}\left(\frac{\xi_i}{\vartheta_{3i}}\right) = \begin{cases} 1 & \xi_i > \vartheta_{3i}, \\ \frac{\xi_i}{\vartheta_{3i}} & |\xi_i| \le \vartheta_{3i}, \\ -1 & \xi_i < -\vartheta_{3i}, \end{cases} \tag{5.29}$$

where $\vartheta_{2i} > 0$ and $\vartheta_{3i} > 0$ are design parameters.

For the simulation, the fractional order $\alpha = 0.95$, the initial conditions of system (5.27) are chosen as $x(0) = [-2, -1]^{\mathrm{T}}$. The control parameters are chosen as $k_{01} = k_{02} = 56$, $k_{11} = k_{12} = 3$, $k_{21} = k_{22} = 5$, $\vartheta_{21} = \vartheta_{22} = 0.3$, and $\vartheta_{31} = \vartheta_{32} = 0.2$. The

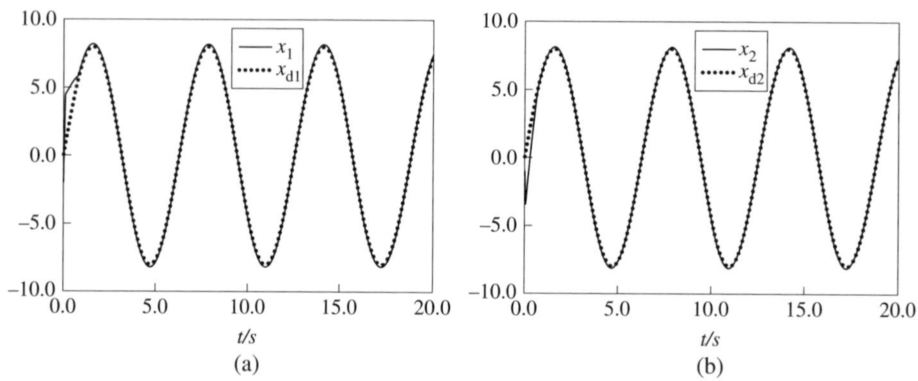

Figure 5.6 Output y of fractional-order system (5.27) follows desired trajectory x_d: (a) x_1 and x_{d1}; (b) x_2 and x_{d2}.

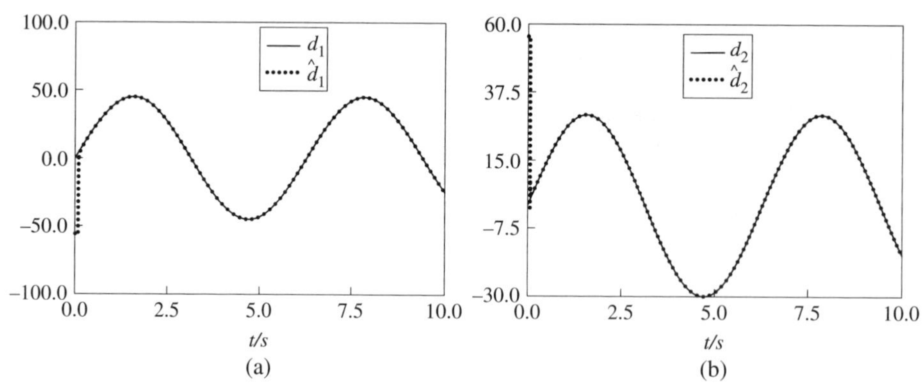

Figure 5.7 Disturbance estimation results for the simulation of fractional-order system (5.27): (a) d_1 and $\hat{d}_1$; (b) d_2 and $\hat{d}_2$.

desired trajectories are chosen as $x_{1d} = x_{2d} = 8\sin(t)$. The external disturbances are assumed as $d_1(t) = 45\sin(t)$ and $d_2(t) = 30\sin(t)$.

On the basis of the proposed sliding-control scheme, the tracking results are shown in Figure 5.6 for the fractional-order nonlinear system (5.27) with external disturbances. We note that satisfactory tracking performance is obtained based on the simulation results. Furthermore, the disturbance estimation results are presented in Figure 5.7 and Figure 5.8. According to Figure 5.7 and Figure 5.8, we can obtain that the disturbance observer is effective and feasible. The control input signals, which are bounded, are shown in Figure 5.9. Therefore, the proposed sliding-mode control technique for the fractional-order nonlinear system using fractional-order SMDO is effective, based on the simulation results.

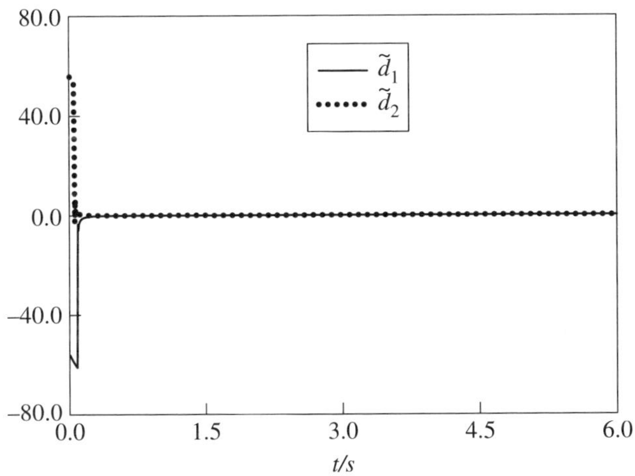

Figure 5.8 Estimation errors $\tilde{d}_1$ and $\tilde{d}_2$ for disturbances d_1 and d_2 in fractional-order system (5.27).

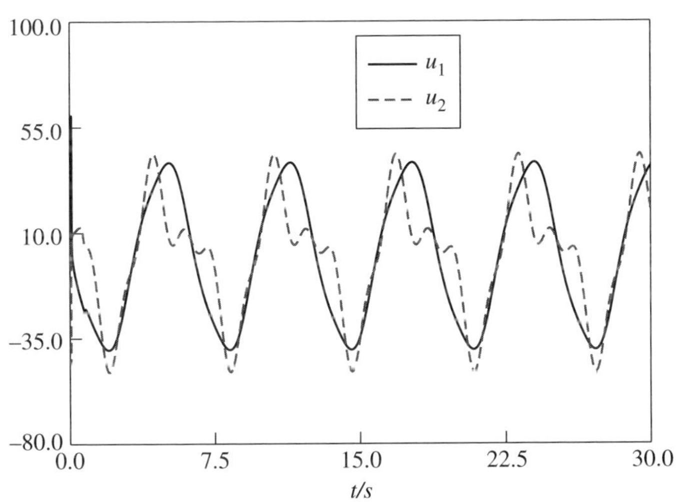

Figure 5.9 Control inputs u_1 and u_2 of fractional-order system (5.27).

5.4 Conclusion

A sliding-mode control scheme has been proposed for a class of fractional-order nonlinear systems using a fractional-order SMDO in this chapter. The SMDO has been designed to guarantee the convergence of the disturbance estimation error. Using the developed disturbance observer, the sliding-mode control scheme has been developed to guarantee the convergence of all closed-loop system signals. Numerical simulation results have been presented to show the effectiveness of the developed fractional-order SMDO-based tracking control scheme under the effect of the unknown disturbances.

6

Disturbance-Observer-Based Neural Control for Uncertain Fractional-Order Rotational Mechanical System

6.1 Problem Statement

On the basis of the Caputo definition of the fractional derivative (2.17), we consider the following fractional-order rotational mechanical system with a centrifugal governor in the following form [226]:

$$\begin{aligned} D^{\alpha}x_1 &= x_2, \\ D^{\alpha}x_2 &= a_0(x_3+a_1)^2\sin(x_1-a_2)\cos(x_1-a_2)-\sin(x_1-a_2)-a_3x_2, \\ D^{\alpha}x_3 &= a_4\cos(x_1-a_2)-a_5, \end{aligned} \tag{6.1}$$

where α is the fractional order with $0<\alpha<1$, $x_1(t)$, $x_2(t)$ and $x_3(t)$ are the state variables of the fractional-order system (6.1), a_0, a_3, a_4, and a_5 are known constants, $\cos a_2 = a_5/a_4$, and $a_1 = a_5/(a_0a_4)$.

From Equation (6.1), the fractional-order nonlinear model of a rotational mechanical system with a centrifugal governor in the presence of unknown uncertainties, external disturbances, and control inputs can be described as follows:

$$\begin{aligned} D^{\alpha}x &= F(x)+\Delta F(x)+Gu(t)+d(t), \\ y &= x, \end{aligned} \tag{6.2}$$

where α is the fractional order with $0<\alpha<1$, $G\in R^{3\times 3}$ is a known control gain matrix and G is an invertible matrix, $x=[x_1,x_2,x_3]^{\top}\in R^3$ is the state vector of the fractional-order system (6.2), $F(x)=[f_1(x),f_2(x),f_3(x)]^{\top}\in R^3$ is the known nonlinear function vector with $f_1(x)=x_2$, $f_2(x)=a_0(x_3+a_1)^2\sin(x_1-a_2)\cos(x_1-a_2)-\sin(x_1-a_2)-a_3x_2$, and $f_3(x)=a_4\cos(x_1-a_2)-a_5$, $\Delta F(x)=[\Delta f_1(x),\Delta f_2(x),\Delta f_3(x)]^{\top}\in R^3$ is the unknown nonlinear uncertainty, $u(t)=[u_1(t),u_2(t),u_3(t)]^{\top}\in R^3$ is the control input, $d(t)=[d_1(t),d_2(t),d_3(t)]^{\top}\in R^3$ is the external disturbance, and $y\in R^3$ is the system output vector.

In this chapter, we design a nonlinear FODO based adaptive neural control scheme to track the desired output of the uncertain FONS (6.2). The radial basis function neural network is used to approximate unknown nonlinear functions in the uncertain FONS (6.2) with external disturbances. On the basis of the proposed control scheme, the signal x could follow a given desired trajectory x_{d} in the presence of system uncertainties and

Robust Adaptive Control for Fractional-Order Systems with Disturbance and Saturation, First Edition.
Mou Chen, Shuyi Shao, and Peng Shi.

unknown external disturbances. The proposed control scheme will be rigorously shown to guarantee that all the signals in the closed-loop system remain bounded.

To proceed with the design of the robust adaptive neural control for the uncertain FONS (6.2) subjected to external disturbances, the following assumptions are required.

Assumption 6.1 *For the given desired trajectory x_d, the Caputo fractional derivative of x_d exists.*

Assumption 6.2 *For the external disturbance $d_j(t)$ with $j = 1, 2, 3$, the Caputo fractional derivative of $d_j(t)$ is bounded, that is $|D^\alpha d_j(t)| \leq b_{3j}$, where $b_{3j} > 0$ is an unknown positive constant.*

6.2 Adaptive Neural Control Design

6.2.1 Design of Fractional-Order Disturbance Observer

Without loss of generality, according to the uncertain FONS (6.2), we have

$$D^\alpha x_j = f_j(x) + \Delta f_j(x) + \overline{u}_j + d_j(t), \tag{6.3}$$

where x_j is the jth element of x, $f_j(x)$ is the jth element of $F(x)$, $\Delta f_j(x)$ is the jth element of $\Delta F(x)$, $\overline{u}_j$ is the jth element of $\overline{u}$ with $\overline{u} = Gu$, $d_j(t)$ is the jth element of $d(t)$, and $j = 1, 2, 3$.

On the basis of Lemma 2.4, the neural network is employed to approximate $\rho_j \Delta f_j(x)$ with $\rho_j < -2$, and we obtain

$$D^\alpha x_j = f_j(x) + \frac{1}{\rho_j}(\theta_j^{*\mathrm{T}} \phi_j(Z) + \varepsilon_j^*) + \overline{u}_j + d_j(t), \tag{6.4}$$

where $Z = x$.

Since the disturbance $d(t)$ in Equation (6.2) is unknown, $d(t)$ cannot be applied to develop robust tracking control for the uncertain FONS (6.2). To overcome this problem, a nonlinear FODO is designed to estimate disturbance.

For the system (6.4), the nonlinear FODO is designed as follows:

$$\begin{aligned} D^\alpha \eta_j &= f_j(x) + \frac{1}{\rho_j}\hat{\theta}_j^{\mathrm{T}} \phi_j(Z) + \overline{u}_j + \hat{d}_j(t), \\ D^\alpha \chi_j &= k_j \hat{\theta}_j^{\mathrm{T}} \phi_j(Z), \\ \hat{d}_j &= -\rho_j(x_j - \eta_j) - \frac{1}{k_j}\chi_j, \end{aligned} \tag{6.5}$$

where η_j and χ_j are the state variables of the nonlinear FODO, $\hat{d}_j(t)$ is the output of the disturbance observer, and the adaptive law $\hat{\theta}_j$ will be described in the next section.

According to Equations (6.4) and (6.5), we have

$$D^\alpha x_j - D^\alpha \eta_j = \frac{1}{\rho_j}\tilde{\theta}_j^{\mathrm{T}} \phi_j(Z) + \frac{1}{\rho_j}\varepsilon_j^* + \tilde{d}_j, \tag{6.6}$$

where $\tilde{\theta}_j = \theta_j^* - \hat{\theta}_j$, $\tilde{d}_j$ is the disturbance estimation error, and $\tilde{d}_j = d_j - \hat{d}_j$.

Differentiating $\hat{d}_j$ and considering Equation (6.5) yields

$$D^\alpha \hat{d}_j = -\rho_j(D^\alpha x_j - D^\alpha \eta_j) - \frac{1}{k_j}D^\alpha \chi_j$$
$$= -\rho_j\left(\frac{1}{\rho_j}\tilde{\theta}_j^\top \phi_j(Z) + \frac{1}{\rho_j}\varepsilon_j^* + \tilde{d}_j\right) - \hat{\theta}_j^\top \phi_j(Z). \tag{6.7}$$

6.2.2 Controller Design and Stability Analysis

This section develops a nonlinear FODO-based adaptive neural tracking control scheme for the uncertain FONS (6.2). The tracking error is defined as

$$e(t) = x(t) - x_\mathrm{d}(t), \tag{6.8}$$

where $e(t) = [e_1(t), e_2(t), e_3(t)]^\top$ is the tracking error vector and $x_\mathrm{d}(t) = [x_{\mathrm{d}1}(t), x_{\mathrm{d}2}(t), x_{\mathrm{d}3}(t)]^\top$ is the desired signal vector.

According to Equation (6.8), the dynamic of the tracking error can be written as

$$D^\alpha e(t) = D^\alpha x(t) - D^\alpha x_\mathrm{d}(t)$$
$$= F(x) + \Delta F(x) + Gu + d(t) - D^\alpha x_\mathrm{d}(t). \tag{6.9}$$

On the basis of these discussions, the tracking controller and the adaptive update law will be designed to ensure that the error system (6.9) is ultimately bounded stable. We first consider the following Lyapunov function candidate:

$$V(t) = \frac{1}{2}e^\top(t)Pe(t) + \sum_{j=1}^{3}\frac{1}{2}\tilde{d}_j^2 + \sum_{j=1}^{3}\frac{1}{2h_j}\tilde{\theta}_j^\top\tilde{\theta}_j, \tag{6.10}$$

where $P \in R^{3\times 3}$ is a symmetric and positive definite constant matrix and $h_j > 0$ is a design constant.

In particular, we have

$$D^\alpha \tilde{\theta}_j = -D^\alpha \hat{\theta}_j. \tag{6.11}$$

Invoking Equations (6.10) and (6.11), Lemma 2.1, and Lemma 2.5, we obtain

$$D^\alpha V(t) \le e^\top(t)PD^\alpha e(t) + \sum_{j=1}^{3}\tilde{d}_j D^\alpha \tilde{d}_j - \sum_{j=1}^{3}\frac{1}{h_j}\tilde{\theta}_j^\top D^\alpha \hat{\theta}_j. \tag{6.12}$$

Substituting Equation (6.9) into Equation (6.12), we have

$$D^\alpha V(t) \le e^\top(t)P(F(x) + \Delta F(x) + Gu + d(t) - D^\alpha x_\mathrm{d}(t))$$
$$+ \sum_{j=1}^{3}\tilde{d}_j D^\alpha \tilde{d}_j - \sum_{j=1}^{3}\frac{1}{h_j}\tilde{\theta}_j^\top D^\alpha \hat{\theta}_j. \tag{6.13}$$

On the basis of Lemma 2.4, the neural network is used to approximate $\rho_j \Delta f_j(x)$ with $\rho_j < -2$, and we obtain

$$D^\alpha V(t) \le e^\top(t)P(F(x) + \frac{1}{\rho}\Theta + \frac{1}{\rho}\varepsilon^* + Gu + d(t) - D^\alpha x_\mathrm{d}(t))$$
$$+ \sum_{j=1}^{3}\tilde{d}_j D^\alpha \tilde{d}_j - \sum_{j=1}^{3}\frac{1}{h_j}\tilde{\theta}_j^\top D^\alpha \hat{\theta}_j, \tag{6.14}$$

where

$$\frac{1}{\rho}\varepsilon^* = \left[\frac{1}{\rho_1}\varepsilon_1^*, \frac{1}{\rho_2}\varepsilon_2^*, \frac{1}{\rho_3}\varepsilon_3^*\right]^{\mathrm{T}},$$

$$\frac{1}{\rho}\Theta = \left[\frac{1}{\rho_1}\theta_1^{*\mathrm{T}}\phi_1(Z), \frac{1}{\rho_2}\theta_2^{*\mathrm{T}}\phi_2(Z), \frac{1}{\rho_3}\theta_3^{*\mathrm{T}}\phi_3(Z)\right]^{\mathrm{T}},$$

and $Z = x$.

The adaptive neural controller is designed as

$$u = -kG^{-1}P^{-1}e(t) - G^{-1}F(x) - G^{-1}\hat{d}(t) - G^{-1}\frac{1}{\rho}\hat{\Theta} + G^{-1}D^\alpha x_{\mathrm{d}}(t) - G^{-1}P^{-1}\|P\|^2 e(t), \tag{6.15}$$

where $k > 0$ is a design constant and

$$\frac{1}{\rho}\hat{\Theta} = \left[\frac{1}{\rho_1}\hat{\theta}_1^{\mathrm{T}}\phi_1(Z), \frac{1}{\rho_2}\hat{\theta}_2^{\mathrm{T}}\phi_2(Z), \frac{1}{\rho_3}\hat{\theta}_3^{\mathrm{T}}\phi_3(Z)\right]^{\mathrm{T}}.$$

Substituting Equation (6.15) into Equation (6.14), we have

$$D^\alpha V(t) \le -ke^{\mathrm{T}}(t)e(t) + e^{\mathrm{T}}(t)P\frac{1}{\rho}\tilde{\Theta} + e^{\mathrm{T}}(t)P\tilde{d}(t) - \|P\|^2\|e(t)\|^2 + e^{\mathrm{T}}(t)P\frac{1}{\rho}\varepsilon^* + \sum_{j=1}^{3}\tilde{d}_j D^\alpha \tilde{d}_j - \sum_{j=1}^{3}\frac{1}{h_j}\tilde{\theta}_j^{\mathrm{T}} D^\alpha \hat{\theta}_j, \tag{6.16}$$

where

$$\frac{1}{\rho}\tilde{\Theta} = \left[\frac{1}{\rho_1}\tilde{\theta}_1^{\mathrm{T}}\phi_1(Z), \frac{1}{\rho_2}\tilde{\theta}_2^{\mathrm{T}}\phi_2(Z), \frac{1}{\rho_3}\tilde{\theta}_3^{\mathrm{T}}\phi_3(Z)\right]^{\mathrm{T}},$$

$\tilde{d}(t) = d(t) - \hat{d}(t)$, and $\tilde{\Theta} = \Theta - \hat{\Theta}$.

Furthermore, the adaptive law for $\hat{\theta}_j$ is chosen as

$$D^\alpha \hat{\theta}_j = \frac{h_j}{\rho_j}(e^{\mathrm{T}}(t)P)_j\phi_j(Z) - h_j m_j \hat{\theta}_j, \tag{6.17}$$

where $h_j > 0$ and $m_j > 0$ are design constants, and $(e^{\mathrm{T}}(t)P)_j$ denotes the jth element of $e^{\mathrm{T}}(t)P$, with $j = 1, 2, 3$.

According to Equation (6.17) and Lemma 2.5, we have

$$\begin{aligned} -\sum_{j=1}^{3}\frac{1}{h_j}\tilde{\theta}_j^{\mathrm{T}} D^\alpha \hat{\theta}_j &\le -\sum_{j=1}^{3}\frac{1}{h_j}h_j(e^{\mathrm{T}}(t)P)_j\frac{1}{\rho_j}\tilde{\theta}_j^{\mathrm{T}}\phi_j(Z) + \sum_{j=1}^{3}\left(\frac{m_j}{2}\left\|\theta_j^*\right\|^2 - \frac{m_j}{2}\left\|\tilde{\theta}_j\right\|^2\right) \\ &= \sum_{j=1}^{3}\left(\frac{m_j}{2}\left\|\theta_j^*\right\|^2 - \frac{m_j}{2}\left\|\tilde{\theta}_j\right\|^2\right) - e^{\mathrm{T}}(t)P\frac{1}{\rho}\tilde{\Theta}. \end{aligned} \tag{6.18}$$

From Equation (6.18), Equation (6.16) can be rewritten as

$$D^\alpha V(t) \le -ke^{\mathrm{T}}(t)e(t) + e^{\mathrm{T}}(t)P\frac{1}{\rho}\varepsilon^* + e^{\mathrm{T}}(t)P\tilde{d}(t) - \|P\|^2\|e(t)\|^2 + \sum_{j=1}^{3}\left(\frac{m_j}{2}\left\|\theta_j^*\right\|^2 - \frac{m_j}{2}\left\|\tilde{\theta}_j\right\|^2\right) + \sum_{j=1}^{3}\tilde{d}_j D^\alpha \tilde{d}_j, \tag{6.19}$$

with

$$e^{\top}(t)P\frac{1}{\rho}\varepsilon^* \leq \|e(t)\|\ \|P\|\ \varphi, \tag{6.20}$$

$$\begin{aligned} e^{\top}(t)P\tilde{d}(t) &\leq \|e(t)\|\ \|P\|\ \left\|\tilde{d}(t)\right\| \\ &\leq \frac{1}{2}\|e(t)\|^2\|P\|^2 + \frac{1}{2}\left\|\tilde{d}(t)\right\|^2, \end{aligned} \tag{6.21}$$

where $\left\|\frac{1}{\rho}\varepsilon^*\right\| \leq \varphi$ and $\varphi > 0$ is an unknown constant.

Substituting Equations (6.20) and (6.21) into Equation (6.19), we have

$$\begin{aligned} D^{\alpha}V(t) \leq{}& -ke^{\top}(t)e(t) + \|e(t)\|\ \|P\|\ \varphi - \frac{1}{2}\|e(t)\|^2\|P\|^2 + \frac{1}{2}\left\|\tilde{d}(t)\right\|^2 \\ &+ \sum_{j=1}^{3}\left(\frac{m_j}{2}\left\|\theta_j^*\right\|^2 - \frac{m_j}{2}\left\|\tilde{\theta}_j\right\|^2\right) + \sum_{j=1}^{3}\tilde{d}_j D^{\alpha}\tilde{d}_j, \end{aligned} \tag{6.22}$$

with

$$\|e(t)\|\ \|P\|\ \varphi - \frac{1}{2}\|P\|^2\|e(t)\|^2 \leq \frac{1}{2}\varphi^2. \tag{6.23}$$

Combining Equations (6.22) and (6.23), we obtain

$$\begin{aligned} D^{\alpha}V(t) \leq{}& -ke^{\top}(t)e(t) + \frac{1}{2}\varphi^2 + \frac{1}{2}\left\|\tilde{d}(t)\right\|^2 \\ &+ \sum_{j=1}^{3}\left(\frac{m_j}{2}\left\|\theta_j^*\right\|^2 - \frac{m_j}{2}\left\|\tilde{\theta}_j\right\|^2\right) + \sum_{j=1}^{3}\tilde{d}_j D^{\alpha}\tilde{d}_j. \end{aligned} \tag{6.24}$$

Invoking Assumption 6.2 and Equation (6.7), Equation (6.24) can be written as

$$\begin{aligned} D^{\alpha}V(t) \leq{}& -ke^{\top}(t)e(t) + \frac{1}{2}\varphi^2 + \frac{1}{2}\left\|\tilde{d}(t)\right\|^2 + \sum_{j=1}^{3}\left(\frac{m_j}{2}\left\|\theta_j^*\right\|^2 - \frac{m_j}{2}\left\|\tilde{\theta}_j\right\|^2\right) \\ &+ \sum_{j=1}^{3}\tilde{d}_j(D^{\alpha}d_j - D^{\alpha}\hat{d}_j) \\ \leq{}& -ke^{\top}(t)e(t) + \frac{1}{2}\varphi^2 + \frac{1}{2}\left\|\tilde{d}(t)\right\|^2 + \sum_{j=1}^{3}\left(\frac{m_j}{2}\left\|\theta_j^*\right\|^2 - \frac{m_j}{2}\left\|\tilde{\theta}_j\right\|^2\right) \\ &+ \sum_{j=1}^{3}\left(\frac{1}{2}\tilde{d}_j^2 + \frac{1}{2}b_{3j}^2 - \tilde{d}_j\left(-\rho_j\left(\frac{1}{\rho_j}\tilde{\theta}_j^{\top}\phi_j(Z) + \frac{1}{\rho_j}\varepsilon_j^* + \tilde{d}_j\right) - \hat{\theta}_j^{\top}\phi_j(Z)\right)\right) \\ ={}& -ke^{\top}(t)e(t) + \frac{1}{2}\varphi^2 + \frac{1}{2}\left\|\tilde{d}(t)\right\|^2 + \sum_{j=1}^{3}\left(\frac{m_j}{2}\left\|\theta_j^*\right\|^2 - \frac{m_j}{2}\left\|\tilde{\theta}_j\right\|^2\right) \\ &+ \sum_{j=1}^{3}\left(\left(\frac{1}{2} + \rho_j\right)\tilde{d}_j^2 + \tilde{d}_j\varepsilon_j^* + \tilde{d}_j\theta_j^{*\top}\phi_j(Z) + \frac{1}{2}b_{3j}^2\right). \end{aligned} \tag{6.25}$$

According to Equation (6.25), we have

$$D^{\alpha}V(t) \leq -ke^{\top}(t)e(t) + \frac{1}{2}\varphi^2 + \frac{1}{2}\left\|\tilde{d}(t)\right\|^2 + \sum_{j=1}^{3}\left(\frac{m_j}{2}\left\|\theta_j^*\right\|^2 - \frac{m_j}{2}\left\|\tilde{\theta}_j\right\|^2\right)$$

$$
\begin{aligned}
&+\sum_{j=1}^{3}\left(\left(\frac{3}{2}+\rho_j\right)\tilde{d}_j^2+\frac{1}{2}\varepsilon_j^{*2}+\frac{1}{2}\left(\theta_j^{*\mathrm{T}}\phi_j(Z)\right)^2+\frac{1}{2}b_{3j}^2\right)\\
&\leq -ke^{\mathrm{T}}(t)e(t)+\frac{1}{2}\varphi^2+\frac{1}{2}\left\|\tilde{d}(t)\right\|^2+\sum_{j=1}^{3}\left(\frac{m_j}{2}\left\|\theta_j^*\right\|^2-\frac{m_j}{2}\left\|\tilde{\theta}_j\right\|^2\right)\\
&+\sum_{j=1}^{3}\left(\left(\frac{3}{2}+\rho_j\right)\tilde{d}_j^2+\frac{1}{2}\bar{\varepsilon}_j^2+\frac{1}{2}(\bar{\theta}_j\bar{\phi})^2+\frac{1}{2}b_{3j}^2\right),
\end{aligned}
\tag{6.26}
$$

where $\left\|\theta_j^*\right\| \leq \bar{\theta}_j$ and $|\phi_j(Z)| \leq \bar{\phi}$.

Furthermore, Equation (6.26) can be rewritten as

$$
\begin{aligned}
D^{\alpha}V(t) &\leq -ke^{\mathrm{T}}(t)e(t)+\sum_{j=1}^{3}(2+\rho_j)\tilde{d}_j^2-\sum_{j=1}^{3}\frac{m_j}{2}\left\|\tilde{\theta}_j\right\|^2+\frac{1}{2}\varphi^2\\
&+\sum_{j=1}^{3}\left(\frac{m_j}{2}\bar{\theta}_j^2\right)+\sum_{j=1}^{3}\left(\frac{1}{2}\bar{\varepsilon}_j^2+\frac{1}{2}(\bar{\theta}_j\bar{\phi})^2+\frac{1}{2}b_{3j}^2\right).
\end{aligned}
\tag{6.27}
$$

From Equation (6.27), we obtain

$$
\begin{aligned}
D^{\alpha}V(t) &\leq -\frac{2k}{\lambda_{\max}(P)}\frac{1}{2}e^{\mathrm{T}}(t)Pe(t)+\sum_{j=1}^{3}(2+\rho_j)\tilde{d}_j^2-\sum_{j=1}^{3}\frac{m_j}{2}\left\|\tilde{\theta}_j\right\|^2\\
&+\frac{1}{2}\varphi^2+\sum_{j=1}^{3}\left(\frac{m_j}{2}\bar{\theta}_j^2\right)+\sum_{j=1}^{3}\left(\frac{1}{2}\bar{\varepsilon}_j^2+\frac{1}{2}(\bar{\theta}_j\bar{\phi})^2+\frac{1}{2}b_{3j}^2\right)\\
&\leq -MV(t)+\overline{M},
\end{aligned}
\tag{6.28}
$$

where $j = 1, 2, 3$,

$$
M=\min\left(\frac{2k}{\lambda_{\max}(P)}, h_jm_j, -2(2+\rho_j)\right),
$$

and

$$
\overline{M}=\frac{1}{2}\varphi^2+\sum_{j=1}^{3}\left(\frac{m_j}{2}\bar{\theta}_j^2\right)+\sum_{j=1}^{3}\left(\frac{1}{2}\bar{\varepsilon}_j^2+\frac{1}{2}(\bar{\theta}_j\bar{\phi})^2+\frac{1}{2}b_{3j}^2\right).
$$

This design procedure can be summarized in the following theorem, which contains the results of the FODO-based adaptive neural control for the uncertain FONS (6.2) with unknown time-varying external disturbance.

Theorem 6.1 *Consider the uncertain FONS (6.4) in the presence of the unknown time-varying external disturbance $d(t)$, which is estimated via the fractional-order nonlinear disturbance observer (6.5). By using the nonlinear FODO (6.5), the disturbance $d(t)$ can be approximated well via appropriately adjusting design parameters. The updated law (6.17) of the neural network parameter is chosen and the robust tracking control scheme is designed according to Equation (6.15). All closed-loop system signals are uniformly stable under the proposed disturbance-observer-based adaptive* *neural tracking control (6.15). Furthermore, the tracking error $e(t)$ and the disturbance approximation error $\tilde{d}(t)$ satisfy*

$$\|e(t)\| \le \sqrt{\frac{4\overline{M}}{M\lambda_{\min}(P)}}$$

and

$$\left\|\tilde{d}(t)\right\| \le \sqrt{\frac{4\overline{M}}{M}},$$

respectively.

Proof 6.1 *By using Lemma 2.3, there exists a constant $t_1 > 0$, such that for all $t \in (t_1, \infty)$, we have*

$$|V(t)| \le \frac{2\overline{M}}{M}. \tag{6.29}$$

Combining Equations (6.10) and (6.29), we have

$$\|e(t)\| \le \sqrt{\frac{4\overline{M}}{M\lambda_{\min}(P)}}, \tag{6.30}$$

$$\left\|\tilde{d}(t)\right\| \le \sqrt{\frac{4\overline{M}}{M}}. \tag{6.31}$$

Equations (6.30) and (6.31) illustrate that the tracking error $e(t)$ is ultimately bounded and that the disturbance estimation error $\tilde{d}(t)$ is also ultimately bounded. Meanwhile, it is also shown that all the signals in the closed-loop system remain bounded. This concludes the proof. □

6.3 Simulation Example

In this section, simulation results are presented to illustrate the effectiveness of the proposed robust adaptive neural control scheme for the uncertain FONS with external disturbances. From Equation (6.1), the fractional-order nonlinear model of a rotational mechanical system with a centrifugal governor in the presence of unknown uncertainties, external disturbances, and control inputs can be described as follows:

$$\begin{aligned}
D^\alpha x_1 &= x_2 + \Delta f_1(x) + d_1(t) + G_1 u(t),\\
D^\alpha x_2 &= a_0(x_3 + a_1)^2 \sin(x_1 - a_2)\cos(x_1 - a_2) - \sin(x_1 - a_2)\\
&\quad - a_3 x_2 + \Delta f_2(x) + d_2(t) + G_2 u(t),\\
D^\alpha x_3 &= a_4 \cos(x_1 - a_2) - a_5 + \Delta f_3(x) + d_3(t) + G_3 u(t),
\end{aligned} \tag{6.32}$$

where $G_j (j = 1, 2, 3)$ is the jth line of G.

In this simulation, the fractional order is chosen as $\alpha = 0.92$, the initial conditions of system (6.32) are chosen as $x(0) = [0.3, 0.4, 0.2]^{\mathrm{T}}$, and the system parameters are selected as $a_0 = 0.25$, $a_1 = 2.6892$, $a_2 = 0.6812$, $a_3 = 0.7$, $a_4 = 2.5$, and $a_5 = 1.942$. The control parameters are chosen as $k = 100$, $k_j = h_j = m_j = 1$, and $\rho_j = -100$, with $j = 1, 2, 3$. The matrices are designed as $G = \mathrm{diag}[1, 1, 1]$ and $P = \mathrm{diag}[1, 1, 1]$. The uncertainty terms are assumed as $\Delta f_1(x) = 3\sin(x_1)$, $\Delta f_2(x) = 4\sin(x_2)$ and $\Delta f_3(x) = 5\sin(x_3)$. The desired trajectories are chosen as $x_{1\mathrm{d}} = \sin(t)$, $x_{2\mathrm{d}} = 1.5\sin(t)$, and $x_{3\mathrm{d}} = 2\sin(t)$. The external disturbances are assumed as $d_1(t) = 100\sin(t)$, $d_2(t) = 110\sin(t)$, and $d_3(t) = 120\sin(t)$. On the basis of the result of Ishtev [225], we have $\lambda_1 D^{\alpha}\sin(\lambda_2 t) = -\lambda_1\frac{1}{2}\mathrm{i}(i\lambda_2)^m t^{m-\alpha}(E_{1,m-\alpha+1}(\mathrm{i}\lambda t) - (-1)^n E_{1,m-\alpha+1}(-\mathrm{i}\lambda t))$, where i denotes the square root of minus one and λ_1 and λ_2 are arbitrary numbers. In this simulation, the parameter is assumed as $m = 1$ and the fractional order is chosen as $\alpha = 0.92$. Thus, $\lambda_1\lambda_2^{\alpha}\sin(\lambda_2 t + (\pi\alpha/2))$ can be applied to approximate $\lambda_1 D^{\alpha}\sin(\lambda_2 t)$. The comparison result is shown in Figure 6.1 for the case of $\lambda_1 = 1$ and $\lambda_2 = 2$. According to Figure 6.1, Assumption 6.1 and Assumption 6.2 are satisfied.

The simulation results of the uncertain FONS (6.32) with external disturbances are shown in Figure 6.2, Figure 6.3, Figure 6.4, Figure 6.5, and Figure 6.6 under the proposed adaptive neural control scheme. The tracking results of output signals and desired signals are given in Figure 6.2a–c. It is shown that the tracking performance is satisfactory. Figure 6.3 shows that the tracking errors $e_1(t)$, $e_2(t)$, and $e_3(t)$ are bounded. Furthermore, the estimate performance of the proposed nonlinear FODO (6.5) is presented in Figure 6.4 and Figure 6.5. It is evident from Figure 6.4 and Figure 6.5 that the disturbance observer is effective and feasible. The control input signals, which are bounded, are shown in Figure 6.6. It is concluded from these simulation results that the proposed adaptive neural control technique is effective for uncertain fractional-order nonlinear systems using FODOs.

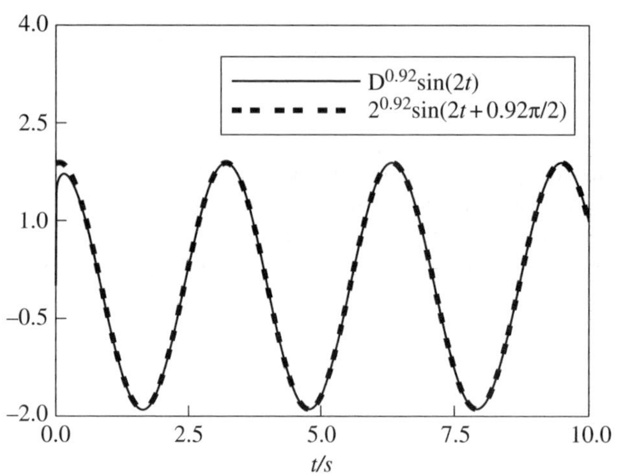

Figure 6.1 Comparison result of $2^{0.92}\sin(2t + (0.92\pi/2))$ and $D^{0.92}\sin(2t)$.

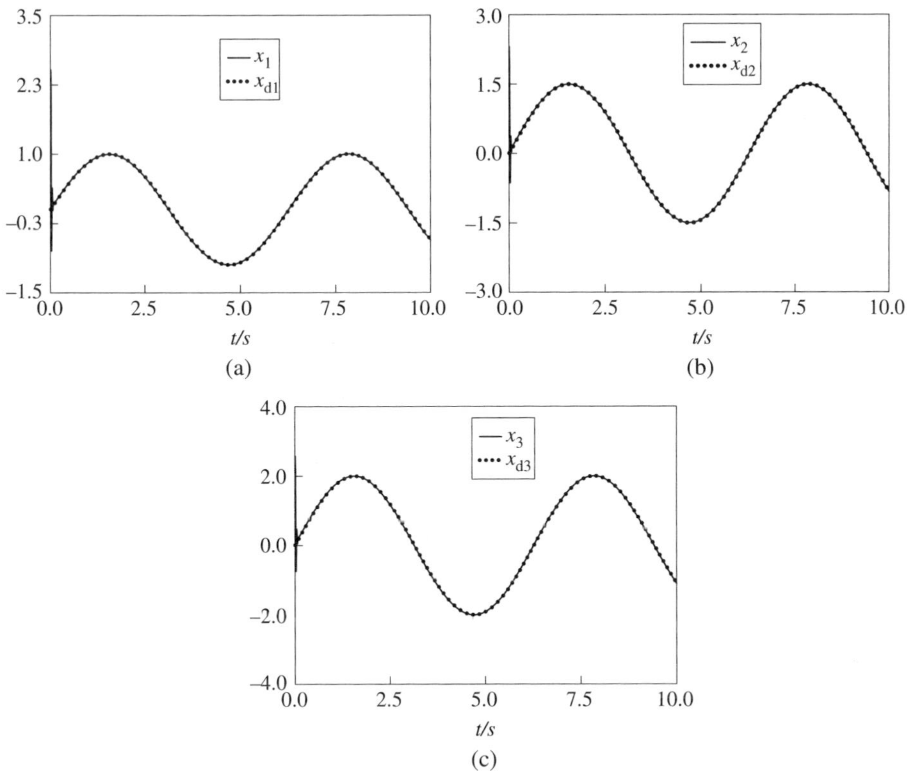

Figure 6.2 Output y of the system (6.32) follows the desired trajectory x_d: (a) x_1 and x_{d1}; (b) x_2 and x_{d2}; (c) x_3 and x_{d3}.

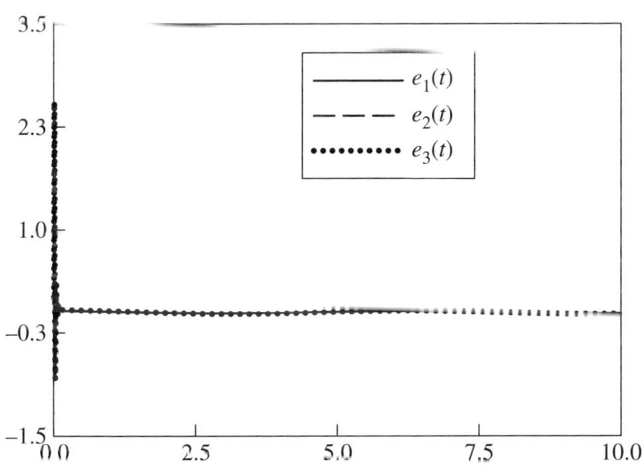

Figure 6.3 Tracking errors $e_1(t)$, $e_2(t)$, and $e_3(t)$ for desired trajectories x_{d1}, x_{d2} and x_{d3}.

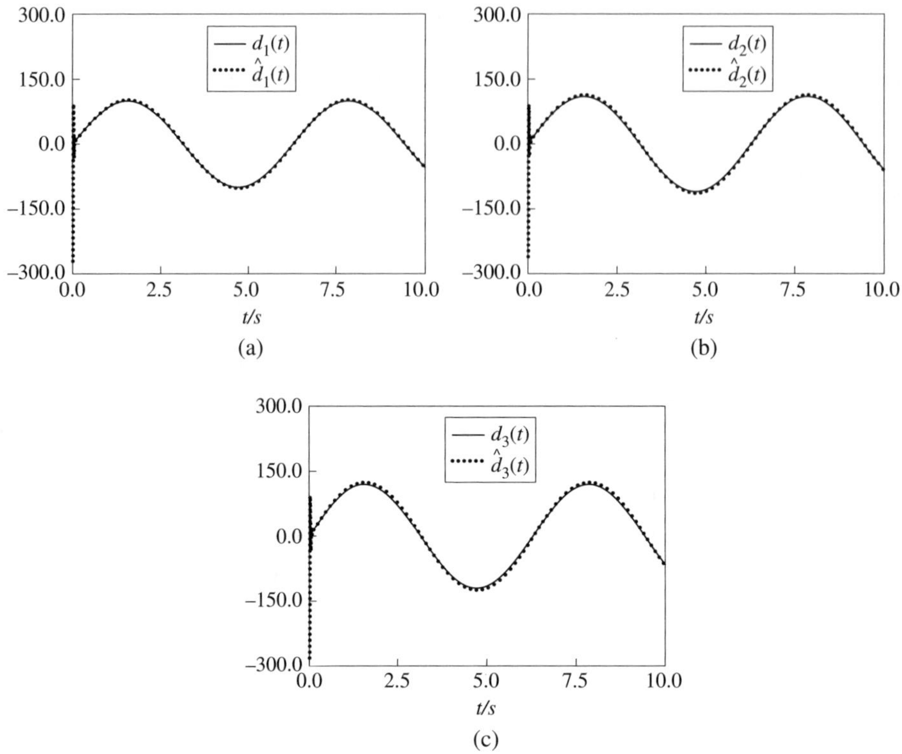

Figure 6.4 Disturbance $d(t)$ and approximation output of $\hat{d}(t)$: (a) $d_1(t)$ and $\hat{d}_1(t)$; (b) $d_2(t)$ and $\hat{d}_2(t)$; (c) $d_3(t)$ and $\hat{d}_3(t)$.

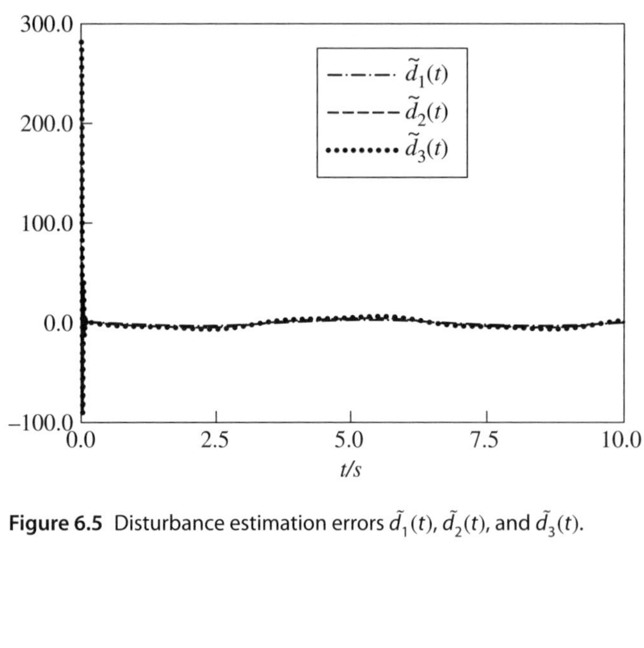

Figure 6.5 Disturbance estimation errors $\tilde{d}_1(t)$, $\tilde{d}_2(t)$, and $\tilde{d}_3(t)$.

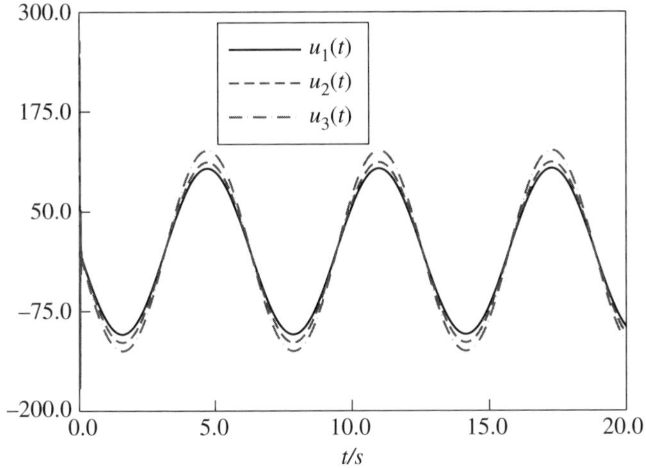

Figure 6.6 Control Inputs $u_1(t)$, $u_2(t)$, and $u_3(t)$ of the system (6.32).

6.4 Conclusion

An adaptive neural tracking control has been proposed for a class of uncertain fractional-order nonlinear systems in this chapter. To improve the ability of disturbance attenuation and the control performance of the FONS subjected to external unknown bounded disturbances and model uncertainties, the fractional-order nonlinear disturbance observer together with neural network approximation has been employed to estimate the disturbance. By using the designed nonlinear FODO and the neural network, the FODO-based adaptive neural network control has been developed for uncertain fractional-order nonlinear systems with external disturbances. The stability of the closed-loop system has been proved based on the fractional-order Lyapunov method. Finally, simulation results have been presented to illustrate the effectiveness of the proposed adaptive neural control scheme.

7

Adaptive Neural Tracking Control for Uncertain Fractional-Order Chaotic Systems Subject to Input Saturation and Disturbance

7.1 Problem Statement

According to the Caputo definition of the fractional derivative (2.17), the following uncertain FOCS in the presence of input saturation and unknown external disturbance is described as follows:

$$\begin{aligned} D^{\alpha}x_i(t) &= x_{i+1}(t), \\ &\vdots \\ D^{\alpha}x_n(t) &= f(x(t)) + \Delta f(x(t)) + \text{sat}(u(t)) + d(t), \\ y &= x_1, \end{aligned} \tag{7.1}$$

where α is the fractional order with $0 < \alpha < 1$, $D^{\alpha}(\cdot)$ is the fractional derivative, $x_i(t) \in R$ $(1 \leq i \leq n)$ are the state variables of the chaotic system, which are measurable, $y \in R$ is the system output, $u(t) \in R$ is the desired control input, $d(t)$ is the unknown time-varying disturbance, $f(x(t))$ is the known nonlinear function with $x(t) = [x_1(t), x_2(t), \cdots, x_n(t)]^{\top} \in R^n$, $\Delta f(x(t))$ is the unknown nonlinear function, and $\text{sat}(u(t))$ is the input saturation function defined as follows:

$$\text{sat}(u(t)) = \begin{cases} u_{\max}\text{sign}(u(t)), & |u(t)| \geq u_{\max}, \\ u(t), & |u(t)| < u_{\max}, \end{cases} \tag{7.2}$$

where $u_{\max}$ is a known bound of $\text{sat}(u(t))$ and $\text{sign}(\cdot)$ is the standard sign function.

The aim of this chapter is to design a SMFODO to approximate the unknown external disturbance, and to propose an adaptive neural control scheme based on the designed SMFODO to control the output signal of the uncertain FOCS (7.1), which could follow a given desired trajectory $x_{1d}(t)$ in the presence of input saturation and unknown external disturbance. Meanwhile, the proposed control scheme will be rigorously shown to guarantee that all the signals in the closed-loop system remain bounded.

To facilitate the design of the neural tracking control for the uncertain FOCS (7.1) subjected to input saturation and external disturbance, the following assumptions are necessary in this study.

Assumption 7.1 *For the fractional-order chaotic system (7.1) in the presence of input saturation and unknown external disturbance, the stable controller exists.*

Robust Adaptive Control for Fractional-Order Systems with Disturbance and Saturation, First Edition.
Mou Chen, Shuyi Shao, and Peng Shi.

Assumption 7.2 *For the given desired trajectory $x_{1d}(t)$, the Caputo fractional derivative of $x_{1d}(t)$ exists and is bounded.*

Assumption 7.3 *For the external disturbance $d(t)$, the external disturbance $d(t)$ is bounded with $|d(t)| \leq \Theta$ and the Caputo fractional derivative of $d(t)$ is bounded with $|D^\alpha d(t)| \leq \delta$, where $\Theta > 0$ and $\delta > 0$ are unknown positive constants.*

7.2 Adaptive Neural Control Design Based on Fractional-Order Disturbance Observer

In this section, an adaptive neural control will be proposed for an uncertain FOCS with unknown external disturbance and input saturation based on the designed SMFODO. To handle the input saturation, the following auxiliary system with the same order as the FOCS (7.1) is constructed to counteract the effect of the input saturation, as follows:

$$\begin{aligned} D^\alpha \varphi_i(t) &= -c_i\varphi_i(t) + \varphi_{i+1}(t), \\ &\vdots \\ D^\alpha \varphi_n(t) &= -c_n\varphi_n(t) + \Delta u(t), \end{aligned} \tag{7.3}$$

where $\varphi_i(t)$ $(i = 1, 2, \dots, n)$ are the state variables of the auxiliary system, $c_i > 0$ are the design constants, and $\Delta u(t) = \text{sat}(u(t)) - u(t)$.

Remark 7.1 *Since the control moment or control force is limited for the practical actuator, the practical system can be controlled via finite control input $u(t)$ in industry. On the basis of the system controllability, we can assume that $\Delta u(t)$ is upper bounded and $\Delta u(t)$ satisfies $|\Delta u(t)| \leq \overline{u}$, where $\overline{u} > 0$ is a positive constant.*

Using the state variables of the auxiliary system (7.3), the adaptive neural control scheme is designed using the backstepping technique. The detailed design process is given as follows.

Step 1 In the first step, we define the error variable as $e_1(t) = x_1(t) - x_{1d}(t) - \varphi_1(t)$ and $e_2(t) = x_2(t) - \beta_1(t) - \varphi_2(t)$, where $\beta_1(t)$ is the virtual control law and will be designed.

Considering Equations (7.1) and (7.3), we obtain

$$\begin{aligned} D^\alpha e_1(t) &= D^\alpha x_1(t) - D^\alpha x_{1d}(t) - D^\alpha \varphi_1(t) \\ &= x_2(t) - D^\alpha x_{1d}(t) + c_1\varphi_1(t) - \varphi_2(t), \end{aligned} \tag{7.4}$$

Invoking the definition of $e_2(t)$, Equation (7.4) can be written as

$$D^\alpha e_1(t) = e_2(t) + \beta_1(t) - D^\alpha x_{1d}(t) + c_1\varphi_1(t). \tag{7.5}$$

Furthermore, the virtual control law $\beta_1(t)$ in the first step is designed as

$$\beta_1(t) = D^\alpha x_{1d}(t) - c_1\varphi_1(t) - l_1 e_1(t), \tag{7.6}$$

where $l_1 > 0$ is a design constant.

Substituting Equation (7.6) into Equation (7.5) yields

$$D^\alpha e_1(t) = -l_1 e_1(t) + e_2(t). \tag{7.7}$$

From Equation (7.7), we have

$$
\begin{aligned}
e_1(t)D^\alpha e_1(t) &= -l_1 e_1^2(t) + e_1(t)e_2(t) \\
&\leq -(l_1 - 0.5)e_1^2(t) + 0.5e_2^2(t).
\end{aligned} \tag{7.8}
$$

According to Equation (7.3), we obtain

$$
\begin{aligned}
\varphi_1(t)D^\alpha \varphi_1(t) &= -c_1\varphi_1^2(t) + \varphi_1(t)\varphi_2(t) \\
&\leq -(c_1 - 0.5)\varphi_1^2(t) + 0.5\varphi_2^2(t).
\end{aligned} \tag{7.9}
$$

Considering the signals $e_1(t)$ and $\varphi_1(t)$, the Lyapunov function candidate is chosen as

$$
V_1(t) = \frac{1}{2}e_1^2(t) + \frac{1}{2}\varphi_1^2(t). \tag{7.10}
$$

On the basis of Lemma 2.1, the Caputo derivative of $V_1(t)$ can be described as

$$
D^\alpha V_1(t) \leq e_1(t)D^\alpha e_1(t) + \varphi_1(t)D^\alpha\varphi_1(t). \tag{7.11}
$$

Invoking Equations (7.8), (7.9), and (7.11), we have

$$
D^\alpha V_1(t) \leq -(l_1 - 0.5)e_1^2(t) + 0.5e_2^2(t) - (c_1 - 0.5)\varphi_1^2(t) + 0.5\varphi_2^2(t), \tag{7.12}
$$

where $0.5e_2^2(t)$ and $0.5\varphi_2^2(t)$ will be handled in the next step.

***Step i* (2 ≤ *i* ≤ *n* − 1)** Define the error variable as $e_i(t) = x_i(t) - \beta_{i-1}(t) - \varphi_i(t)$ and $e_{i+1}(t) = x_{i+1}(t) - \beta_i(t) - \varphi_{i+1}(t)$, where $\beta_{i-1}(t)$ and $\beta_i(t)$ are the virtual control laws designed in the $i-1$th step and the ith step, respectively.

Considering Equations (7.1) and (7.3), we have

$$
\begin{aligned}
D^\alpha e_i(t) &= D^\alpha x_i(t) - D^\alpha \beta_{i-1}(t) - D^\alpha \varphi_i(t) \\
&= x_{i+1}(t) - D^\alpha \beta_{i-1}(t) + c_i\varphi_i(t) - \varphi_{i+1}(t).
\end{aligned} \tag{7.13}
$$

Invoking the definition of $e_{i+1}(t)$, Equation (7.13) can be written as

$$
D^\alpha e_i(t) = e_{i+1}(t) + \beta_i(t) - D^\alpha \beta_{i-1}(t) + c_i\varphi_i(t). \tag{7.14}
$$

To eliminate the tedious analytic computations of fractional derivatives of the virtual control law $\beta_{i-1}(t)$, the differentiator is employed to obtain the fractional derivatives of the virtual control law $\beta_{i-1}(t)$. According to Lemma 2.6, we have

$$
\begin{aligned}
D^\alpha w_{1,i-1}(t) &= w_{2,i-1}(t), \\
D^\alpha w_{2,i-1}(t) &= -R_{i-1}\tanh\Bigg(w_{1,i-1}(t) - \beta_{i-1}(t) \\
&\qquad + \frac{w_{2,i-1}(t)|w_{2,i-1}(t)|}{R_{i-1}}\left[1 - \frac{\Gamma^2(\alpha+1)}{\Gamma(2\alpha+1)}\right]\Bigg),
\end{aligned} \tag{7.15}
$$

where R_{i-1} is a design positive constant.

Invoking Equation (7.15) and Lemma 2.6, we obtain

$$
D^\alpha \beta_{i-1}(t) = \xi_{2,i-1}(t) + w_{2,i-1}(t), \tag{7.16}
$$

where $\xi_{2,i-1}(t)$ is the estimation error of the differentiator. From Lemma 2.6, we know $|\xi_{2,i-1}(t)| \leq \bar{\xi}_{2,i-1}$, with $\bar{\xi}_{2,i-1} > 0$.

Considering Equations (7.14) and (7.16) yields

$$D^\alpha e_i(t) = e_{i+1}(t) + \beta_i(t) - \xi_{2,i-1}(t) - w_{2,i-1}(t) + c_i\varphi_i(t). \tag{7.17}$$

Furthermore, the virtual control law $\beta_i(t)$ is designed as

$$\beta_i(t) = w_{2,i-1}(t) - c_i\varphi_i(t) - l_i e_i(t), \tag{7.18}$$

where $l_i > 0$ is a design constant.

Substituting Equation (7.18) into Equation (7.17) yields

$$D^\alpha e_i(t) = -l_i e_i(t) + e_{i+1}(t) - \xi_{2,i-1}(t). \tag{7.19}$$

From Equation (7.19) and Lemma 2.6, we have

$$\begin{aligned} e_i(t)D^\alpha e_i(t) &= -l_i e_i^2(t) + e_i(t)e_{i+1}(t) - e_i(t)\xi_{2,i-1}(t) \\ &\leq -(l_i - 1)e_i^2(t) + 0.5e_{i+1}^2(t) + 0.5\bar{\xi}_{2,i-1}^2. \end{aligned} \tag{7.20}$$

According to Equation (7.3), we obtain

$$\begin{aligned} \varphi_i(t)D^\alpha \varphi_i(t) &= -c_i\varphi_i^2(t) + \varphi_i(t)\varphi_{i+1}(t) \\ &\leq -(c_i - 0.5)\varphi_i^2(t) + 0.5\varphi_{i+1}^2(t). \end{aligned} \tag{7.21}$$

Considering the signals $e_i(t)$ and $\varphi_i(t)$, the Lyapunov function candidate is chosen as

$$V_i(t) = \frac{1}{2}e_i^2(t) + \frac{1}{2}\varphi_i^2(t). \tag{7.22}$$

On the basis of Lemma 2.1, the Caputo derivative of $V_i(t)$ can be described as

$$D^\alpha V_i(t) \leq e_i(t)D^\alpha e_i(t) + \varphi_i(t)D^\alpha \varphi_i(t). \tag{7.23}$$

Invoking Equations (7.20), (7.21), and (7.23), we have

$$D^\alpha V_i(t) \leq -(l_i - 1)e_i^2(t) + 0.5e_{i+1}^2(t) + 0.5\bar{\xi}_{2,i-1}^2 - (c_i - 0.5)\varphi_i^2(t) + 0.5\varphi_{i+1}^2(t), \tag{7.24}$$

where $0.5e_{i+1}^2(t)$ and $0.5\varphi_{i+1}^2(t)$ will be handled in the next step.

Step n Considering the system (7.1), the Caputo derivative of $e_n(t) = x_n(t) - \beta_{n-1}(t) - \varphi_n(t)$ can be written as

$$D^\alpha e_n(t) = f(x(t)) + \Delta f(x(t)) + \text{sat}(u(t)) + d(t) - D^\alpha \beta_{n-1}(t) - D^\alpha \varphi_n(t). \tag{7.25}$$

On the basis of Lemma 2.4, a neural network is used to approximate the unknown nonlinear function $\Delta f(x(t))$; we obtain

$$D^\alpha e_n(t) = f(x(t)) + \theta^{*\mathrm{T}}\phi(Z) + \varepsilon^* + \text{sat}(u(t)) + d(t) - D^\alpha \beta_{n-1}(t) - D^\alpha \varphi_n(t). \tag{7.26}$$

Invoking Equation (7.3), we have

$$D^\alpha e_n(t) = f(x(t)) + \theta^{*\mathrm{T}}\phi(Z) + \varepsilon^* + u(t) + d(t) - D^\alpha \beta_{n-1}(t) + c_n\varphi_n(t). \tag{7.27}$$

To eliminate tedious analytic computations of fractional derivatives of the virtual control law $\beta_{n-1}(t)$, the differentiator is employed to obtain the fractional derivatives of the virtual control law $\beta_{n-1}(t)$. According to Lemma 2.6, we obtain

$$D^\alpha w_{1,n-1}(t) = w_{2,n-1}(t),$$

$$D^\alpha w_{2,n-1}(t) = -R_{n-1} \tanh\left(w_{1,n-1}(t) - \beta_{n-1}(t) + \frac{w_{2,n-1}(t)|w_{2,n-1}(t)|}{R_{n-1}} \left[1 - \frac{\Gamma^2(\alpha+1)}{\Gamma(2\alpha+1)}\right]\right), \tag{7.28}$$

where R_{n-1} is the design positive constant.

Considering Equation (7.28) and Lemma 2.6, we obtain

$$D^\alpha \beta_{n-1}(t) = \xi_{2,n-1}(t) + w_{2,n-1}(t), \tag{7.29}$$

where $\xi_{2,n-1}(t)$ is the estimation error of the differentiator. From Lemma 2.6, we know $|\xi_{2,n-1}(t)| \leq \bar{\xi}_{2,n-1}$ with $\bar{\xi}_{2,n-1} > 0$.

Combining Equations (7.27) and (7.29) yields

$$D^\alpha e_n(t) = f(x(t)) + \theta^{*\top}\phi(Z) + \varepsilon^* + u(t) + d(t) - w_{2,n-1}(t) - \xi_{2,n-1}(t) + c_n\varphi_n(t). \tag{7.30}$$

To compensate for the effect of the external disturbance $d(t)$, the SMFODO is designed to estimate it. To develop the SMFODO, the following auxiliary variable is defined as

$$z(t) = \sigma(t) - e_n(t) \tag{7.31}$$

and the intermedial variable $\sigma(t)$ is given by

$$D^\alpha \sigma(t) = -k_1 z(t) + f(x(t)) - \hat{\Theta}\text{sign}(z(t)) + \hat{\theta}^\top\phi(Z) + u(t) - w_{2,n-1}(t) + c_n\varphi_n(t), \tag{7.32}$$

where $k_1 > 0$ is a design constant, $\hat{\Theta}$ is the estimate of the unknown constant Θ, and $\hat{\theta}$ is the estimate of θ^*.

According to Equations (7.30) and (7.32), the Caputo derivative of Equation (7.31) yields

$$\begin{aligned} D^\alpha z(t) &= D^\alpha \sigma(t) - D^\alpha e_n(t) \\ &= -k_1 z(t) - \hat{\Theta}\text{sign}(z(t)) + \tilde{\theta}^\top\phi(Z) - \varepsilon^* - d(t) + \xi_{2,n-1}(t), \end{aligned} \tag{7.33}$$

where $\tilde{\theta} = \hat{\theta} - \theta^*$.

Considering Assumption 7.3 and Equation (7.33), we have

$$\begin{aligned} z(t)D^\alpha z(t) &= -k_1 z^2(t) - z(t)\hat{\Theta}\text{sign}(z(t)) + z(t)\tilde{\theta}^\top\phi(Z) \\ &\quad - z(t)\varepsilon^* - z(t)d(t) + z(t)\xi_{2,n-1}(t) \\ &\leq -(k_1 - 0.5)z^2(t) + |z(t)|\tilde{\Theta} + z(t)\tilde{\theta}^\top\phi(Z) + 0.5(\bar{\varepsilon} + \bar{\xi}_{2,n-1})^2, \end{aligned} \tag{7.34}$$

where $\tilde{\Theta} = \Theta - \hat{\Theta}$ and $|\varepsilon^*| \leq \bar{\varepsilon}$.

The SMFODO is designed as

$$\begin{aligned} \hat{d}(t) &= k_2(\chi(t) - z(t)), \\ D^\alpha \chi(t) &= -k_1 z(t) - \hat{\Theta}\text{sign}(z(t)) - \hat{d}(t) + \hat{\theta}^\top\phi(Z), \end{aligned} \tag{7.35}$$

where $\hat{d}(t)$ is the estimate of the disturbance $d(t)$, and $k_2 > 0$ is a design constant of the SMFODO.

Invoking Equations (7.33) and (7.35), we obtain

$$\begin{aligned} D^\alpha \hat{d}(t) &= k_2(D^\alpha \chi(t) - D^\alpha z(t)) \\ &= k_2(\theta^{*\top}\phi(Z) + \varepsilon^* + \tilde{d}(t) - \xi_{2,n-1}(t)), \end{aligned} \tag{7.36}$$

where $\tilde{d}(t) = d(t) - \hat{d}(t)$.

Consideration of Assumption 7.3 and the definition of $\tilde{d}(t)$ yields

$$\begin{aligned} \frac{1}{k_2}\tilde{d}(t)D^\alpha \tilde{d}(t) &= \frac{1}{k_2}\tilde{d}(t)(D^\alpha d(t) - D^\alpha \hat{d}(t)) \\ &= \frac{1}{k_2}\tilde{d}(t)D^\alpha d(t) - \tilde{d}(t)\tilde{d}(t) - \tilde{d}(t)(\theta^{*\top}\phi(Z) + \varepsilon^* - \xi_{2,n-1}(t)) \\ &\le -\left(0.5 - 0.5\frac{1}{k_2^2}\right)\tilde{d}^2(t) + 0.5\delta^2 + 0.5(\overline{\theta}\,\overline{\phi} + \overline{\varepsilon} + \overline{\xi}_{2,n-1})^2, \end{aligned} \tag{7.37}$$

where $\|\theta^*\| \le \overline{\theta}$ and $\|\phi(Z)\| \le \overline{\phi}$.

Define $V_{\mathrm{d}}(t) = \frac{1}{2k_2}\tilde{d}^2(t)$. Thus, we have

$$D^\alpha V_{\mathrm{d}}(t) \le -\left(k_2 - \frac{1}{k_2}\right)V_{\mathrm{d}}(t) + 0.5\delta^2 + 0.5(\overline{\theta}\,\overline{\phi} + \overline{\varepsilon} + \overline{\xi}_{2,n-1})^2. \tag{7.38}$$

From Lemma 2.3 and Equation (7.38), we know that the disturbance estimation error $\tilde{d}(t)$ is bounded when $k_2 - (1/k_2) > 0$. Therefore, we can assume $|\tilde{d}(t)| \le \varsigma$ where $\varsigma > 0$ is an unknown constant.

Furthermore, the controller $u(t)$ is designed as

$$u(t) = -k_3 e_n(t) - f(x(t)) - \hat{\theta}^\top\phi(Z) - \hat{d}(t) - e_n(t)\hat{\varsigma} + w_{2,n-1}(t) - c_n\varphi_n(t), \tag{7.39}$$

where $k_3 > 0$ is a design constant and $\hat{\varsigma}$ is the estimate of the unknown constant ς.

Substituting Equation (7.39) into Equation (7.30) yields

$$D^\alpha e_n(t) = -k_3 e_n(t) - \tilde{\theta}^\top\phi(Z) + \tilde{d}(t) - e_n(t)\hat{\varsigma} - \xi_{2,n-1}(t) + \varepsilon^*. \tag{7.40}$$

From Equation (7.40), we have

$$\begin{aligned} e_n(t)D^\alpha e_n(t) &= -k_3 e_n^2(t) - e_n(t)\tilde{\theta}^\top\phi(Z) + e_n(t)\tilde{d}(t) \\ &\quad - e_n^2(t)\hat{\varsigma} - e_n(t)\xi_{2,n-1}(t) + e_n(t)\varepsilon^* \\ &\le -k_3 e_n^2(t) - e_n(t)\tilde{\theta}^\top\phi(Z) + |e_n(t)\tilde{d}(t)| \\ &\quad - e_n^2(t)\hat{\varsigma} + 0.5e_n^2(t) + 0.5(\overline{\xi}_{2,n-1} + \overline{\varepsilon})^2 \\ &\le -(k_3 - 0.5)e_n^2(t) - e_n(t)\tilde{\theta}^\top\phi(Z) \\ &\quad - e_n^2(t)\hat{\varsigma} + |e_n(t)|\varsigma + 0.5(\overline{\xi}_{2,n-1} + \overline{\varepsilon})^2. \end{aligned} \tag{7.41}$$

According to Equation (7.3), we obtain

$$\begin{aligned} \varphi_n(t)D^\alpha\varphi_n(t) &= -c_n\varphi_n^2(t) + \varphi_n(t)\Delta u(t) \\ &\le -(c_n - 0.5)\varphi_n^2(t) + 0.5\overline{u}^2. \end{aligned} \tag{7.42}$$

Considering the signals $e_n(t)$, $z(t)$, $\tilde{d}(t)$, $\tilde{\Theta}$, $\tilde{\theta}$, $\tilde{\varsigma}$, and $\varphi_n(t)$, the function candidate is chosen as

$$V_n = \frac{1}{2}e_n^2(t) + \frac{1}{2}z^2(t) + \frac{1}{2k_2}\tilde{d}^2(t) + \frac{1}{2k_4}\tilde{\Theta}^2 + \frac{1}{2k_5}\tilde{\theta}^\top\tilde{\theta} + \frac{1}{2k_6}\tilde{\varsigma}^2 + \frac{1}{2}\varphi_n^2(t), \tag{7.43}$$

where $k_4 > 0$, $k_5 > 0$, and $k_6 > 0$ are design constants and $\tilde{\varsigma} = \varsigma - \hat{\varsigma}$.

On the basis of Lemma 2.1, the Caputo derivative of $V_n(t)$ can be described as

$$D^\alpha V_n \leq e_n(t)D^\alpha e_n(t) + z(t)D^\alpha z(t) + \frac{1}{k_2}\tilde{d}(t)D^\alpha \tilde{d}(t) + \frac{1}{k_4}\tilde{\Theta}D^\alpha\tilde{\Theta} + \frac{1}{k_5}\tilde{\theta}^\top D^\alpha\tilde{\theta} + \frac{1}{k_6}\tilde{\varsigma}D^\alpha\tilde{\varsigma} + \varphi_n(t)D^\alpha\varphi_n(t). \quad (7.44)$$

Invoking Equations (7.34), (7.37), (7.41), and (7.42), Equation (7.44) can be written as

$$\begin{aligned} D^\alpha V_n &\leq e_n(t)D^\alpha e_n(t) + z(t)D^\alpha z(t) + \frac{1}{k_2}\tilde{d}(t)D^\alpha \tilde{d}(t) \\ &\quad - \frac{1}{k_4}\tilde{\Theta}D^\alpha\hat{\Theta} + \frac{1}{k_5}\tilde{\theta}^\top D^\alpha\hat{\theta} - \frac{1}{k_6}\tilde{\varsigma}D^\alpha\hat{\varsigma} + \varphi_n(t)D^\alpha\varphi_n(t) \\ &\leq -(k_3 - 0.5)e_n^2(t) - e_n(t)\tilde{\theta}^\top\phi(Z) - e_n^2(t)\hat{\varsigma} \\ &\quad + |e_n(t)|\varsigma + 0.5(\overline{\xi}_{2,n-1} + \overline{\varepsilon})^2 - (k_1 - 0.5)z^2 \\ &\quad + |z(t)|\tilde{\Theta} + z(t)\tilde{\theta}^\top\phi(Z) + 0.5(\overline{\varepsilon} + \xi)^2 + 0.5\delta^2 \\ &\quad - \left(0.5 - 0.5\frac{1}{k_2^2}\right)\tilde{d}^2(t) + 0.5(\overline{\theta}\,\overline{\phi} + \overline{\varepsilon} + \overline{\xi})^2 - (c_n - 0.5)\varphi_n^2(t) \\ &\quad + 0.5\overline{u}^2 - \frac{1}{k_4}\tilde{\Theta}D^\alpha\hat{\Theta} + \frac{1}{k_5}\tilde{\theta}^\top D^\alpha\hat{\theta} - \varsigma\frac{1}{k_6}D^\alpha\hat{\varsigma}. \end{aligned} \quad (7.45)$$

Consider the adaptive laws for $\hat{\Theta}$, $\hat{\theta}$, and $\hat{\varsigma}$ as follows:

$$D^\alpha\hat{\Theta} = k_4(|z(t)| - \hat{\Theta}), \quad (7.46)$$

$$D^\alpha\hat{\theta} = k_5((e_n(t) - z(t))\phi(Z) - \hat{\theta}), \quad (7.47)$$

$$D^\alpha\hat{\varsigma} = k_6\left(e_n^2(t) - \hat{\varsigma}\right). \quad (7.48)$$

Substituting Equations (7.46), (7.47), and (7.48) into Equation (7.45), and considering the following facts:

$$\begin{aligned} -\frac{1}{k_4}\tilde{\Theta}D^\alpha\hat{\Theta} &= -|z(t)|\tilde{\Theta} + \tilde{\Theta}\hat{\Theta} \\ &\leq -|z(t)|\tilde{\Theta} - 0.5\tilde{\Theta}^2 + 0.5\Theta^2, \end{aligned} \quad (7.49)$$

$$\begin{aligned} \frac{1}{k_5}\tilde{\theta}^\top D^\alpha\hat{\theta} &= \tilde{\theta}^\top(e_n(t) - z(t))\phi(Z) - \tilde{\theta}^\top\hat{\theta} \\ &\leq \tilde{\theta}(e_n(t) - z(t))\phi(Z) - 0.5\|\tilde{\theta}\|^2 + 0.5\|\theta\|^2, \end{aligned} \quad (7.50)$$

$$\begin{aligned} -\frac{1}{k_6}\tilde{\varsigma}D^\alpha\hat{\varsigma} &= -\tilde{\varsigma}e_n^2(t) + \tilde{\varsigma}\hat{\varsigma} \\ &\leq -\tilde{\varsigma}e_n^2(t) - 0.5\tilde{\varsigma}^2 + 0.5\varsigma^2, \end{aligned} \quad (7.51)$$

$$-e_n^2(t)\hat{\varsigma} - \tilde{\varsigma}e_n^2(t) = -\varsigma e_n^2(t), \quad (7.52)$$

$$|e_n(t)|\varsigma - \varsigma e_n^2(t) \leq \frac{1}{4}\varsigma, \quad (7.53)$$

we have

$$D^\alpha V_n \leq -(k_3 - 0.5)e_n^2(t) - (k_1 - 0.5)z^2(t) - 0.5\tilde{\Theta}^2$$

$$
\begin{aligned}
&-\left(0.5-0.5\frac{1}{k_2^2}\right)\tilde{d}^2(t)-0.5\tilde{\varsigma}^2-(c_n-0.5)\varphi_n^2(t)\\
&-0.5\|\tilde{\theta}\|^2+(\bar{\xi}_{2,n-1}+\bar{\varepsilon})^2+0.5\Theta^2+0.5\|\theta\|^2\\
&+0.5\delta^2+0.5(\bar{\theta}\,\bar{\phi}+\bar{\varepsilon}+\bar{\xi}_{2,n-1})^2+0.5\bar{u}^2+0.5\varsigma^2+\frac{1}{4}\varsigma.
\end{aligned}
\tag{7.54}
$$

This design procedure of the FODO-based adaptive neural control can be summarized in the following theorem, which contains the results of adaptive control for uncertain FOCS (7.1) in the presence of input saturation and external disturbance.

Theorem 7.1 *Consider the uncertain FOCS (7.1) in the presence of an unknown external disturbance and an input saturation that satisfies Assumptions 7.2 and 7.3. If the full state information is available and all initial conditions lie in a given compact set, the adaptive control law is proposed as Equation (7.39), parameter updated laws are chosen as Equations (7.46), (7.47), and (7.48), and the SMFODO is designed as Equation (7.35). Then, there exist appropriate design parameters $l_i > 0$, $c_i > 0$, and $k_j > 0$, such that the overall closed-loop control system is uniformly stable, in the sense that all of the closed-loop signals $e_i(t)$, $z(t)$, $\tilde{d}(t)$, $\tilde{\Theta}$, $\tilde{\theta}$, $\tilde{\varsigma}$, and $\varphi_i(t)$ are bounded, where $i = 1, \ldots, n$ and $j = 1, \ldots, 6$. Furthermore, on the basis of the integrated effect of system uncertainties and external disturbances, the tracking error signal $e_1(t)$ ultimately remains within the compact set $\Omega_{e_1(t)}$ as $t \to \infty$, where $\Omega_{e_1(t)}$ is defined by $\Omega_{e_1(t)} := \{e_1(t) \in R \,|\, |e_1(t)| \le \sqrt{\Omega_0}\}$, and where Ω_0 will be given later.*

Proof 7.1 *The Lyapunov function candidate for the whole closed-loop system is chosen as*

$$
V(t)=\sum_{i=1}^{n}V_i(t).
\tag{7.55}
$$

Considering Equations (7.12), (7.24), and (7.54), the Caputo derivative of $V(t)$ is

$$
\begin{aligned}
D^{\alpha}V(t)\le&-(l_1-0.5)e_1^2(t)-\sum_{\eta_1=2}^{n-1}(l_{\eta_1}-1.5)e_{\eta_1}^2(t)-(k_3-1)e_n^2(t)\\
&-(c_1-0.5)\varphi_1^2(t)-\sum_{\eta_2=2}^{n}(c_{\eta_2}-1)\varphi_{\eta_2}^2(t)-(k_1-0.5)z^2(t)\\
&-\left(0.5-0.5\frac{1}{k_2^2}\right)\tilde{d}^2(t)-0.5\tilde{\Theta}^2-0.5\|\tilde{\theta}\|^2-0.5\tilde{\varsigma}^2\\
&+\sum_{\eta_3=2}^{n-1}0.5\bar{\xi}_{2,\eta_3-1}^2+(\bar{\xi}_{2,n-1}+\bar{\varepsilon})^2+0.5\Theta^2+0.5\|\theta\|^2\\
&+0.5\delta^2+0.5\varsigma^2+\frac{1}{4}\varsigma+0.5(\bar{\theta}\,\bar{\phi}+\bar{\varepsilon}+\bar{\xi}_{2,n-1})^2+0.5\bar{u}^2\\
\le&-\kappa V(t)+C,
\end{aligned}
\tag{7.56}
$$

where

$$
\kappa := \min\begin{pmatrix} 2l_1-1, 2l_{\eta_1}-3, 2k_3-2, 2c_1-1, k_4,\\ 2c_{\eta_2}-2, 2k_1-1, k_2-\frac{1}{k_2}, k_5, k_6\end{pmatrix},
\tag{7.57}
$$

$$C := \sum_{\eta_3=2}^{n-1} 0.5\overline{\xi}_{2,\eta_3-1}^2 + (\overline{\xi}_{2,n-1} + \overline{\varepsilon})^2 + 0.5\Theta^2 + 0.5\|\theta\|^2 + 0.5\delta^2$$
$$+ 0.5(\overline{\theta}\,\overline{\phi} + \overline{\varepsilon} + \overline{\xi}_{2,n-1})^2 + 0.5\overline{u}^2 + 0.5\varsigma^2 + \frac{1}{4}\varsigma. \tag{7.58}$$

To ensure that $\kappa > 0$, *the design parameters* $l_i > 0$, $c_i > 0$, *and* $k_j > 0$ *must make* $2l_1 - 1 > 0$, $2l_{\eta_1} - 3 > 0$, $2k_3 - 2 > 0$, $2c_1 - 1 > 0$, $2c_{\eta_2} - 2 > 0$, $2k_1 - 1 > 0$, *and* $k_2 - \frac{1}{k_2} > 0$, *where* $i = 1, \ldots, n$, $j = 1, \ldots, 6$, $\eta_1 = 2, \ldots, n-1$, $\eta_2 = 2, \ldots, n$, *and* $\eta_3 = 2, \ldots, n-1$.

By using Lemma 2.3, there exists a constant $t_1 > 0$, *such that for all* $t \in (t_1, \infty)$, *we have*

$$|V(t)| \le \frac{2C}{\kappa}. \tag{7.59}$$

Combining (7.55) and (7.59), we have

$$|e_1(t)| \le \sqrt{\frac{4C}{\kappa}}. \tag{7.60}$$

From Equation (7.60), we obtain that $\Omega_0 = 4C/\kappa$. *Conversely, on the basis of Lemma 2.3 and Equation (7.58), we can conclude that the auxiliary design variables* $z(t)$ *and* $\varphi_i(t)$ *and the error signals* $e_i(t)$, $\tilde{d}(t)$, $\tilde{\Theta}$, $\tilde{\theta}$, $\tilde{\varsigma}$ *converge to a compact set, where* $i = 1, \ldots, n$. *This concludes the proof.* □

Remark 7.2 *In this section, the SMFODO-based adaptive neural tracking control scheme is proposed for a class of uncertain fractional-order chaotic systems with input saturation and unknown external disturbance. To restrain the effect of the unknown external disturbance, the SMFODO is designed to estimate the unknown disturbance. At the same time, to tackle the input saturation, an auxiliary design system (7.3) is introduced to compensate for the effect of the saturation constraint. Meanwhile, the output* $\hat{d}(t)$ *of the SMFODO and the auxiliary system state* $\varphi_n(t)$ *are used to design the adaptive neural controller. Therefore, the integrated effect of the unknown external disturbance and input saturation has been explicitly considered in the developed tracking control scheme for uncertain fractional-order chaotic systems. According to Theorem 7.1, it is evident that the input saturation* $sat(u(t))$ *produced by the designed control command* $u(t)$ *can guarantee that the closed-loop system is bounded stable.*

Remark 7.3 *For the uncertain FOCS (7.1), to solve tedious analytic computations of Caputo derivatives of the virtual control laws in the backstepping procedure, the fractional-order differentiator is employed to approximate the derivatives of the virtual control laws in each step. Meanwhile, the approximation errors of the fractional-order differentiator are considered in the stability analysis of the whole closed-loop system.*

7.3 Simulation Examples

To illustrate the effectiveness of the proposed SMFODO-based adaptive neural control scheme for the uncertain FOCS with external disturbance and input saturation, the numerical simulations of two fractional-order chaotic systems will be studied.

7.3.1 Fractional-Order Chaotic Electronic Oscillator

Consider the following fractional-order chaotic electronic oscillator model [227]:

$$\begin{aligned}
D^{\alpha}x_1(t) &= x_2(t),\\
D^{\alpha}x_2(t) &= x_3(t),\\
D^{\alpha}x_3(t) &= -\hbar_1(x_1(t)+x_2(t)+x_3(t)-\psi(x_1(t))),
\end{aligned} \tag{7.61}$$

where $x_1(t)$, $x_2(t)$, and $x_3(t)$ are system state variables, $\hbar_1$ is a designed parameter, and the function $\psi(x_1(t))$ satisfies

$$\psi(x_1(t)) = \begin{cases} 1, & x_1(t) \geq 0,\\ -1, & x_1(t) < 0. \end{cases} \tag{7.62}$$

The fractional order is chosen as $\alpha = 0.98$ and the parameter is set as $\hbar_1 = 0.8$; the system (7.61) is a chaotic system based on the theory of Tavazoei and Haeri [228]. Furthermore, the chaotic behavior of Equation (7.61) is shown in Figure 7.1, for initial conditions of system (7.61) chosen as $x(0) = [0.3, 0.4, 0.2]^{\top}$.

According to these simulation results, it can be seen that the system (7.61) is chaotic without control input. To facilitate the output signal $x_1(t)$ of the system (7.61) to track the desired signal $x_{1\mathrm{d}}(t)$, the input control is considered in Equation (7.61). At the same time, the system uncertainty and the unknown external disturbance are also considered for the system (7.61).

From Equations (7.1) and (7.61), we have the following:

$$\begin{aligned}
D^{\alpha}x_1(t) &= x_2(t),\\
D^{\alpha}x_2(t) &= x_3(t),\\
D^{\alpha}x_3(t) &= f(x(t)) + \Delta f(x(t)) + d(t) + \mathrm{sat}(u(t)),
\end{aligned} \tag{7.63}$$

where $f(x(t)) = -\hbar_1(x_1(t)+x_2(t)+x_3(t)-\psi(x_1(t)))$, with $x(t) = [x_1(t), x_2(t), x_3(t)]^{\top}$, $u(t)$ is the desired control input, $d(t)$ is the unknown time-varying disturbance, and $\Delta f(x(t))$ is the unknown nonlinear function.

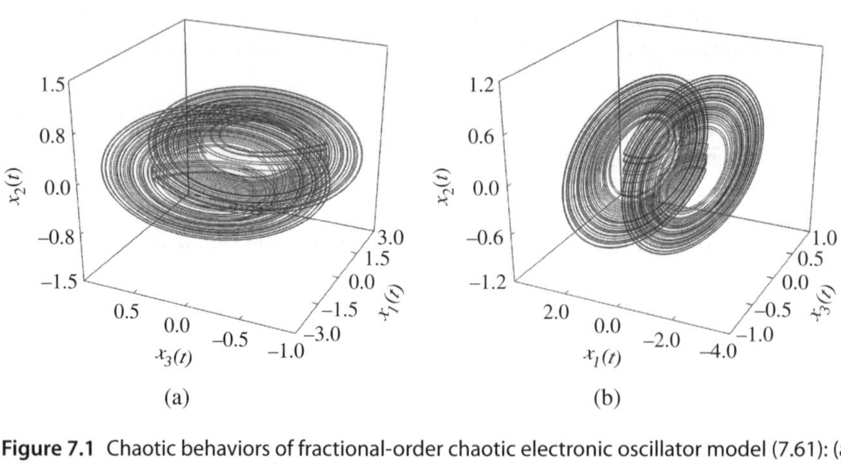

Figure 7.1 Chaotic behaviors of fractional-order chaotic electronic oscillator model (7.61): (a) $x_1(t)$–$x_3(t)$–$x_2(t)$ space; (b) $x_3(t)$–$x_1(t)$–$x_2(t)$ space.

Theorem 7.1 is applied to the uncertain fractional-order electronic oscillator model (7.63) to render the output signal $x_1(t)$ to track the reference signal $x_{1d}(t) = \sin(t)$. In the simulation studies, for fractional order $\alpha = 0.98$, the initial conditions are chosen as $x(0) = [0.3, 0.4, 0.2]^\top$, $\varphi(0) = [0.1, 0.1, 0.1]^\top$, $w_{1,1}(0) = 0.1$, $w_{1,2}(0) = 0.1$, $w_{2,1}(0) = 0.1$, $w_{2,2}(0) = 0.1$, $\sigma(0) = -2$, $\chi(0) = -10$, $\hat{\Theta}_0 = 0.1$, and $\hat{\zeta}_0 = 0.1$. The system parameters are selected as $\hbar_1 = 1.1$ and $\hbar_2 = 0.38$. The control parameters are chosen as $R_1 = R_2 = 10$, $c_1 = c_2 = 1$, $c_3 = 20$, $l_1 = 7$, $l_2 = 10$, $k_1 = k_3 = k_4 = k_5 = k_6 = 20$, and $k_2 = 80$. The uncertainty is assumed as $\Delta f(x) = 0.01\sin(x_1 x_2 x_3)$. The external disturbance is assumed as $d(t) = 20\sin(t)$. Furthermore, we define $e(t) = x_1(t) - x_{1d}(t)$. On the basis of the result of Ishteva [225], we have $\lambda_1 D^\alpha \sin(\lambda_2 t) = -\lambda_1 \frac{1}{2} \mathrm{i}(\mathrm{i}\lambda_2)^m t^{m-\alpha}(E_{1,m-\alpha+1}(\mathrm{i}\lambda t) - (-1)^n E_{1,m-\alpha+1}(-\mathrm{i}\lambda t))$, where i denotes the square root of minus one and λ_1 and λ_2 are arbitrary numbers. Therefore, Assumptions 7.2 and 7.3 are satisfied.

Simulation results of the uncertain fractional-order electronic oscillator model (7.63) are shown in Figure 7.2, Figure 7.3, and Figure 7.4. Tracking results of the uncertain fractional-order electronic oscillator model with external disturbance and input saturation under the proposed adaptive neural control scheme are shown in Figure 7.2.

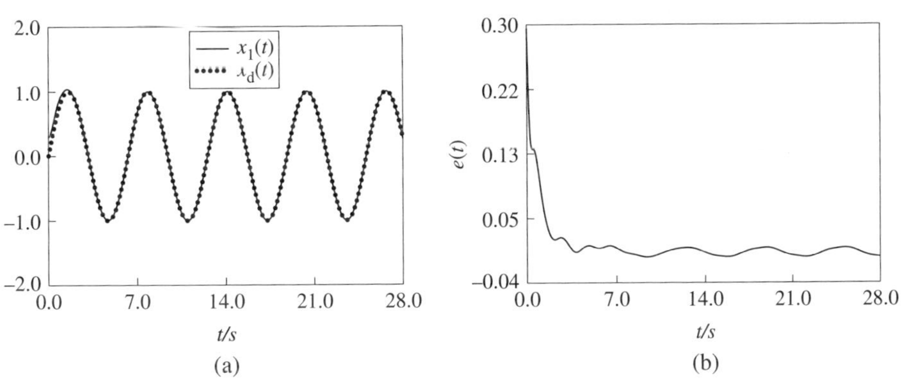

Figure 7.2 Tracking control results of the fractional-order chaotic electronic oscillator (7.61) (a) Output $x_1(t)$ follows desired trajectory $x_d(t)$; (b) tracking error $e(t)$.

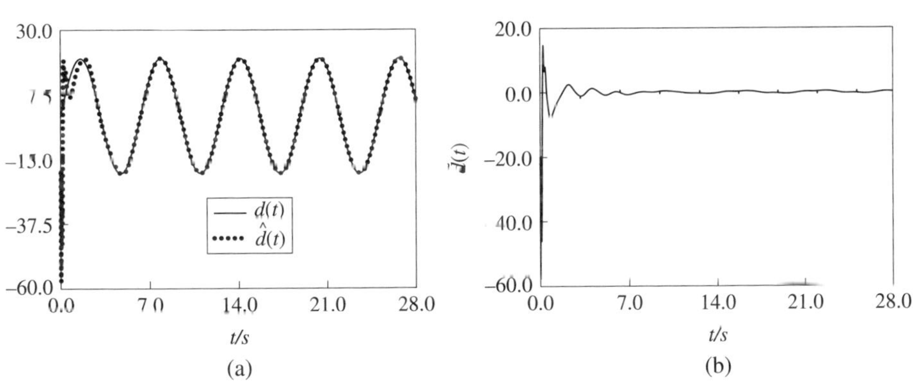

Figure 7.3 Disturbance estimation results of the fractional-order chaotic electronic oscillator (7.61) (a) Disturbance $d(t)$ and approximation output of $\hat{d}(t)$; (b) observation error $\tilde{d}(t)$.

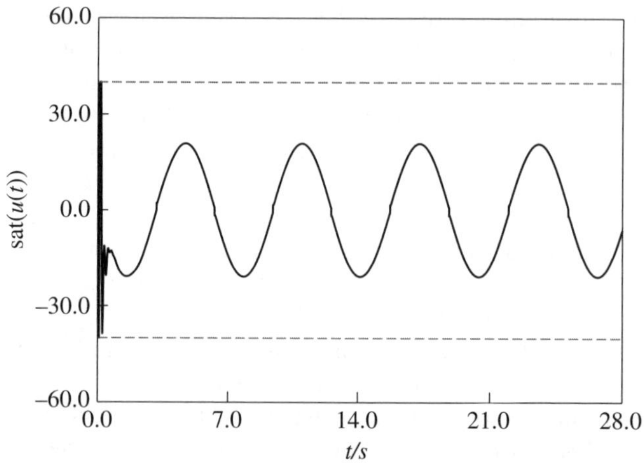

Figure 7.4 Control input sat(u(t)) of the fractional-order chaotic electronic oscillator (7.61).

According to Figure 7.2a, the tracking performance of the uncertain FOCS (7.63) is satisfactory. The tracking error $e(t)$ between the output signal $x_1(t)$ and the desired signal $x_{1d}(t)$ is bounded, as shown in Figure 7.2b. Furthermore, the observation performance of the proposed SMFODO (7.35) is presented in Figure 7.3. It is evident from Figure 7.3 that the disturbance observer could approximate the unknown external disturbance well. The control input signal, which is bounded, is shown in Figure 7.4. It is concluded from these simulation results that the proposed adaptive neural control scheme for uncertain fractional-order chaotic systems based on the SMFODO is effective.

7.3.2 Fractional-Order Modified Jerk System

The fractional-order modified jerk system is given as follows [229]:

$$\begin{aligned} D^\alpha x_1(t) &= x_2(t), \\ D^\alpha x_2(t) &= x_3(t), \\ D^\alpha x_3(t) &= -\hbar_1 x_3(t) - x_2(t) + |x_1(t)| - 2, \end{aligned} \tag{7.64}$$

where $x_1(t)$, $x_2(t)$, and $x_3(t)$ are system state variables, and $\hbar_1$ is a design parameter.

The fractional order is set as $\alpha = 0.99$ and the parameter is chosen as $\hbar_1 = 0.54$; the system (7.64) is a chaotic system based on the theory of Tavazoei and Haeri [228]. Furthermore, if the initial condition of the system (7.64) is chosen as $x(0) = [0.1, 0.2, 0.2]^\mathrm{T}$, the chaotic behavior of the system (7.64) is given in Figure 7.5.

From these simulation results, it can be seen that the system (7.64) is chaotic without input control. To make the output signal $x_1(t)$ of the system (7.64) track the desired signal $x_{1d}(t)$, the input control is considered in Equation (7.64). Meanwhile, the unknown uncertainty and the unknown external disturbance are also considered in Equation (7.64). From Equations (7.1) and (7.64), we have the following:

$$\begin{aligned} D^\alpha x_1(t) &= x_2(t), \\ D^\alpha x_2(t) &= x_3(t), \\ D^\alpha x_3(t) &= f(x(t)) + \Delta f(x(t)) + d(t) + \text{sat}(u(t)), \end{aligned} \tag{7.65}$$

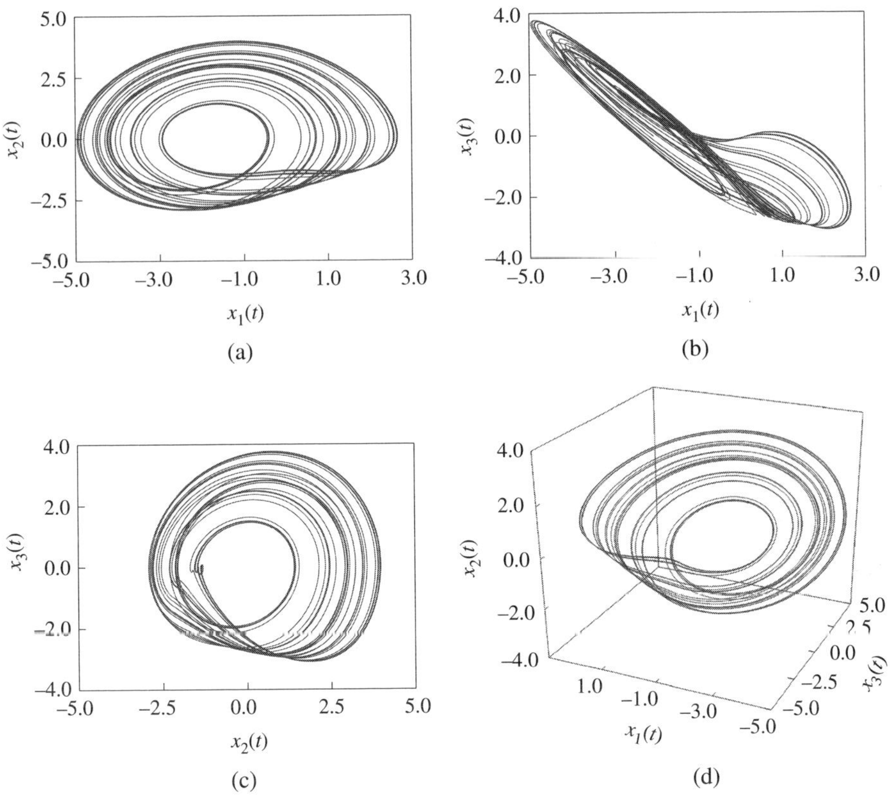

Figure 7.5 Chaotic behavior of (7.64): (a) $x_1(t)$–$x_2(t)$ plane; (b) $x_1(t)$–$x_3(t)$ plane; (c) $x_2(t)$–$x_3(t)$ plane; (d) $x_3(t)$–$x_1(t)$–$x_2(t)$ space.

where $f(x(t)) = -\hbar_1 x_3(t) - x_2(t) + |x_1(t)| - 2$ with $x(t) = [x_1(t), x_2(t), x_3(t)]^{\mathrm{T}}$, $u(t)$ is the desired control input, $d(t)$ is the unknown time-varying disturbance, and $\Delta f(x(t))$ is the unknown nonlinear function.

According to Theorem 7.1, the output signal $x_1(t)$ of the uncertain fractional-order modified Jerk system (7.65) is guaranteed to track the reference signal $x_{1\mathrm{d}}(t) = \sin(t)$. To realize this simulation, the fractional order is chosen as $\alpha = 0.99$, the initial conditions are chosen as $x(0) = [0.1, 0.2, 0.2]^{\mathrm{T}}$, $\varphi(0) = [0.1, 0.1, 0.1]^{\mathrm{T}}$, $w_{1,1}(0) = 0.1$, $w_{1,2}(0) = 0.1$, $w_{2,1}(0) = 0.1$, $w_{2,2}(0) = 0.1$, $\sigma(0) = -2$, $\chi(0) = -10$, $\dot{\Theta}_0 = 0.1$, and $\dot{\zeta}_0 = 0.1$. The system parameter is selected as $\hbar_1 = 0.45$. The control parameters are chosen as $R_1 = R_2 = 10$, $c_1 = c_2 = 1$, $c_3 = 20$, $l_1 = 7$, $l_2 = 10$, $k_1 = k_3 = k_4 = k_5 = k_6 = 20$, and $k_2 = 10$. The uncertainty is assumed as $\Delta f(x) = 0.01\sin(x_1 x_2 x_3)$. The external disturbance is assumed as $d(t) = 40\sin(t)$. Furthermore, we define $e(t) = x_1(t) - x_{1\mathrm{d}}(t)$.

Simulation results are shown in Figure 7.6, Figure 7.7, and Figure 7.8 for the uncertain fractional-order modified Jerk system (7.65). According to the proposed adaptive neural control scheme, the tracking results are shown in Figure 7.6 for the uncertain fractional-order modified Jerk system with external disturbance and input saturation. From Figure 7.6, we note that satisfactory tracking performance is obtained. Figure 7.6b shows that the tracking error $e(t)$ is bounded. In addition, the disturbance estimation

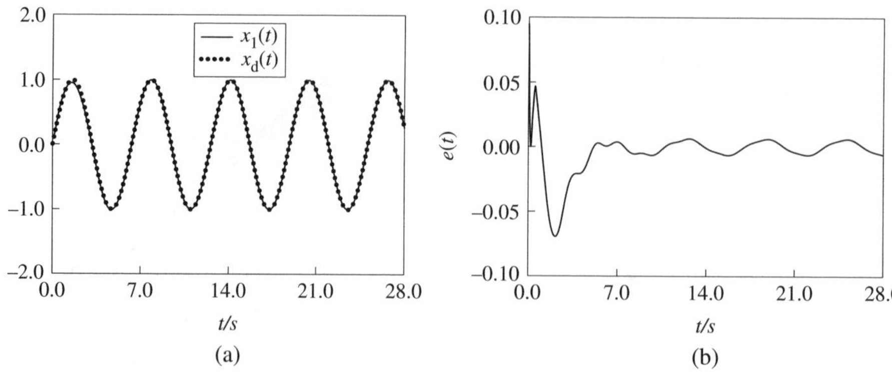

Figure 7.6 Tracking control results of the fractional-order modified jerk system (7.64) (a) Output $x_1(t)$ follows desired trajectory $x_d(t)$; (b) tracking error $e(t)$.

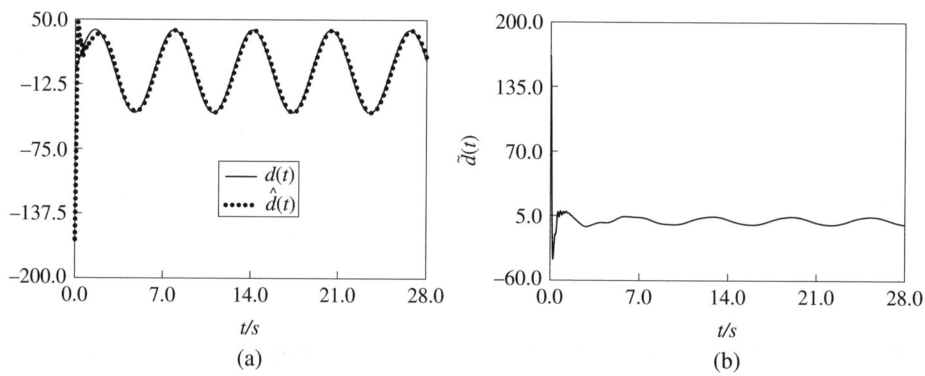

Figure 7.7 Disturbance estimation results of the fractional-order modified jerk system (7.64) (a) Disturbance $d(t)$ and approximation output of $\hat{d}(t)$; (b) observation error $\tilde{d}(t)$.

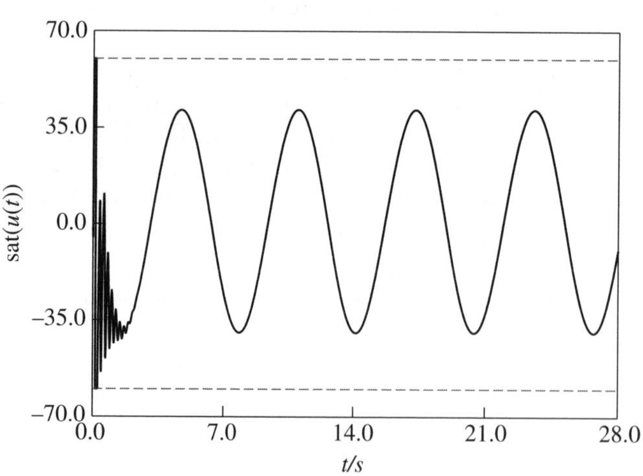

Figure 7.8 Control input sat(u(t)) of the fractional-order modified jerk system (7.64).

performance of the proposed SMFODO (7.35) is shown in Figure 7.7. It is evident, based on Figure 7.7, that the disturbance observer could approximate the unknown external disturbance well. The bounded control input signal is shown in Figure 7.8. According to these simulation results, we can conclude that the proposed SMFODO-based adaptive neural control scheme is viable for uncertain fractional-order chaotic systems.

7.4 Conclusion

An adaptive neural tracking control scheme has been proposed for a class of uncertain fractional-order chaotic systems subjected to unknown disturbance and input saturation in this chapter. An auxiliary design system has been used to compensate for the effect of the input saturation. At the same time, an SMFODO has been designed to guarantee the convergence of the disturbance estimation error. On the basis of the radial basis function neural network, the auxiliary system, and the SMFODO, an adaptive neural control scheme has been presented for fractional-order chaotic systems with unknown disturbance and input saturation. Under the integrated effect of unknown external disturbance and unknown uncertainty, the bounded convergence of all closed-loop signals has been guaranteed. Numerical simulation results have been given to show the effectiveness of the developed control scheme.

An adaptive [illegible] tracking control scheme has been proposed for [illegible].

8

Stabilization Control of Continuous-Time Fractional Positive Systems Based on Disturbance Observer

8.1 Problem Statement

8.1.1 Notation and Definitions

- R^n denotes the n-dimensional real space.
- N^{T} denotes the transpose of a matrix N.
- A matrix $N_0 \in R^{n\times n}$ is called a Metzler matrix if its off diagonal elements are non-negative.
- A matrix $N_1 \in R^{n\times n}$ can be seen as a non-negative matrix if all its elements are non-negative, e.g., $N_1 \geq 0$.
- A vector Φ can be seen as a non-negative vector if all its elements are non-negative, e.g., $\Phi \geq 0$. It is said to be positive if all its components are positive ($\Phi > 0$). Furthermore, it is said to be negative if all its components are negative ($\Phi < 0$).

8.1.2 Preliminaries

Based on the Caputo definition of the fractional derivative, we consider the continuous-time fractional-order system with unknown non-negative constant disturbance as follows:

$$\begin{aligned} D^{\alpha}x(t) &= Ax(t) + Bu(t) + Bd, \\ x(0) &= x_0 \geq 0, \end{aligned} \tag{8.1}$$

where $A \in R^{n\times n}$ and $B \in R^{n\times n}$ are constant matrices, $x(t) = [x_1(t), x_2(t), \cdots, x_n(t)]^{\mathrm{T}} \in R^n$ is a state vector, $u(t) = [u_1(t), u_2(t), \cdots, u_n(t)]^{\mathrm{T}} \in R^n$ is a control input vector, $d \in R^n$ is an unknown non-negative constant disturbance vector, $x_0 \in R^n$ is an initial condition vector, and $0 < \alpha < 1$. This chapter aims to develop a sufficient condition based on the state feedback control method and a FODO to ensure that:

1) The fractional-order system (8.1) is positive.
2) The fractional-order system is asymptotically stable, that is, $\lim_{t\to\infty} x(t) = 0$.

To obtain these conclusions, the following assumptions, definitions and lemmas are required.

Robust Adaptive Control for Fractional-Order Systems with Disturbance and Saturation, First Edition.
Mou Chen, Shuyi Shao, and Peng Shi.

Assumption 8.1 *The matrix B is an invertible matrix.*

Assumption 8.2 *The external disturbance d is a non-negative constant vector, and d is bounded.*

Lemma 8.1 **([156])** *Considering the system (8.1) without the control input $u(t)$ and the disturbance d, we have the following:*

$$\begin{aligned} D^{\alpha}x(t) &= Ax(t), \\ x(0) &= x_0 \geq 0. \end{aligned} \tag{8.2}$$

The solution of the fractional-order system (8.2) is

$$x(t) = E_{\alpha}(At^{\alpha})x_0, \tag{8.3}$$

where $E_{\alpha}(\cdot)$ denotes the ML function defined by

$$E_{\alpha}(At^{\alpha}) = \sum_{l=0}^{\infty} \frac{(At^{\alpha})^l}{\Gamma(l\alpha + 1)}. \tag{8.4}$$

Lemma 8.2 **([158])** *Considering the continuous-time fractional-order linear system (8.2), when A is a Metzler matrix, we obtain $0 \leq x(t) \leq \overline{\kappa}$ with $\overline{\kappa} > 0$, if and only if $A\overline{\kappa} \leq 0$ for any initial condition $0 \leq x_0 \leq \overline{\kappa}$.*

Definition 8.1 **([156])** *If $x(t)$ is non-negative for all $t \geq 0$, the system (8.2) can be regarded as a fractional positive system.*

According to Definition 8.1, the following results are given.

Lemma 8.3 **([161])** *The continuous-time fractional-order linear system (8.2) is positive if and only if the matrix A is a Metzler matrix.*

Lemma 8.4 **([161])** *The fractional-order system (8.2) is asymptotically stable if and only if*

$$|\arg(\lambda_i(A))| > \frac{\alpha\pi}{2}, \qquad i = 1, 2, \ldots, n, \tag{8.5}$$

where $\lambda_i(A)$ stands for the ith eigenvalue of matrix A and the $\arg(\cdot)$ *denotes the argument.*

Lemma 8.5 **([230])** *The fractional-order system (8.2) is asymptotically stable if there exists a Lyapunov function $V(x(t))$ and class-κ functions φ_i, $i = 1, 2, 3$, satisfying*

$$\begin{aligned} &\varphi_1(\|x(t)\|) \leq V(x(t)) \leq \varphi_2(\|x(t)\|), \\ &D^{\alpha}V(x(t)) \leq -\varphi_3(\|x(t)\|). \end{aligned} \tag{8.6}$$

Lemma 8.6 *The system (8.2) is positive and asymptotically stable if there exists a vector $\eta \in R^n$, $\eta > 0$, and the matrix A is a Metzler matrix such that*

$$A\eta < 0. \tag{8.7}$$

Proof 8.1 *On the basis of Lemma 8.3, we obtain that the system (8.2) is positive. Since $A\eta < 0$, we have that $-A\eta > 0$. If the matrix A is a Metzler matrix, the off-diagonal elements of $-A$ are non-positive. On the basis of the results of Araki and Kondo [231], we obtain that there always exists a positive constant vector $\rho \in R^n$ with $\rho > 0$, such that $(-A)^\top \rho > 0$, which can be further written as $\rho^\top A < 0$. Since the system (8.2) is positive, the state variable satisfies $x(t) \geq 0$. In addition, we define the Lyapunov function $V(x(t)) = \rho^\top x(t)$. Combining $V(x(t)) = \rho^\top x(t)$ and the positive system (8.2), we have*

$$D^\alpha V(x(t)) = \rho^\top D^\alpha x(t) = \rho^\top A x(t). \tag{8.8}$$

According to $\rho^\top A < 0$ and Equation (8.8), we obtain

$$D^\alpha V(x(t)) < 0. \tag{8.9}$$

From Lemma 8.5 and Equation (8.9), the positive system (8.2) is asymptotically stable. According to Araki and Kondo [231], $A\eta < 0$ and $\rho^\top A < 0$ are equivalent. This concludes the proof. □

Lemma 8.7 *Consider a fractional-order system as follows:*

$$\begin{aligned} D^\alpha x(t) &= (A_0 + B_0 K_0) x(t), \\ x(0) &= x_0 \geq 0, \end{aligned} \tag{8.10}$$

where $A_0 \in R^{n\times n}$, $B_0 \in R^{n\times n}$ and $K_0 \in R^{n\times n}$ are matrices. For the system (8.10), the vectors $h = [h_1, h_2, \cdots, h_n] \in R^n$ and $\beta_j \in R^n (j = 1, 2, \ldots, n)$, if the following conditions are satisfied in the form

$$\begin{aligned} A_0 h + B_0 \sum_{j=1}^{n} \beta_j &< 0, \\ h &> 0, \\ a_{0ij} h_j + b_{0i} \beta_j &\geq 0, \qquad i \neq j, \end{aligned} \tag{8.11}$$

where $i = 1, 2, \ldots, n$, h_j are the elements of h, a_{0ij} are the elements of matrix A_0, and b_{0i} are the row vectors of matrix B_0.

Then, the system (8.10) is positive and asymptotically stable for any initial condition $x_0 \geq 0$, with

$$K_0 = \begin{bmatrix} \frac{\beta_1}{h_1} & \frac{\beta_2}{h_2} & \cdots & \frac{\beta_n}{h_n} \end{bmatrix}. \tag{8.12}$$

Proof 8.2 *Assume that the condition (8.11) is satisfied with K_0 given by Equation (8.12). By applying Lemma 8.6 to the system (8.10), we have*

$$A_0 h + B_0 K_0 h < 0. \tag{8.13}$$

Furthermore, Equation (8.12) can be rewritten as

$$K_0 h = \sum_{j=1}^{n} \beta_j. \tag{8.14}$$

Therefore, Equation (8.13) can be equivalently written as the first equation of system (8.11). In addition, the system (8.10) is positive if matrix $A_0 + B_0 K_0$ is Metzler, that is

$(A_0 + B_0K_0)_{ij} \geq 0$ for $i \neq j$. By replacing K_0 with its expression (8.12), we obtain

$$a_{0ij} + \frac{b_{0i}\beta_j}{h_j} \geq 0, \qquad i \neq j, \tag{8.15}$$

which is equivalent to the third equation of system (8.11). This concludes the proof. □

8.2 Main Results

In this section, a FODO is introduced to estimate the external unknown non-negative constant disturbance of a continuous-time fractional-order linear system (8.1). Furthermore, the stability of the closed-loop fractional positive system is investigated, based on a FODO.

8.2.1 Fractional Disturbance Observer

According to the integer-order disturbance observer design technique presented in Chen *et al.* [105], the FODO will be introduced for the fractional-order system (8.1) in this section. On the basis of the property of the Caputo definition, we obtain that the Caputo derivative of a constant is zero. Thus, for a constant disturbance, we have

$$D^{\alpha} d = O, \tag{8.16}$$

where $O \in R^n$ denotes a zero vector.

From the system (8.1), d can be written as

$$Bd = D^{\alpha}x(t) - Ax(t) - Bu(t). \tag{8.17}$$

The FODO is described as

$$D^{\alpha}\hat{d} = -GB\hat{d} + GD^{\alpha}x(t) - G(Ax(t) + Bu(t)), \tag{8.18}$$

where G is a design matrix of the FODO (8.18). The disturbance estimation error is defined as

$$\tilde{d} = d - \hat{d}. \tag{8.19}$$

According to Equations (8.16), (8.17), (8.18), and (8.19), we obtain

$$D^{\alpha}\tilde{d} = GB\hat{d} - GBd = -GB\tilde{d}, \tag{8.20}$$

if the design matrix G is chosen such that the disturbance estimation error system

$$D^{\alpha}\tilde{d} + GB\tilde{d} = O \tag{8.21}$$

tends to stability. Then the disturbance estimate output $\hat{d}$ can approximate to d, if $t \to \infty$.

Since $D^{\alpha}x(t)$ is not available, the FODO (8.18) cannot be realized. To handle the aforementioned issue, an auxiliary design system is defined as

$$\psi(t) = \hat{d} - Gx(t). \tag{8.22}$$

Considering Equations (8.18) and (8.22), we have

$$D^{\alpha}\psi(t) = D^{\alpha}\hat{d} - GD^{\alpha}x(t)$$

$$= -GB(\psi(t) + Gx(t)) - G(Ax(t) + Bu(t))$$
$$= -GB\psi(t) - GBGx(t) - G(Ax(t) + Bu(t)). \tag{8.23}$$

From Equation (8.22), the disturbance estimate output $\hat{d}$ can be written as

$$\hat{d} = \psi(t) + Gx(t). \tag{8.24}$$

The designed FODO consists of Equations (8.23) and (8.24). Invoking Equations (8.19), (8.23), and (8.24), the disturbance estimation error is given by

$$\begin{aligned} D^\alpha \tilde{d} &= D^\alpha d - D^\alpha \hat{d} \\ &= -D^\alpha \psi(t) - GD^\alpha x(t) \\ &= GB\psi(t) + G(BGx(t) - Bd + D^\alpha x(t)) - GD^\alpha x(t) \\ &= GB(\hat{d} - d) = -F\tilde{d}, \end{aligned} \tag{8.25}$$

where $F = GB$.

From Equation (8.25), we know that the disturbance estimate output $\hat{d}$ can approximate the disturbance d well if the matrix $-F$ satisfies $|\arg(\lambda_i(-F))| > \alpha\pi/2$, $i = 1, 2, \ldots, n$. Thus, $\tilde{d}$ is asymptotically stable based on Lemma 8.4.

According to this analysis, we know that the disturbance estimation error $\tilde{d}$ is upper bounded.

Remark 8.1 *If $-F$ is a Metzler matrix and the matrix $-F$ satisfies $|\arg(\lambda_i(-F))| > \alpha\pi/2$, $i = 1, 2, \ldots, n$, the system (8.25) is positive and asymptotically stable for any initial condition $\tilde{d}(0) \geq 0$.*

Remark 8.2 *Disturbance observers can estimate external disturbances via known information of the controlled plants, such as system states, control input, or control output, and the output of disturbance observers can be used to design the control law. According to the designed FODO (8.23 and 8.24), we know that the auxiliary variable ψ can be obtained by appropriately choosing the control matrix G and designing the control law u. Therefore, $\hat{d}$ can be realized using known state variables x and ψ. Furthermore, the disturbance estimation errors are convergent if the matrix G is chosen appropriately based on the stability theory of fractional-order systems [164].*

To ensure that the closed-loop system is positive and asymptotically stable, the initial conditions must satisfy $x(0) \geq 0$ and $\tilde{d}(0) \geq 0$ for the fractional-order system (8.1). Thus, to guarantee the initial value $\tilde{d}(0) \geq 0$ of the disturbance estimation error, the following theorem is given.

Theorem 8.1 *The initial condition $\tilde{d}(0)$ can satisfy $\tilde{d}(0) \geq 0$ if there exists an initial condition $\psi(0) \leq 0$, such that*

$$\psi(0) + Gx(0) \leq 0. \tag{8.26}$$

Proof 8.3 *According to Equation (8.24), we have*

$$\hat{d}(0) = \psi(0) + Gx(0). \tag{8.27}$$

Combining Equations (8.26) and (8.27), we obtain

$$\hat{d}(0) \leq 0. \tag{8.28}$$

When d is an unknown non-negative constant disturbance, one has

$$d(0) \geq 0. \tag{8.29}$$

On the basis of this analysis, we obtain that $\tilde{d}(0) = d(0) - \hat{d}(0) \geq 0$. □

8.2.2 Stabilization Control of Fractional Positive System

Consider the continuous-time fractional-order system given as Equation (8.1). Using the state-feedback control method and the designed disturbance observer (8.23 and 8.24), the desired control input is designed as

$$u(t) = Kx(t) - \hat{d}, \tag{8.30}$$

where K is a design matrix. On the basis of Equations (8.1), (8.23), (8.24), and (8.30), the closed-loop control system is illustrated in Figure 8.1. Substituting Equation (8.30) into Equation (8.1), we obtain

$$\begin{aligned} D^{\alpha}x(t) &= (A + BK)x(t) + B\tilde{d}, \\ x(0) &= x_0 \geq 0. \end{aligned} \tag{8.31}$$

To analyze the stability of the closed-loop system (8.31), an augmented system based on Equations (8.25) and (8.31) is given as follows:

$$\begin{aligned} D^{\alpha}\tilde{d} &= -F\tilde{d}, \\ D^{\alpha}x(t) &= (A + BK)x(t) + B\tilde{d}, \\ \tilde{d}(0) &= \tilde{d}_0 \geq 0, \\ x(0) &= x_0 \geq 0. \end{aligned} \tag{8.32}$$

On the basis of Equation (8.32), we have

$$\begin{aligned} D^{\alpha}z(t) &= Qz(t), \\ z(0) &= z_0 \geq 0, \end{aligned} \tag{8.33}$$

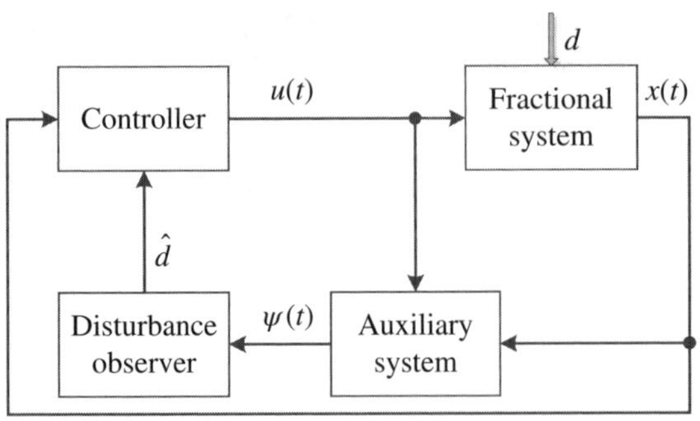

Figure 8.1 Whole closed-loop control system.

where

$$z(t) = \begin{bmatrix} \tilde{d} \\ x(t) \end{bmatrix}, \qquad Q = \begin{bmatrix} -F & O_1 \\ B & A + BK \end{bmatrix},$$

and $O_1 \in R^{n\times n}$ denotes a zero matrix.

Remark 8.3 *If Q is a Metzler matrix and satisfies $|\arg(\lambda_i(Q))| > \alpha\pi/2$, $i = 1, 2, \ldots, n$, the system (8.33) is positive and asymptotically stable for any initial condition $z(0) \geq 0$.*

Furthermore, the following theorem is given based on Lemma 8.7 for the fractional-order system (8.33).

Theorem 8.2 *For the vectors $\Upsilon = [\Upsilon^*, \Upsilon^{**}]^\top \in R^{2n}$, $\Upsilon^* = [\Upsilon_{11}, \Upsilon_{12}, \cdots, \Upsilon_{1n}]^\top \in R^n$, $\Upsilon^{**} = [\Upsilon_{21}, \Upsilon_{22}, \cdots, \Upsilon_{2n}]^\top \in R^n$, the matrices $\varepsilon_1 = [\varepsilon_{11}, \varepsilon_{12}, \cdots, \varepsilon_{1n}] \in R^{n\times n}$, and $\varepsilon_2 = [\varepsilon_{21}, \varepsilon_{22}, \cdots, \varepsilon_{2n)}] \in R^{n\times n}$, if the following conditions are satisfied:*

$$\begin{aligned} \sum_{i_1=1}^{n} \varepsilon_{1i_1} &< 0, \\ \frac{\varepsilon_{1i_1j_1}}{\Upsilon_{1j_1}} &\geq 0, \qquad i_1 \neq j_1, \\ B\Upsilon^* + A\Upsilon^{**} + B\sum_{i_1=1}^{n} \varepsilon_{2i_1} &< 0, \\ \Upsilon &> 0, \\ a_{i_1j_1}\Upsilon_{2j_1} + b_{i_1}\varepsilon_{2j_1} &\geq 0, \qquad i_1 \neq j_1, \end{aligned} \tag{8.34}$$

where $i_1 = 1, 2, \ldots, n$, $j_1 = 1, 2, \ldots, n$, Υ_{1j_1} are the j_1th elements of Υ^*, Υ_{2j_1} are the j_1th elements of Υ^{**}, $a_{i_1j_1}$ are the elements of the ith rows and jth columns of matrix A, $\varepsilon_{1i_1j_1}$ are the elements of the i_1th rows and j_1th columns of ε_1, b_{i_1} are the i_1th row vectors of matrix B, and

$$-F = \left[\frac{\varepsilon_{11}}{\Upsilon_{11}}, \frac{\varepsilon_{12}}{\Upsilon_{12}}, \cdots, \frac{\varepsilon_{1n}}{\Upsilon_{1n}}\right], \qquad K = \left[\frac{\varepsilon_{21}}{\Upsilon_{21}}, \frac{\varepsilon_{22}}{\Upsilon_{22}}, \quad , \frac{\varepsilon_{2n}}{\Upsilon_{2n}}\right], \tag{8.35}$$

then the augmented system (8.33) is positive and asymptotically stable for any initial condition $z_0 \geq 0$.

Proof 8.4 *Assume that the condition (8.34) is satisfied with $-F$ and K given by Equation (8.35). By applying Lemma 8.6 to the augmented system (8.33), we have*

$$\begin{aligned} \Upsilon &> 0, \\ -F\Upsilon^* &< 0, \\ B\Upsilon^* + A\Upsilon^{**} + BK\Upsilon^{**} &< 0. \end{aligned} \tag{8.36}$$

Furthermore, we obtain

$$\begin{aligned} -F\Upsilon^* &- \sum_{i_1=1}^{n} \varepsilon_{1i_1}, \\ K\Upsilon^{**} &= \sum_{i_1=1}^{n} \varepsilon_{2i_1}. \end{aligned} \tag{8.37}$$

Hence, $-F\Upsilon^* < 0$ *and* $B\Upsilon^* + A\Upsilon^{**} + BK\Upsilon^{**} < 0$ *can be written as*

$$\sum_{i_1=1}^{n} \varepsilon_{1i_1} < 0$$

and

$$B\Upsilon^* + A\Upsilon^{**} + B\sum_{i_1=1}^{n} \varepsilon_{2i_1} < 0,$$

respectively. Moreover, the augmented system (8.33) is positive if Q *is a Metzler matrix, that is* $(-F)_{i_1j_1} \geq 0$ *for* $i_1 \neq j_1$*, and* $(A+BK)_{i_1j_1} \geq 0$ *for* $i_1 \neq j_1$*. By considering the expressions of* $-F$ *and* K*, we have*

$$\begin{aligned} \frac{\varepsilon_{1i_1j_1}}{\Upsilon_{1j_1}} &\geq 0, \qquad i_1 \neq j_1, \\ a_{i_1j_1} + b_{i_1}\frac{\varepsilon_{2j_1}}{\Upsilon_{2j_1}} &\geq 0, \qquad i_1 \neq j_1. \end{aligned} \tag{8.38}$$

According to this analysis, we can know that the augmented system (8.33) is positive and asymptotically stable for any initial condition $z_0 \geq 0$ *if the condition (8.34) is satisfied with* $-F$ *and* K *given by Equation (8.35). This concludes the proof.* □

8.2.3 Simulation of Fractional Positive System

Considering a continuous-time fractional-order linear system with unknown constant disturbance described by Equation (8.1), the following system matrices A and B are given as

$$A = \begin{bmatrix} -1.1 & 0.5 \\ -0.3 & -0.7 \end{bmatrix}, \qquad B = \begin{bmatrix} 1 & 0 \\ 0 & 1 \end{bmatrix}. \tag{8.39}$$

This section aims at designing a controller $u(t) = Kx(t) - \hat{d}$ to render the closed-loop system states non-negative and asymptotically stable. Thus, the conditions in Theorem 8.2 must be satisfied. According to the linear programming method, the control gain matrices F and K are obtained as

$$F = \begin{bmatrix} 2.0055 & -0.9899 \\ -0.9312 & 2.1318 \end{bmatrix}, \qquad K = \begin{bmatrix} -0.9549 & 0.1806 \\ 0.8603 & -1.0801 \end{bmatrix}. \tag{8.40}$$

For the augmented system (8.33), we have

$$Q = \begin{bmatrix} -2.0055 & 0.9899 & 0 & 0 \\ 0.9312 & -2.1318 & 0 & 0 \\ 1 & 0 & -2.0549 & 0.6806 \\ 0 & 1 & 0.5603 & -1.7801 \end{bmatrix}. \tag{8.41}$$

From Equation (8.41), the eigenvalues of Q are given as $\lambda_1(Q) = -1.1065$, $\lambda_2(Q) = -1.2849$, $\lambda_3(Q) = -3.0308$, and $\lambda_4(Q) = -2.5501$. Thus, the eigenvalues of Q are located in the left half of the complex plane. Then the augmented system (8.33) is asymptotically stable. Furthermore, we obtain that Q is a Metzler matrix. Therefore, the augmented

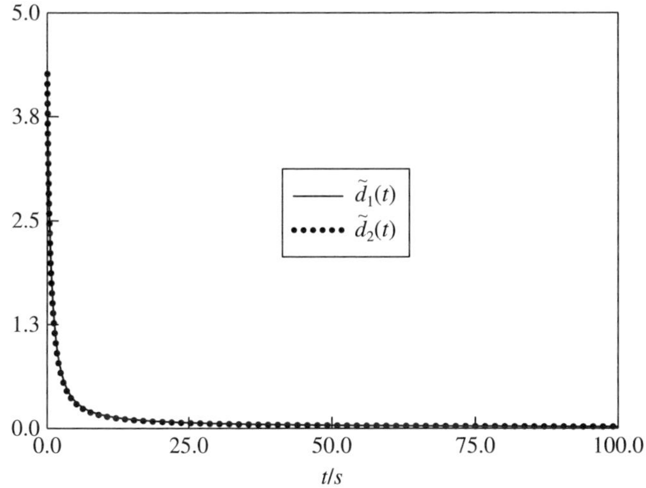

Figure 8.2 Disturbance estimation errors $\tilde{d}_1$ and $\tilde{d}_2$ for the simulation of fractional-order system (8.1) and (8.39).

system (8.33) is positive for any initial condition $z_0 \succeq 0$. From this analysis, we conclude that the augmented system (8.33) is positive and asymptotically stable under conditions (8.40) and (8.41).

For the numerical simulation, we choose the fractional order as $\alpha = 0.8$; the constant disturbances are assumed as $d_1 = d_2 = 4$. The initial conditions are chosen as $x_1(0) = 0.1, x_2(0) = 0.2, \psi_1(0) = -0.3$, and $\psi_2(0) = -0.6$. Then we obtain that $\tilde{d}_1(0) = 4.3124$ and $\tilde{d}_2(0) = 4.2199$. On the basis of the numerical simulation method for fractional-order systems [164], the estimation performance of the FODO (8.23) and (8.24) is shown in Figure 8.2, which shows that good estimation performance is achieved. In addition, if we use Equations (8.3) and (8.4) to realize the numerical simulation of the augmented system (8.33), the ML function will converge slowly for the case of the large variable.

Using the same numerical simulation method, the simulation results of augmented system (8.33) are presented in Figure 8.3. From Figure 8.3, the state variables $x_1(t)$ and $x_2(t)$ are positive and asymptotically stable. According to these simulation results, we conclude that the FODO (8.23) and (8.24) is effective for constant disturbances, and that the designed controller $u(t) = Kx(t) - \hat{d}$ can guarantee that the closed-loop system states are non-negative and asymptotically stable. Furthermore, the simulation results are shown in Figure 8.4 for the case of the closed-loop control system without FODO. From Figure 8.4, the state variables $x_1(t)$ and $x_2(t)$ are not be controlled to zero. However, the state variables $x_1(t)$ and $x_2(t)$ are controlled to zero for the case of the closed-loop control system with FODO based on Figure 8.3. Thus, the developed FODO-based control scheme can improve the performance of the closed-loop control system.

8.2.4 Stabilization Control of Fractional Bounded Positive System

In this section, a sufficient condition is given to guarantee that a closed-loop fractional bounded positive system is asymptotically stable. Invoking the system (8.1), the following continuous-time fractional-order system with disturbance is described by

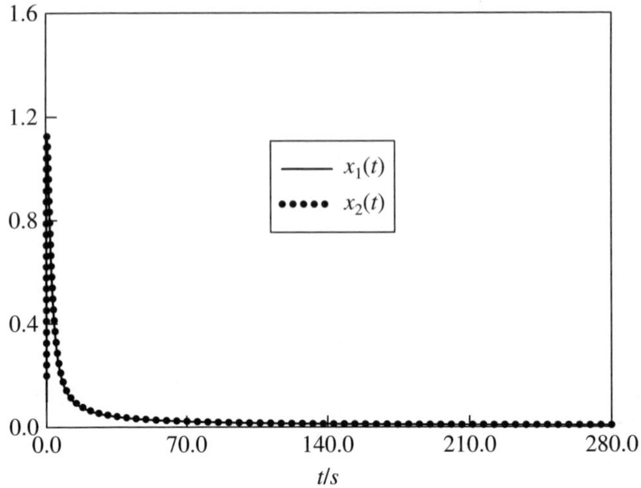

Figure 8.3 Responses of state variables $x_1(t)$ and $x_2(t)$ for the fractional-order system (8.1) and (8.39) with FODO.

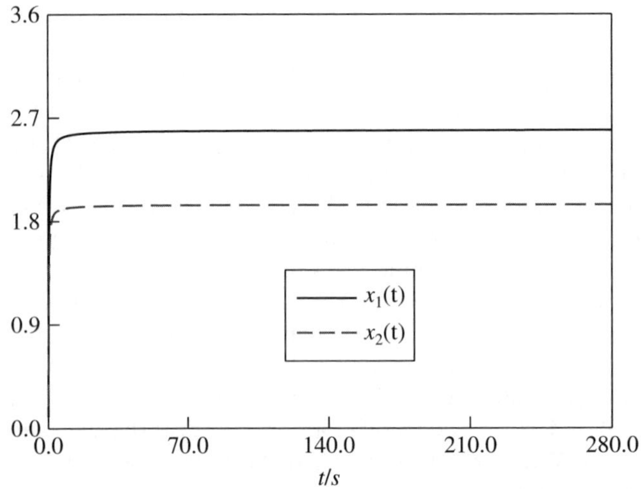

Figure 8.4 Responses of state variables $x_1(t)$ and $x_2(t)$ for the fractional-order system (8.1) and (8.39) without FODO.

the following:

$$
\begin{aligned}
D^{\alpha}x(t) &= Ax(t) + Bu(t) + Bd, \\
0 &< \alpha < 1, \\
0 \leq x(0) &= x_0 \leq \overline{x},
\end{aligned}
\tag{8.42}
$$

where $\overline{x} = [\overline{x}_1, \overline{x}_2, \cdots, \overline{x}_n] \in R^n$ is the bound of $x(t)$.

Combining Equations (8.33) and (8.42), we obtain

$$\begin{aligned} D^{\alpha}z(t) &= Qz(t), \\ 0 < \alpha &< 1, \\ 0 \le z(0) = z_0 &\le \bar{z}, \end{aligned} \tag{8.43}$$

where $\bar{z}$ is the bound of $z(t)$. We define $\bar{z} = [\bar{z}^*, \bar{z}^{**}]^\top \in R^{2n}$, $\bar{z}^* = [\bar{z}_{11}, \bar{z}_{12}, \cdots, \bar{z}_{1n}]^\top \in R^n$, and $\bar{z}^{**} = [\bar{z}_{21}, \bar{z}_{22}, \cdots, \bar{z}_{2n}]^\top \in R^n$, with $\bar{z}^{**} = \bar{x}$.

Considering Theorem 8.2 and the system (8.43), we have the following theorem.

Theorem 8.3 *For the matrices* $\vartheta_1 = [\vartheta_{11}, \vartheta_{12}, \cdots, \vartheta_{1n}] \in R^{n\times n}$ *and* $\vartheta_2 = [\vartheta_{21}, \vartheta_{22}, \cdots, \vartheta_{2n}] \in R^{n\times n}$, *if the following conditions are satisfied:*

$$\begin{aligned} \sum_{i_1=1}^{n} \vartheta_{1i_1} &< 0, \\ \frac{\vartheta_{1i_1j_1}}{\bar{z}_{1j_1}} &\ge 0, \qquad i_1 \ne j_1, \\ B\bar{z}^* + A\bar{z}^{**} + B\sum_{i_1=1}^{n} \vartheta_{2i_1} &< 0, \\ \bar{z} &> 0, \\ a_{i_1j_1}\bar{z}_{2j_1} + b_{i_1}\vartheta_{2j_1} &\ge 0, \qquad i_1 \ne j_1, \end{aligned} \tag{8.44}$$

where $i_1 = 1, 2, \ldots, n$, $j_1 = 1, 2, \ldots, n$, $\bar{z}_{1j_1}$ *are the* j_1*th elements of* $\bar{z}^*$, $\bar{z}_{2j_1}$ *are the* j_1*th elements of* $\bar{z}^{**}$, $a_{i_1j_1}$ *are the elements of matrix A*, $\vartheta_{1i_1j_1}$ *are the elements of* i_1*th rows and* j_1*th columns of* ϑ_1, b_{i_1} *are the* i_1*th row vectors of matrix B, and*

$$-F = \left[\frac{\vartheta_{11}}{\bar{z}_{11}}, \quad \frac{\vartheta_{12}}{\bar{z}_{12}}, \quad \cdots, \quad \frac{\vartheta_{1n}}{\bar{z}_{1n}}\right], \qquad K = \left[\frac{\vartheta_{21}}{\bar{z}_{21}}, \quad \frac{\vartheta_{22}}{\bar{z}_{22}}, \quad \cdots, \quad \frac{\vartheta_{2n}}{\bar{z}_{2n}}\right]. \tag{8.45}$$

Then, the state is bounded for the system (8.43), and the system (8.43) is positive and asymptotically stable for any initial condition $0 \le z(0) = z_0 \le \bar{z}$.

Proof 8.5 *Invoking Theorem 8.2 and Equation (8.44), we have that Q is a Metzler matrix. Since* $\bar{z} > 0$, *we obtain that the augmented system (8.43) is asymptotically stable. In addition, on the basis of Lemma 8.2, the state variables of the positive system (8.43) are bounded such that* $0 \le z(t) \le \bar{z}$ *for any initial condition* $0 \le z(0) = z_0 \le \bar{z}$ □

According to Theorem 8.3, we can conclude that $0 \le x(t) \le \bar{z}^{**}$. Since $\bar{z}^{**} = \bar{x}$, we have that $0 \le x(t) \le \bar{x}$ for any initial condition $0 \le x(0) = x_0 \le \bar{x}$. Therefore, the state variables of the closed-loop fractional-order system (8.43) are bounded, and the system (8.43) is positive and asymptotically stable for any initial condition $0 \le x(0) = x_0 \le \bar{x}$.

8.2.5 Simulation of Fractional Bounded Positive System

In this section, a fractional electrical circuit [162] is employed to illustrate the effectiveness of the proposed control scheme. The fractional electrical circuit is shown in Figure 8.5, where x_1 and x_2 are capacitor voltages, u_1 and u_2 are source voltages, V_1 and

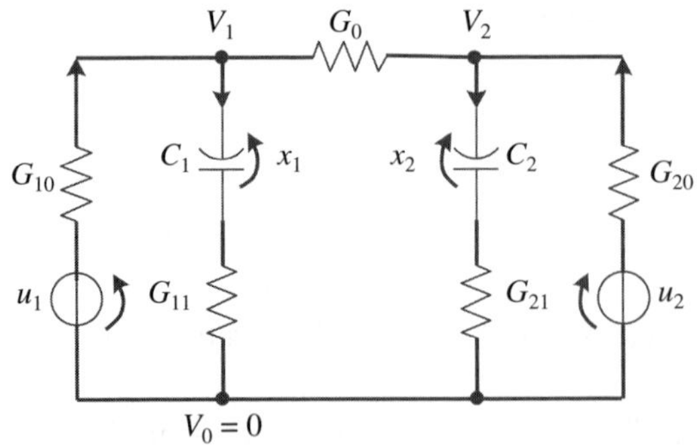

Figure 8.5 Fractional electrical circuit.

V_2 are node voltages, C_1 and C_2 are ultracapacitors with fractional order $\alpha = 0.8$, and $G_0 = 0$, G_{10}, G_{11}, G_{20}, and G_{21} are conductances. According to Figure 8.5, we have

$$\begin{aligned} C_1 D^\alpha x_1 &= G_{11}(V_1 - x_1), \\ C_2 D^\alpha x_2 &= G_{21}(V_2 - x_2). \end{aligned} \tag{8.46}$$

Furthermore, we obtain

$$\begin{aligned} (G_{10} + G_{11})V_1 &= G_{11}x_1 + G_{10}u_1, \\ (G_{20} + G_{21})V_2 &= G_{21}x_2 + G_{20}u_2. \end{aligned} \tag{8.47}$$

From Equation (8.47), this yields

$$\begin{aligned} V_1 &= \frac{G_{11}}{G_{10} + G_{11}}x_1 + \frac{G_{10}}{G_{10} + G_{11}}u_1, \\ V_2 &= \frac{G_{21}}{G_{20} + G_{21}}x_2 + \frac{G_{20}}{G_{20} + G_{21}}u_2. \end{aligned} \tag{8.48}$$

According to Equation (8.46), one gets

$$\begin{bmatrix} D^\alpha x_1 \\ D^\alpha x_2 \end{bmatrix} = \begin{bmatrix} -\frac{G_{11}}{C_1} & 0 \\ 0 & -\frac{G_{21}}{C_2} \end{bmatrix} \begin{bmatrix} x_1 \\ x_2 \end{bmatrix} + \begin{bmatrix} \frac{G_{11}}{C_1} & 0 \\ 0 & \frac{G_{21}}{C_2} \end{bmatrix} \begin{bmatrix} V_1 \\ V_2 \end{bmatrix}. \tag{8.49}$$

Combining Equations (8.48) and (8.49), we have

$$\begin{bmatrix} D^\alpha x_1 \\ D^\alpha x_2 \end{bmatrix} = A \begin{bmatrix} x_1 \\ x_2 \end{bmatrix} + B \begin{bmatrix} u_1 \\ u_2 \end{bmatrix}, \tag{8.50}$$

where

$$A = \begin{bmatrix} \frac{G_{11}}{C_1}\left(\frac{G_{11}}{G_{10}+G_{11}} - 1\right) & 0 \\ 0 & \frac{G_{21}}{C_2}\left(\frac{G_{21}}{G_{20}+G_{21}} - 1\right) \end{bmatrix},$$

$$B = \begin{bmatrix} \frac{G_{10}G_{11}}{C_1(G_{10}+G_{11})} & 0 \\ 0 & \frac{G_{20}G_{21}}{C_2(G_{20}+G_{21})} \end{bmatrix}.$$

According to this analysis, the fractional-order circuit always has perfect status. However, in the real world, many uncertain factors exist in circuits, such as consumption of internal resistance and wires, and so on. Thus, the disturbance is considered in Equation (8.50); it can be written as

$$\begin{bmatrix} D^\alpha x_1 \\ D^\alpha x_2 \end{bmatrix} = A \begin{bmatrix} x_1 \\ x_2 \end{bmatrix} + B \begin{bmatrix} u_1 \\ u_2 \end{bmatrix} + B \begin{bmatrix} d_1 \\ d_2 \end{bmatrix}. \tag{8.51}$$

Without loss of generality, to simplify the notation, we choose the values of C_1, C_2, α, G_0, G_{10}, G_{11}, G_{20}, and G_{21} as $C_1 = 1$, $C_2 = 2$, $\alpha = 0.8$, $G_0 = 0$, $G_{10} = 4$, $G_{11} = 4$, $G_{20} = 3$, and $G_{21} = 6$. Thus, the system matrices A and B are described as follows:

$$A = \begin{bmatrix} -2 & 0 \\ 0 & -1 \end{bmatrix}, \qquad B = \begin{bmatrix} 2 & 0 \\ 0 & 1 \end{bmatrix}. \tag{8.52}$$

According to the state feedback control method, a controller $u(t) = Kx(t) - \hat{d}$ is designed to stabilize the fractional-order system, so that the closed-loop system is non-negative and asymptotically stable. Therefore, the conditions in Theorem 8.3 must be satisfied. Based on the linear programming method, we obtain the control gain matrices F and K as follows:

$$F = \begin{bmatrix} 1.8364 & -0.9339 \\ -0.8524 & 2.0119 \end{bmatrix}, \qquad K = \begin{bmatrix} -0.6634 & 0.4165 \\ 0.6514 & -0.8937 \end{bmatrix}. \tag{8.53}$$

Furthermore, the bound of $z(t)$ is given by

$$\bar{z} = \begin{bmatrix} 67.3028 & 61.4325 & 94.3070 & 103.1118 \end{bmatrix}^{\mathrm{T}}. \tag{8.54}$$

On the basis of the augmented system (8.43), we obtain

$$Q = \begin{bmatrix} -1.8364 & 0.9339 & 0 & 0 \\ 0.8524 & -2.0119 & 0 & 0 \\ 2 & 0 & -2.6634 & 0.4165 \\ 0 & 1 & 0.6514 & -1.8973 \end{bmatrix}. \tag{8.55}$$

According to Equation (8.55), the eigenvalues of Q are given by $\lambda_1(Q) = -2.9262$, $\lambda_2(Q) = -2.8207$, $\lambda_3(Q) = -1.0276$, and $\lambda_4(Q) = -1.6309$. Therefore, the eigenvalues of Q are located in the left half of the complex plane. Then we obtain that the augmented system (8.43) is asymptotically stable. In addition, we obtain that Q is a Metzler matrix. Thus, for the initial condition $0 \leq z(0) = z_0 \leq \bar{z}$, the augmented system (8.43) will be positive. On the basis of this analysis, we obtain that the augmented system (8.43) is positive and asymptotically stable based on the conditions (8.53) and (8.55).

In this numerical simulation, the fractional order is chosen as $\alpha = 0.8$, and the constant disturbances are assumed as $d_1(t) = d_2(t) = 5$. Furthermore, we choose the initial conditions $x_1(0) = 0.1$, $x_2(0) = 0.2$, $\psi_1(0) = -0.2$, and $\psi_2(0) = -0.5$. Then, $\tilde{d}_1(0) = 5.1985$ and $\tilde{d}_2(0) = 5.1483$ are obtained. According to the numerical simulation method for fractional-order systems [164], Figure 8.6 shows the estimation performance of the FODO (8.23) and (8.24). It can be seen that good estimation performance is realized. For the numerical simulation of Equation (8.43), the simulation results are presented in Figure 8.7. From Figure 8.7, the state variables $x_1(t)$ and $x_2(t)$ are bounded, positive, and asymptotically stable. From these simulation results, we conclude that the FODO

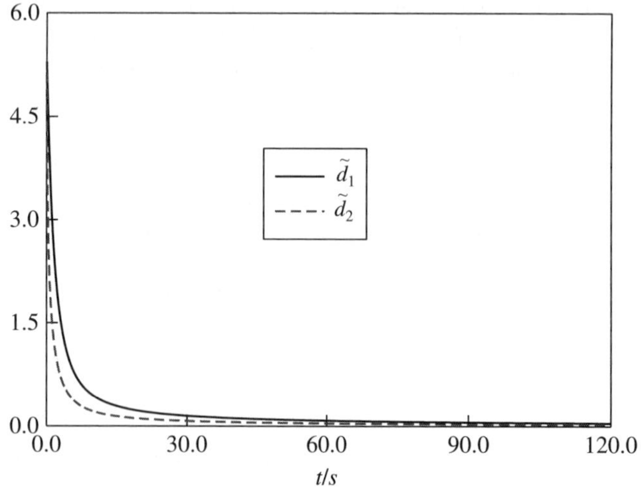

Figure 8.6 Disturbance estimation errors $\tilde{d}_1(t)$ and $\tilde{d}_2(t)$ for the simulation of the fractional electrical circuit (8.46).

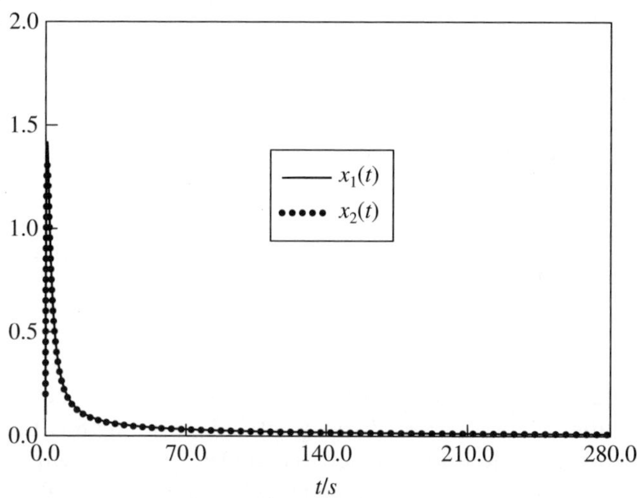

Figure 8.7 Responses of state variables $x_1(t)$ and $x_2(t)$ for the fractional electrical circuit (8.46) with FODO.

(8.23) and (8.24) is effective for constant disturbances, and the designed controller $u(t) = Kx(t) - \hat{d}$ can guarantee that the closed-loop system states are positive and asymptotically stable.

Conversely, the simulation results of the closed-loop control system without FODO are given in Figure 8.8. According to Figure 8.8, we know that the state variables $x_1(t)$ and $x_2(t)$ cannot be controlled to zero. However, the state variables $x_1(t)$ and $x_2(t)$ can converge to zero when the FODO is considered in the closed-loop control system, as

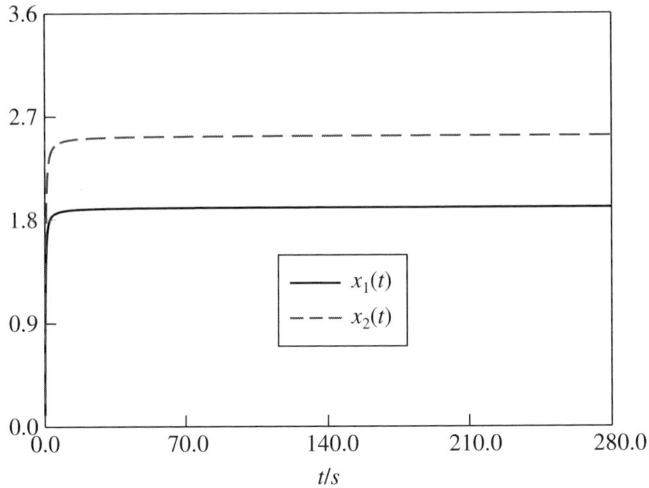

Figure 8.8 Responses of state variables $x_1(t)$ and $x_2(t)$ for the fractional electrical circuit (8.46) without FODO.

shown in Figure 8.8. As a result, the developed disturbance estimation technique can reject the effect of external disturbances. Therefore, the performance of the closed-loop system can be improved by the developed control scheme.

8.3 Conclusion

In this chapter, the FODO-based stabilization control scheme has been studied for a continuous-time fractional-order linear positive system in the presence of unknown constant disturbance. To improve the disturbance attenuation, a FODO has been employed to approximate the unknown disturbances. By using the developed FODO and the state feedback control method, a stabilization controller has been designed to guarantee that the closed-loop system states are positive and asymptotically stable. Furthermore, a sufficient condition of stabilization has been given for the case of constrained states of fractional-order positive systems with constant disturbances. At the same time, two numerical simulations have been shown to illustrate the effectiveness of the developed control scheme.

Conclusion

9

Sliding-Mode Synchronization Control for Fractional-Order Chaotic Systems with Disturbance

9.1 Problem Statement

On the basis of the Caputo definition of the fractional derivative (2.17), the FOCS will be introduced. Considering the following FOCS as the drive system:

$$D^{\alpha}x(t) = Ax(t) + f(x(t)), \tag{9.1}$$

where the fractional order satisfies $0 < \alpha < 1$, $A \in R^{n\times n}$ denotes a constant matrix, $x(t) = [x_1(t), x_2(t), \cdots, x_n(t)]^{\mathrm{T}} \in R^n$ is a state vector, and $f(x(t)) = [f_1(x(t)), f_2(x(t)), \cdots, f_n(x(t))]^{\mathrm{T}} \in R^n$ is the nonlinear function vector.

The response system is defined as

$$D^{\alpha}y(t) = Ay(t) + f(y(t)) + d(t) + u(t), \tag{9.2}$$

where $y(t) = [y_1(t), y_2(t), \cdots, y_n(t)]^{\mathrm{T}} \in R^n$ is the state vector, $f(y(t)) = [f_1(y(t)), f_2(y(t)), \cdots, f_n(y(t))]^{\mathrm{T}} \in R^n$ is the nonlinear function vector, $d(t) = [d_1(t), d_2(t), \cdots, d_n(t)]^{\mathrm{T}} \subset R^n$ is the unknown bounded disturbance, and $u(t) = [u_1(t), u_2(t), \cdots, u_n(t)]^{\mathrm{T}} \in R^n$ is the control input.

This chapter aims to develop a FODO-based adaptive sliding-mode synchronization control scheme, in which synchronization is realized between two identical fractional-order chaotic systems in the presence of unknown external disturbances. On the basis of the designed sliding-mode controller, the response system can well synchronize the drive system under the proper condition. To obtain the main results, the following assumption is introduced.

Assumption 9.1 *For the external disturbance $d_i(t)$ with $i = 1, 2, \ldots, n$, the Caputo fractional derivative of $d_i(t)$ is bounded, that is $|D^{\alpha}d_i(t)| \leq \zeta_i$, where $\zeta_i > 0$ is an unknown positive constant.*

9.2 Design of Fractional-Order Disturbance Observer

In this section, a FODO will be designed to approximate the external disturbance in the response system (9.2). Without loss of generality, according to the response system

Robust Adaptive Control for Fractional-Order Systems with Disturbance and Saturation, First Edition.
Mou Chen, Shuyi Shao, and Peng Shi.

(9.2), we have

$$D^{\alpha} y_i(t) = \theta_i + f_i(y(t)) + u_i(t) + d_i(t), \tag{9.3}$$

where θ_i is the ith element of $Ay(t)$, $y_i(t)$ is the ith element of $y(t)$, $f_i(y(t))$ is the ith element of $f(y(t))$, $u_i(t)$ is the ith element of $u(t)$, $d_i(t)$ is the ith element of $d(t)$, and $i = 1, 2, \ldots, n$.

Since $d(t)$ in Equation (9.2) is unknown, $d(t)$ cannot be applied to develop the synchronization control scheme for the drive system (9.1) and the response system (9.2). To overcome this problem, a fractional-order nonlinear disturbance observer is designed to estimate the disturbance $d_i(t)$.

To design the FODO, an auxiliary variable is introduced based on the design technique of the integer-order disturbance observer as follows [105]:

$$\phi_i(t) = d_i(t) - \sigma_i y_i(t), \tag{9.4}$$

where $\sigma_i > 0$ is a constant to be determined.

Considering Equations (9.3) and (9.4), the Caputo fractional derivative of $\phi_i(t)$ can be written as

$$\begin{aligned} D^{\alpha}\phi_i(t) &= D^{\alpha} d_i(t) - \sigma_i D^{\alpha} y_i(t) \\ &= -\sigma_i(\theta_i + f_i(y(t)) + d_i(t)) - \sigma_i u_i(t) + D^{\alpha} d_i(t) \\ &= -\sigma_i(\theta_i + f_i(y(t)) + \phi_i(t) + \sigma_i y_i(t)) - \sigma_i u_i(t) + D^{\alpha} d_i(t). \end{aligned} \tag{9.5}$$

To obtain the disturbance estimate, the estimate of the intermediate variable $\phi_i(t)$ is described as

$$D^{\alpha}\hat{\phi}_i = -\sigma_i(\theta_i + f_i(y(t)) + \sigma_i y_i(t)) - \sigma_i\hat{\phi}_i(t) - \sigma_i u_i(t), \tag{9.6}$$

where $\hat{\phi}_i$ is the estimate of ϕ_i.

According to Equation (9.4), the disturbance $d_i(t)$ can be estimated as

$$\hat{d}_i(t) = \hat{\phi}_i(t) + \sigma_i y_i(t). \tag{9.7}$$

Define $\tilde{d}_i(t) = d_i(t) - \hat{d}_i(t)$. Considering Equations (9.4) and (9.7), we have

$$\tilde{\phi}_i(t) = \phi_i(t) - \hat{\phi}_i(t) = d_i(t) - \hat{d}_i(t) = \tilde{d}_i(t). \tag{9.8}$$

Consider Equations (9.5) and (9.6); the Caputo fractional derivative of $\tilde{\phi}_i(t)$ can be written as

$$D^{\alpha}\tilde{\phi}_i(t) = -\sigma_i\tilde{\phi}_i(t) + D^{\alpha} d_i(t). \tag{9.9}$$

On the basis of these discussions, to analyze the convergence of disturbance estimation error $\tilde{d}_i(t)$, a Lyapunov function candidate can be chosen as

$$V_{\mathrm{d}} = \frac{1}{2}\tilde{d}_i^2(t) = \frac{1}{2}\tilde{\phi}_i^2(t). \tag{9.10}$$

Invoking Equation (9.10) and Lemma 2.1, the Caputo fractional derivative of V_{d} is described as

$$D^{\alpha} V_{\mathrm{d}}(t) \le \tilde{\phi}_i(t) D^{\alpha}\tilde{\phi}_i(t). \tag{9.11}$$

Substituting Equation (9.9) into Equation (9.11), we obtain

$$D^{\alpha} V_{\mathrm{d}}(t) \le \tilde{\phi}_i(t)(-\sigma_i\tilde{\phi}_i(t) + D^{\alpha} d_i(t)). \tag{9.12}$$

According to Equation (9.12) and Assumption 9.1, we have

$$\begin{aligned} D^{\alpha} V_{\mathrm{d}}(t) &\leq -\sigma_i \tilde{\phi}_i^2(t) + \frac{1}{2}\tilde{\phi}_i^2(t) + \frac{1}{2}\zeta_i^2 \\ &= -\left(\sigma_i - \frac{1}{2}\right)\tilde{\phi}_i^2(t) + \frac{1}{2}\zeta_i^2 \\ &= -B_0 V_{\mathrm{d}}(t) + B_1, \end{aligned} \tag{9.13}$$

where $B_0 = 2\sigma_i - 1$ and $B_1 = \frac{1}{2}\zeta_i^2$. To ensure that the estimation error is bounded, the design parameter σ_i of the FODO should be chosen to make $\sigma_i > 0.5$. The conclusion that the signals $\tilde{\phi}_i(t)$ and $\tilde{d}_i(t)$ are bounded can be drawn from Equation (9.13) and Lemma 2.3.

On the basis of Lemma 2.3 and Equation (9.13), we obtain

$$|V_{\mathrm{d}}(t)| \leq \frac{\zeta_i^2}{2(\sigma_i - 0.5)}, \tag{9.14}$$

which means that

$$|\tilde{d}_i(t)| \leq \sqrt{\frac{\zeta_i^2}{(\sigma_i - 0.5)}}, \tag{9.15}$$

which demonstrates that the disturbance estimation error $\tilde{d}_i$ is upper bounded. For the external disturbance $d_i(t)$, $i = 1, 2, \ldots, n$, the disturbance approximation error $\tilde{d}_i(t) = d_i(t) - \hat{d}_i(t)$ satisfies $|\tilde{d}_i(t)| \leq \kappa_i$, where $\kappa_i > 0$ is an unknown positive constant. In actual applications, the upper bound of $|\tilde{d}_i(t)|$ is difficult to determine. Thus, the estimated value $\hat{\kappa}_i$ of κ_i is introduced, where $i = 1, 2, \ldots, n$.

This design procedure of FODO can be summarized in the following theorem.

Theorem 9.1 *Consider the response system (9.2) and the FODO designed as Equations (9.6) and (9.7). The disturbance estimation error of the proposed FODO is bounded.*

On the basis of these analyses, Theorem 9.1 can be easily proven.

9.3 Disturbance-Observer-Based Synchronization Control of Fractional-Order Chaotic Systems

In this section, a FODO-based adaptive sliding-mode control scheme will be proposed to guarantee that the trajectories of drive system (9.1) and response system (9.2) are ultimately bounded synchronized. To design the synchronization control scheme, we first define the error variable $e(t) = y(t) - x(t)(e(t) = [e_1(t), e_2(t), \cdots, e_n(t)]^{\mathrm{T}} \in R^n)$. From Equations (9.1) and (9.2), the corresponding synchronization error system is given as

$$D^{\alpha} e(t) = Ae(t) + f(y(t)) - f(x(t)) + d(t) + u(t). \tag{9.16}$$

To investigate the stabilization of fractional-order synchronization error system (9.16), a simple sliding-mode surface is defined as

$$s_i(t) = e_i(t), \tag{9.17}$$

where $i = 1, 2, \ldots, n$.

From Equation (9.17), we have

$$\begin{aligned} D^{\alpha} s_i(t) &= D^{\alpha} e_i(t) \\ &= A_i e(t) + f_i(y(t)) - f_i(x(t)) + d_i(t) + u_i(t), \end{aligned} \tag{9.18}$$

where A_i denotes the ith row of A, $f_i(x(t))$ denotes the ith element of $f(x(t))$.

Using the adaptive sliding control approach, the desired synchronization control input is designed as

$$u_i(t) = -A_i e(t) - (f_i(y(t)) - f_i(x(t))) - \eta_i s_i(t) - \hat{\kappa}_i \mathrm{sign}(s_i(t)) - \hat{d}_i(t), \tag{9.19}$$

where sign$(\cdot)$ is the sign function and $\eta_i > 0$ is a design constant. Meanwhile, the estimated value $\hat{\kappa}_i$ is updated by

$$D^{\alpha} \hat{\kappa}_i = \gamma_i(|s_i(t)| - \hat{\kappa}_i), \tag{9.20}$$

where $\gamma_i > 0$ is a design constant.

If the synchronization control scheme is designed as Equation (9.19) for the fractional-order synchronization error system (9.16), the sliding-mode surface $s_i(t)$ satisfies

$$|s_i(t)| \leq a, \tag{9.21}$$

where $a > 0$ is an unknown constant.

From Equations (9.17) and (9.21), one obtains

$$|e_i(t)| \leq a. \tag{9.22}$$

According to this discussion, if the sliding surface $s_i(t)$ is bounded, then the synchronization error $e_i(t)$ is also bounded. Therefore, the FODO-based adaptive sliding-mode synchronization control scheme for fractional-order chaotic systems with unknown disturbances can be summarized in the following theorem, which will be proved using the fractional-order Lyapunov method.

Theorem 9.2 *For the synchronization error system (9.16) with $0 < \alpha < 1$, the sliding-mode surface is designed according to Equation (9.17). The unknown bounded external disturbance is estimated using the designed FODO (9.6) and (9.7). Then, the synchronization error $e(t)$ is ultimately bounded stable under the adaptive sliding control scheme (9.19) and (9.20).*

Proof 9.1 *To analyze the convergence of synchronization error $e(t)$, we consider the Lyapunov candidate function as*

$$V(t) = \sum_{i=1}^{n} \frac{1}{2} s_i^2(t) + \sum_{i=1}^{n} \frac{1}{2} \tilde{d}_i^2(t) + \sum_{i=1}^{n} \frac{1}{2} \left(\frac{1}{\sqrt{\gamma_i}} (\hat{\kappa}_i - \kappa_i) \right)^2. \tag{9.23}$$

According to Equation (9.23), we have

$$D^{\alpha} V(t) = \sum_{i=1}^{n} \frac{1}{2} D^{\alpha} s_i^2(t) + \sum_{i=1}^{n} \frac{1}{2} D^{\alpha} \left(\frac{1}{\sqrt{\gamma_i}} \tilde{\kappa}_i \right)^2 + \sum_{i=1}^{n} \frac{1}{2} D^{\alpha} \tilde{d}_i^2(t), \tag{9.24}$$

where $\tilde{\kappa}_i = \hat{\kappa}_i - \kappa_i$.

From Lemma 2.1, Equation (9.24) can be written as

$$D^\alpha V(t) \leq \sum_{i=1}^{n} s_i(t) D^\alpha s_i(t) + \sum_{i=1}^{n} \frac{1}{\sqrt{\gamma_i}} \tilde{\kappa}_i D^\alpha \left(\frac{1}{\sqrt{\gamma_i}} \tilde{\kappa}_i \right) + \sum_{i=1}^{n} \frac{1}{2} D^\alpha \tilde{d}_i^2(t). \tag{9.25}$$

Furthermore, Equation (9.25) is equivalent to

$$D^\alpha V(t) \leq \sum_{i=1}^{n} s_i(t) D^\alpha s_i(t) + \sum_{i=1}^{n} \frac{1}{\gamma_i} \tilde{\kappa}_i D^\alpha \tilde{\kappa}_i + \sum_{i=1}^{n} \frac{1}{2} D^\alpha \tilde{d}_i^2(t). \tag{9.26}$$

On the basis of Equation (9.18), one has

$$\begin{aligned} D^\alpha V(t) \leq & \sum_{i=1}^{n} s_i(t)(A_i e(t) + f_i(y(t)) - f_i(x(t))) + \sum_{i=1}^{n} \frac{1}{2} D^\alpha \tilde{d}_i^2(t) \\ & + \sum_{i=1}^{n} s_i(t)(d_i(t) + u_i(t)) + \sum_{i=1}^{n} \frac{1}{\gamma_i} \tilde{\kappa}_i D^\alpha \tilde{\kappa}_i. \end{aligned} \tag{9.27}$$

Referring to $\tilde{\kappa}_i = \hat{\kappa}_i - \kappa_i$, we obtain

$$D^\alpha \tilde{\kappa}_i = D^\alpha \hat{\kappa}_i. \tag{9.28}$$

According to Equations (9.20) and (9.28), we have

$$\begin{aligned} \sum_{i=1}^{n} \frac{1}{\gamma_i} \tilde{\kappa}_i D^\alpha \tilde{\kappa}_i &= \sum_{i=1}^{n} \tilde{\kappa}_i (|s_i(t)| - \hat{\kappa}_i) \\ &= \sum_{i=1}^{n} \tilde{\kappa}_i |s_i(t)| - \sum_{i=1}^{n} \tilde{\kappa}_i \hat{\kappa}_i \\ &\leq \sum_{i=1}^{n} \tilde{\kappa}_i |s_i(t)| - \sum_{i=1}^{n} \frac{1}{2} \tilde{\kappa}_i^2 + \sum_{i=1}^{n} \frac{1}{2} \kappa_i^2. \end{aligned} \tag{9.29}$$

Invoking Equations (9.27) and (9.29), we obtain

$$\begin{aligned} D^\alpha V(t) \leq & \sum_{i=1}^{n} s_i(t)(A_i e(t) + f_i(y(t)) - f_i(x(t))) + \sum_{i=1}^{n} s_i(t)(d_i(t) + u_i(t)) \\ & + \sum_{i=1}^{n} \tilde{\kappa}_i |s_i(t)| - \sum_{i=1}^{n} \frac{1}{2} \tilde{\kappa}_i^2 + \sum_{i=1}^{n} \frac{1}{2} \kappa_i^2 + \sum_{i=1}^{n} \frac{1}{2} D^\alpha \tilde{d}_i^2(t). \end{aligned} \tag{9.30}$$

Substituting Equation (9.19) into Equation (9.30) yields

$$\begin{aligned} D^\alpha V(t) \leq & \sum_{i=1}^{n} s_i(t)(-\eta_i s_i(t) + \tilde{d}_i(t) - \hat{\kappa}_i \operatorname{sign}(s_i(t))) \\ & + \sum_{i=1}^{n} \tilde{\kappa}_i |s_i(t)| - \sum_{i=1}^{n} \frac{1}{2} \tilde{\kappa}_i^2 + \sum_{i=1}^{n} \frac{1}{2} \kappa_i^2 + \sum_{i=1}^{n} \frac{1}{2} D^\alpha \tilde{d}_i^2(t). \end{aligned} \tag{9.31}$$

Furthermore, Equation (9.31) can be rewritten as

$$\begin{aligned} D^\alpha V(t) \leq & \sum_{i=1}^{n} -\eta_i s_i^2(t) + \sum_{i=1}^{n} |s_i(t)||\tilde{d}_i(t)| + \sum_{i=1}^{n} \tilde{\kappa}_i |s_i(t)| - \sum_{i=1}^{n} \frac{1}{2} \tilde{\kappa}_i^2 + \sum_{i=1}^{n} \frac{1}{2} \kappa_i^2 \\ & - \sum_{i=1}^{n} \hat{\kappa}_i |s_i(t)| + \sum_{i=1}^{n} \frac{1}{2} D^\alpha \tilde{d}_i^2(t), \end{aligned} \tag{9.32}$$

with

$$\sum_{i=1}^{n} \tilde{\kappa}_i |s_i(t)| - \sum_{i=1}^{n} \hat{\kappa}_i |s_i(t)| = -\sum_{i=1}^{n} \kappa_i |s_i(t)|. \tag{9.33}$$

According to Equation (9.33) and $|\tilde{d}_i(t)| \le \kappa_i$, this yields

$$D^\alpha V(t) \le \sum_{i=1}^{n} -\eta_i s_i^2(t) - \sum_{i=1}^{n} \frac{1}{2}\tilde{\kappa}_i^2 + \sum_{i=1}^{n} \frac{1}{2}\kappa_i^2 + \sum_{i=1}^{n} \frac{1}{2} D^\alpha \tilde{d}_i^2(t). \tag{9.34}$$

Considering Equations (9.13) and (9.34), we have

$$\begin{aligned} D^\alpha V(t) &\le \sum_{i=1}^{n} -\eta_i s_i^2(t) - \sum_{i=1}^{n} \frac{1}{2}\tilde{\kappa}_i^2 + \sum_{i=1}^{n} \frac{1}{2}\kappa_i^2 \sum_{i=1}^{n} -\left(\sigma_i - \frac{1}{2}\right)\tilde{d}_i^2(t) + \sum_{i=1}^{n} \frac{1}{2}\zeta_i^2 \\ &\le -B_2 V(t) + B_3, \end{aligned} \tag{9.35}$$

where $B_2 = \min(2\eta_i, 1, 2\sigma_i - 1)$ *and*

$$B_3 = \sum_{i=1}^{n} \frac{1}{2}\zeta_i^2 + \sum_{i=1}^{n} \frac{1}{2}\kappa_i^2.$$

To ensure that the synchronization error is bounded, the corresponding design parameters η_i and σ_i should be chosen to make $\eta_i > 0$ and $\sigma_i > 0.5$. According to Equation (9.35) and Lemma 2.3, it may directly show that the signals $s(t)$, $e(t)$, and $\tilde{d}_i(t)$ are ultimately bounded. From Lemma 2.3 and Equation (9.35), we obtain

$$|V(t)| \le \frac{\sum_{i=1}^{n} \zeta_i^2 + \sum_{i=1}^{n} \kappa_i^2}{B_2}, \tag{9.36}$$

which implies that

$$\|s(t)\| \le \sqrt{\frac{2\left(\sum_{i=1}^{n} \zeta_i^2 + \sum_{i=1}^{n} \kappa_i^2\right)}{B_2}}. \tag{9.37}$$

From the inequality (9.37), we can obtain that the synchronization error $e(t)$ and $s(t)$ will be ultimately bounded as $t \to \infty$. Therefore, the synchronization error system (9.16) is bounded stable based on Lemma 2.3. The bounded synchronization of the drive system (9.1) and the response system (9.2) is achieved. This concludes the proof. □

9.4 Simulation Examples

9.4.1 Synchronization Control of Modified Fractional-Order Jerk System

Yu [232] investigated a new chaotic generator by constructing a three-segment piecewise-linear odd function with variable breakpoint. From the differential equation of chaotic generator [232], the modified fractional-order jerk system is given as

Table 9.1 Equilibrium points of the modified fractional-order jerk system.

Linear region	$f_3(x(t))$	Equilibrium points
$x_1(t) > 1$	$-0.5x_1(t) - 2$	$Q_+ = (2, 0, 0)$
$-1 \leq x_1(t) \leq 1$	$-2.5x_1(t)$	$Q_0 = (0, 0, 0)$
$x_1(t) < -1$	$-0.5x_1(t) + 2$	$Q_+ = (-2, 0, 0)$

follows:

$$\begin{aligned} D^\alpha x_1(t) &= x_2(t), \\ D^\alpha x_2(t) &= x_3(t), \\ D^\alpha x_3(t) &= -\varepsilon_1 x_1(t) - x_2(t) - \varepsilon_2 x_3(t) - f_3(x(t)), \end{aligned} \tag{9.38}$$

where $x_1(t)$, $x_2(t)$, and $x_3(t)$ are system state variables, the parameters $\varepsilon_1 = 1.5$ and $\varepsilon_2 = 0.35$, and $f_3(x(t))$ is a piecewise linear function defined by

$$f_3(x(t)) = \frac{1}{2}(\vartheta_0 - \vartheta_1)(|x_1(t) + 1| - |x_1(t) - 1|) + \vartheta_1 x_1(t), \tag{9.39}$$

where $\vartheta_0 < -1 < \vartheta_1 < 0$ and $\vartheta_0 = -2.5$, $\vartheta_1 = -0.5$.

According to the system (9.38) and the piecewise linear function (9.39), the three equilibrium points of the modified fractional-order jerk system are given in Table 9.1. The Jacobian matrix for the system (9.38) can be written as

$$J = \begin{bmatrix} 0 & 1 & 0 \\ 0 & 0 & 1 \\ -1.5 - \frac{\partial f_3(x(t))}{\partial x_1(t)} & -1 & -0.35 \end{bmatrix}. \tag{9.40}$$

On the basis of Table 9.1, the corresponding eigenvalues for equilibrium point Q_0 are $\lambda_1 = 0.6228$ and $\lambda_{2,3} = -0.4864 \pm 1.1701i$. For equilibrium points Q_+ and Q_-, the eigenvalues are $\lambda_1 = -0.7614$ and $\lambda_{2,3} = 0.2057 \pm 1.1274i$. When the fractional order $\alpha = 0.98$ is chosen, we obtain the following characteristic equation of the equilibrium points Q_+ and Q_-:

$$\lambda^{294} + 0.35\lambda^{196} + \lambda^{98} + 1 = 0, \tag{9.41}$$

with unstable $\lambda_{1,2} = 1.0013 \pm 0.0142i$ and $|\arg(\lambda_{1,2})| = 0.0142 < \pi/(2v) = 0.0157$, in which $\vartheta = 100$ (ϑ is the lowest common multiple of the fractional-order denominator). Thus, the modified fractional-order jerk system (9.38) has chaotic dynamic behaviors based on the literature [164]. When the initial values are chosen as $(1, 1, 1)^{\mathrm{T}}$ and the fractional order $\alpha = 0.98$, the fractional-order modified jerk system exhibits chaotic behaviors, as shown in Figure 9.1.

In this section, to illustrate the effectiveness of the proposed synchronization controller, the synchronization of the modified fractional-order jerk system (9.38) is investigated. Consider the FOCS (9.38) as a drive system. From Equation (9.2), the response system is defined as follows:

$$\begin{aligned} D^\alpha y_1(t) &= y_2(t) + d_1(t) + u_1(t), \\ D^\alpha y_2(t) &= y_3(t) + d_2(t) + u_2(t), \\ D^\alpha y_3(t) &= -\varepsilon_1 y_1(t) - y_2(t) - \varepsilon_2 y_3(t) - f_3(y(t)) + d_3(t) + u_3(t), \end{aligned} \tag{9.42}$$

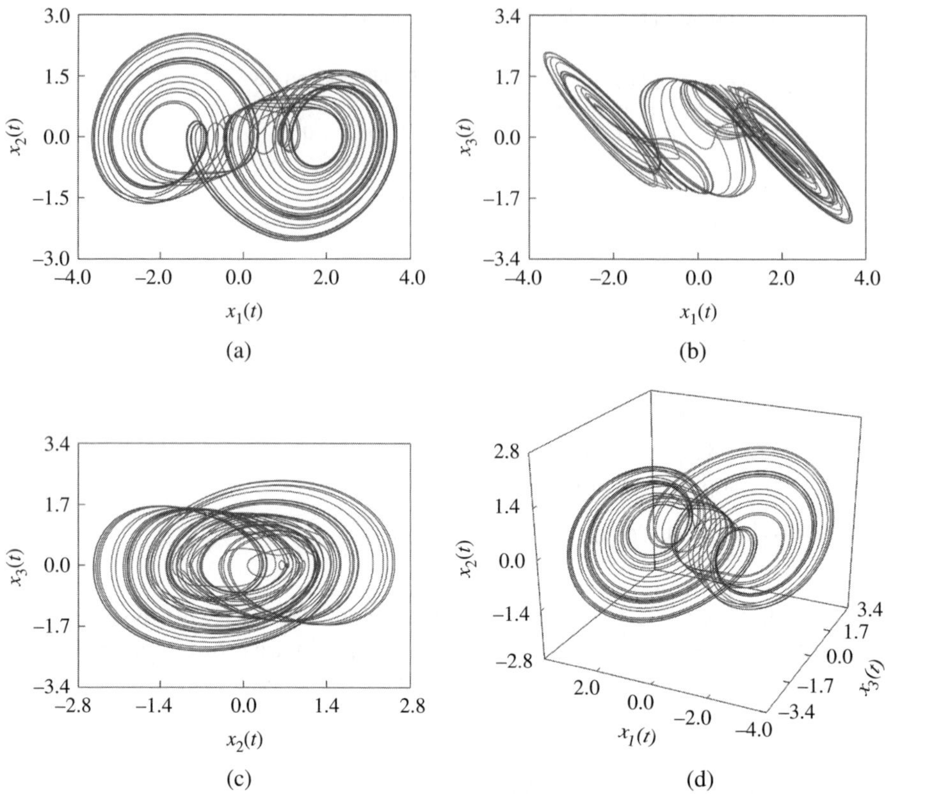

Figure 9.1 Chaotic behaviors of modified fractional-order jerk system: (a) $x_1(t)$–$x_2(t)$ plane; (b) $x_1(t)$–$x_3(t)$ plane; (c) $x_2(t)$–$x_3(t)$ plane; (d) $x_3(t)$–$x_1(t)$–$x_2(t)$ space.

where $y_1(t)$, $y_2(t)$, and $y_3(t)$ are system state variables, $d_1(t)$, $d_2(t)$, and $d_3(t)$ are unknown bounded disturbances, $u_1(t)$, $u_2(t)$, and $u_3(t)$ are designed synchronization control inputs, and $f_3(y(t))$ is defined as

$$f_3(y(t)) = \frac{1}{2}(\vartheta_0 - \vartheta_1)(|y_1(t) + 1| - |y_1(t) - 1|) + \vartheta_1 y_1(t). \tag{9.43}$$

According to Equations (9.38) and (9.42), the synchronization error system can be written as follows:

$$\begin{aligned} D^\alpha e_1(t) &= e_2(t) + d_1(t) + u_1(t), \\ D^\alpha e_2(t) &= e_3(t) + d_2(t) + u_2(t), \\ D^\alpha e_3(t) &= -\varepsilon_1 e_1(t) - e_2(t) - \varepsilon_2 e_3(t) - f_3(y(t)) + f_3(x(t)) + d_3(t) + u_3(t), \end{aligned} \tag{9.44}$$

where $e_1(t)$, $e_2(t)$, and $e_3(t)$ are synchronization error variables.

Referring to the designed controller (9.19), the synchronization controller can be written as follows:

$$\begin{aligned} u_1(t) &= -e_2(t) - \eta_1 s_1(t) - \hat{\kappa}_1 \mathrm{sign}(s_1(t)) - \hat{d}_1(t), \\ u_2(t) &= -e_3(t) - \eta_2 s_2(t) - \hat{\kappa}_2 \mathrm{sign}(s_2(t)) - \hat{d}_2(t), \end{aligned}$$

$$u_3(t) = \varepsilon_1 e_1(t) + e_2(t) + \varepsilon_2 e_3(t) + f_3(y(t)) - f_3(x(t)) - \eta_3 s_3(t)$$
$$- \hat{\kappa}_3 \text{sign}(s_3(t)) - \hat{d}_3(t). \tag{9.45}$$

Substituting Equation (9.45) into Equation (9.44), we have the following:

$$D^\alpha e_1(t) = -\eta_1 s_1(t) - \hat{\kappa}_1 \text{sign}(s_1(t)) + \tilde{d}_1(t),$$
$$D^\alpha e_2(t) = -\eta_2 s_2(t) - \hat{\kappa}_2 \text{sign}(s_2(t)) + \tilde{d}_2(t),$$
$$D^\alpha e_3(t) = -\eta_3 s_3(t) - \hat{\kappa}_3 \text{sign}(s_3(t)) + \tilde{d}_3(t), \tag{9.46}$$

where $D^\alpha \hat{\kappa}_i = \gamma_i(|s_i(t)| - \hat{\kappa}_i)$, with $\gamma_i > 0$ and $i = 1, 2, 3$.

To demonstrate the effectiveness of the proposed FODO-based adaptive sliding-mode synchronization control scheme, numerical simulation results are presented for the modified fractional-order jerk system under the following conditions: initial conditions are chosen as $x_0(t) = [1, 1, 1]^\top$, $y_0(t) = [1.2, 0.6, 0.5]^\top$, $\hat{\kappa}_0 = [0.1, 0.1, 0.1]^\top$, and $\phi_0(t) = [0.1, 0.1, 0.1]^\top$, design parameters are chosen as $\alpha = 0.98$, $\sigma_1 = \sigma_2 = \sigma_3 = 50$, $\gamma_1 = \gamma_2 = \gamma_3 = 0.1$, and $\eta_1 = \eta_2 = \eta_3 = 50$. Disturbances are assumed as $d_1(t) = \cos(2t)$, $d_2(t) = \cos(2t)$, and $d_3(t) = \cos(2t)$. On the basis of the result of Ishteva [225], we have $\rho_1 D^\alpha \cos(\rho_2 t) = \rho_1 \frac{1}{2} (\text{i}\rho_2)^m t^{m-\alpha} (E_{1,m-\alpha+1}(\text{i}\rho_2 t) + (-1)^n E_{1,m-\alpha+1}(-\text{i}\rho_2 t))$, where i denotes the square root of minus one, and ρ_1 and ρ_2 are arbitrary numbers. In this simulation, the parameter $m = 1$ and the fractional order $\alpha = 0.98$. Thus, $\rho_1 \rho_2^\alpha \cos(\rho_2 t + (\pi\alpha/2))$ can be used to approximate $\rho_1 D^\alpha \cos(\rho_2 t)$. The comparison result is shown in Figure 9.2 for $\rho_1 = 1$ and $\rho_2 = 2$. According to Figure 9.2, Assumption 9.1 is satisfied.

Numerical results are shown in Figure 9.3 and Figure 9.4 under the proposed FODO-based adaptive sliding-mode control scheme. State synchronization results of the drive system (9.38) and the response system (9.42) are given in Figure 9.3a–c. It is shown that good synchronization performance is achieved. Figure 9.3d shows that the synchronization errors $e_1(t)$, $e_2(t)$, and $e_3(t)$ are convergent. Furthermore, the disturbance observation performance of the proposed FODO (9.6) and (9.7) is presented in Figure 9.4. It is evident from Figure 9.4 that the disturbance observer is effective and feasible. According to the simulation results, the drive system (9.38) and the response

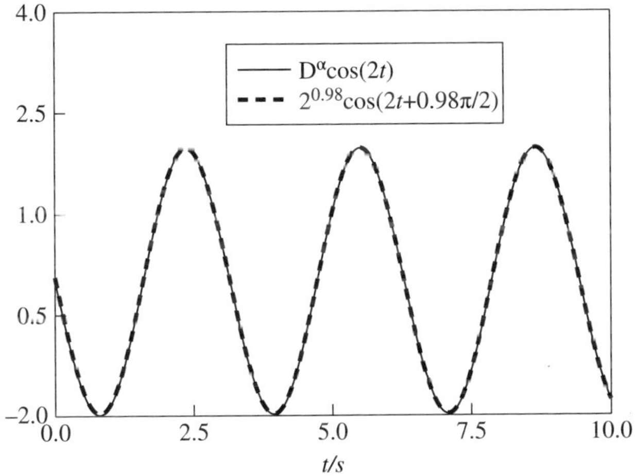

Figure 9.2 Comparison result of $2^{0.98} \cos(2t + (0.98\pi/2))$ and $D^\alpha \cos(2t)$.

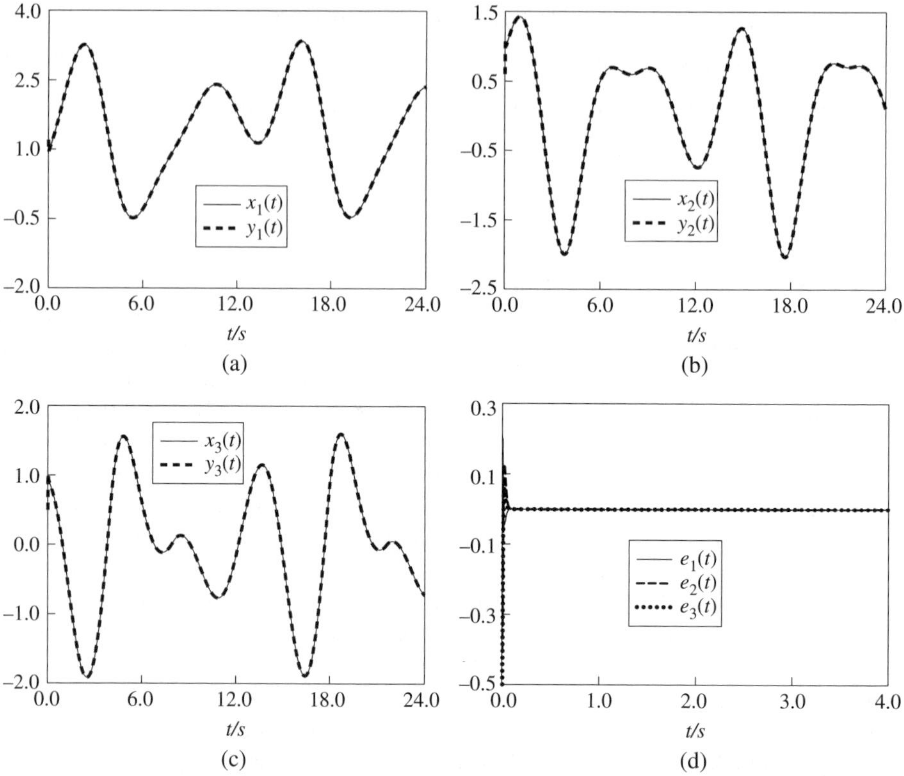

Figure 9.3 Synchronization control results of modified fractional-order jerk system: (a) $x_1(t)$ and $y_1(t)$; (b) $x_2(t)$ and $y_2(t)$; (c) $x_3(t)$ and $y_3(t)$; (d) synchronization errors $e_1(t)$, $e_2(t)$, and $e_3(t)$.

system (9.42) are bounded synchronized under the designed sliding-mode controller (9.19) and the adaptive update law (9.20). Therefore, the proposed FODO-based adaptive sliding-mode synchronization control scheme is valid for fractional-order chaotic systems with external disturbance.

9.4.2 Synchronization Control of Fractional-Order Liu System

To further illustrate the effectiveness of the proposed synchronization controller, synchronization of the fractional-order Liu system [233] is studied in this section. The fractional-order Liu system is given as follows:

$$\begin{aligned} D^{\alpha}x_1(t) &= -\hbar_1 x_1(t) - x_2^2(t), \\ D^{\alpha}x_2(t) &= \hbar_2 x_2(t) - 4x_1(t)x_3(t), \\ D^{\alpha}x_3(t) &= -\hbar_3 x_3(t) + 4x_1(t)x_2(t), \end{aligned} \tag{9.47}$$

where α is the fractional order, $x_1(t)$, $x_2(t)$, and $x_3(t)$ are system state variables, and $\hbar_1$, $\hbar_2$, and $\hbar_3$ are system parameters. The fractional order is chosen as $\alpha = 0.99$, the system parameters are set as $\hbar_1 = 1$, $\hbar_2 = 2.5$, and $\hbar_3 = 5$ and the initial conditions are chosen as $x_1(0) = x_2(0) = x_3(0) = 0.1$. The simulation results of the fractional-order Liu system are shown in Figure 9.5.

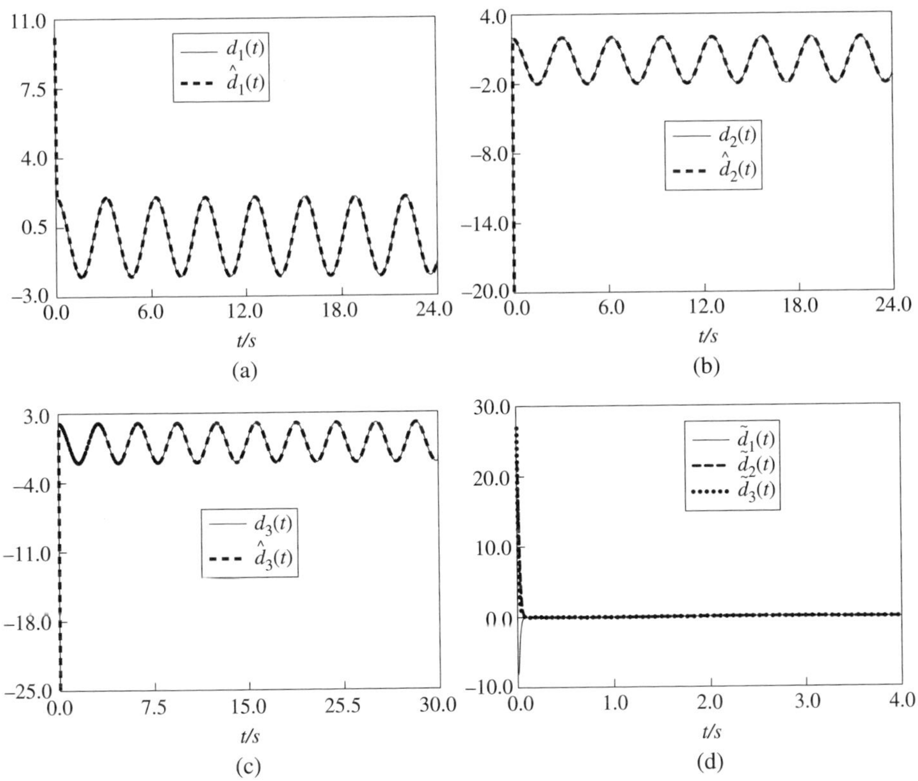

Figure 9.4 Disturbance observer results of the modified fractional-order jerk system: (a) $d_1(t)$ and $\hat{d}_1(t)$; (b) $d_2(t)$ and $\hat{d}_2(t)$; (c) $d_3(t)$ and $\hat{d}_3(t)$; (d) observation errors $\tilde{d}_1(t)$, $\tilde{d}_2(t)$, and $\tilde{d}_3(t)$.

To develop the synchronization control scheme, the fractional-order Liu system (9.47) is taken as the drive system, and the response system is constructed as follows.

$$\begin{aligned} D^{\alpha}y_1(t) &= -\hbar_1 y_1(t) - y_2^2(t) + d_1(t) + u_1(t), \\ D^{\alpha}y_2(t) &= \hbar_2 y_2(t) - 4y_1(t)y_3(t) + d_2(t) + u_2(t), \\ D^{\alpha}y_3(t) &= -\hbar_3 y_3(t) + 4y_1(t)y_2(t) + d_3(t) + u_3(t), \end{aligned} \tag{9.48}$$

where $y_1(t)$, $y_2(t)$, and $y_3(t)$ are system state variables, $d_1(t)$, $d_2(t)$, and $d_3(t)$ are unknown bounded disturbances, and $u_1(t)$, $u_2(t)$, and $u_3(t)$ are designed synchronization control inputs.

According to Equations (9.47) and (9.48), the synchronization error system can be written as follows:

$$\begin{aligned} D^{\alpha}e_1(t) &= -\hbar_1 e_1(t) - y_2^2(t) + x_2^2(t) + d_1(t) + u_1(t), \\ D^{\alpha}e_2(t) &= \hbar_2 e_2(t) - 4y_1(t)y_3(t) + 4x_1(t)x_3(t) + d_2(t) + u_2(t), \\ D^{\alpha}e_3(t) &= -\hbar_3 e_3(t) + 4y_1(t)y_2(t) - 4x_1(t)x_2(t) + d_3(t) + u_3(t), \end{aligned} \tag{9.49}$$

where $e_1(t)$, $e_2(t)$, and $e_3(t)$ are synchronization error variables.

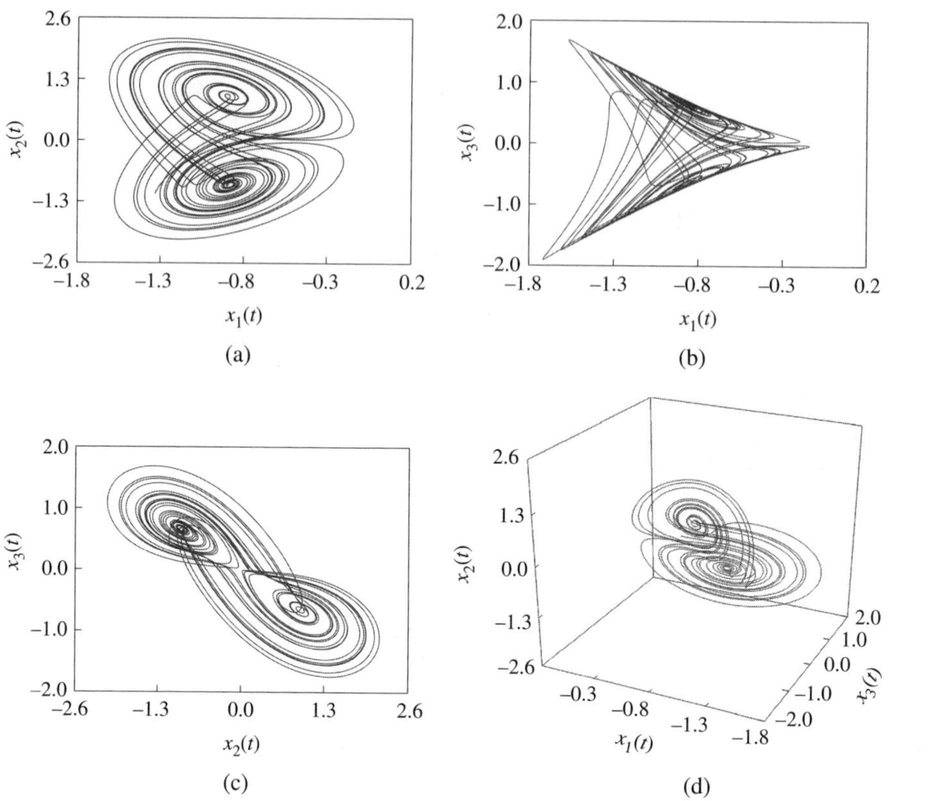

Figure 9.5 Dynamic behaviors of fractional-order Liu system: (a) $x_1(t)$–$x_2(t)$ plane; (b) $x_1(t)$–$x_3(t)$ plane; (c) $x_2(t)$–$x_3(t)$ plane; (d) $x_3(t)$–$x_1(t)$–$x_2(t)$ space.

Invoking the designed controller (9.19), the synchronization controller is given by the following:

$$\begin{aligned}
u_1(t) &= \hbar_1 e_1(t) + y_2^2(t) - x_2^2(t) - \eta_1 s_1(t) - \hat{\kappa}_1 \mathrm{sign}(s_1(t)) - \hat{d}_1(t),\\
u_2(t) &= -\hbar_2 e_2(t) + 4y_1(t)y_3(t) - 4x_1(t)x_3(t) - \eta_2 s_2(t) - \hat{\kappa}_2 \mathrm{sign}(s_2(t)) - \hat{d}_2(t),\\
u_3(t) &= \hbar_3 e_3(t) - 4y_1(t)y_2(t) + 4x_1(t)x_2(t) - \eta_3 s_3(t) - \hat{\kappa}_3 \mathrm{sign}(s_3(t)) - \hat{d}_3(t).
\end{aligned} \tag{9.50}$$

Substituting Equation (9.50) into Equation (9.49), we have the following:

$$\begin{aligned}
D^\alpha e_1(t) &= -\eta_1 s_1(t) - \hat{\kappa}_1 \mathrm{sign}(s_1(t)) + \tilde{d}_1(t),\\
D^\alpha e_2(t) &= -\eta_2 s_2(t) - \hat{\kappa}_2 \mathrm{sign}(s_2(t)) + \tilde{d}_2(t),\\
D^\alpha e_3(t) &= -\eta_3 s_3(t) - \hat{\kappa}_3 \mathrm{sign}(s_3(t)) + \tilde{d}_3(t),
\end{aligned} \tag{9.51}$$

where $D^\alpha \hat{\kappa}_i = \gamma_i(|s_i(t)| - \hat{\kappa}_i)$, with $\gamma_i > 0$ and $i = 1, 2, 3$.

For the numerical simulation, we choose the fractional order as $\alpha = 0.99$; the disturbances are assumed as $d_1(t) = d_2(t) = d_3(t) = 2\sin(2t)$. The initial conditions are chosen as $x_0(t) = [0.5, 0.5, 0.5]^\mathrm{T}$, $y_0(t) = [1, 1.1, 1.5]^\mathrm{T}$, $\hat{\kappa}_0 = [0.2, 0.2, 0.2]^\mathrm{T}$, and $\phi_0(t) = [0.2, 0.2, 0.2]^\mathrm{T}$. The control parameters are designed as $\sigma_1 = \sigma_2 = \sigma_3 = 40$, $\gamma_1 = \gamma_2 = \gamma_3 = 0.2$, and $\eta_1 = \eta_2 = \eta_3 = 60$.

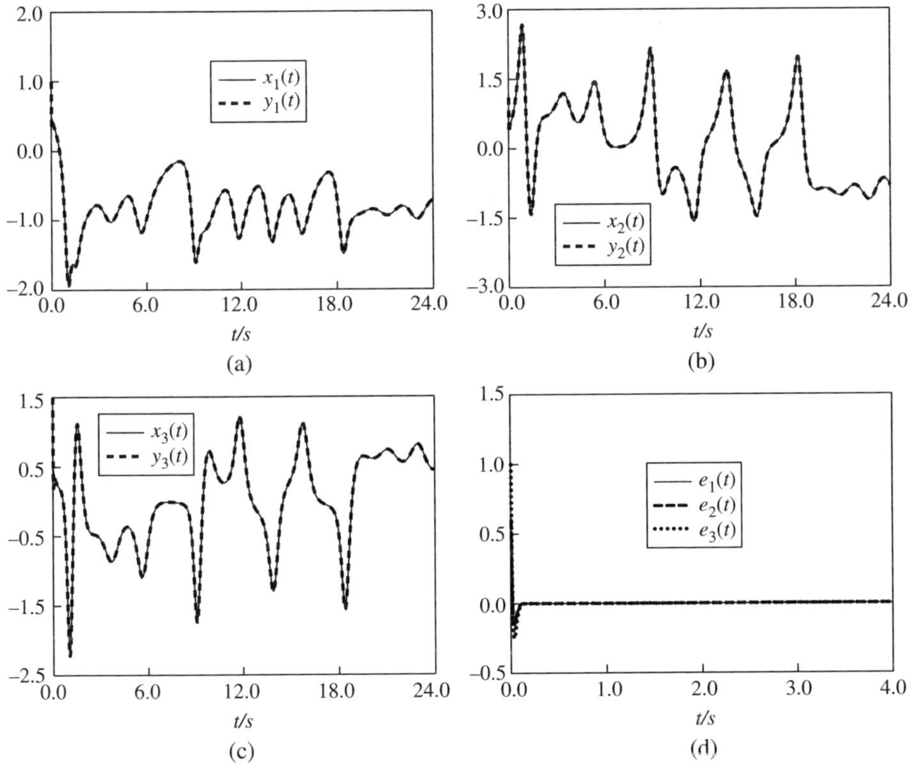

Figure 9.6 Synchronization control results of fractional-order Liu system: (a) $x_1(t)$ and $y_1(t)$; (b) $x_2(t)$ and $y_2(t)$; (c) $x_3(t)$ and $y_3(t)$; (d) synchronization errors $e_1(t)$, $e_2(t)$, and $e_3(t)$.

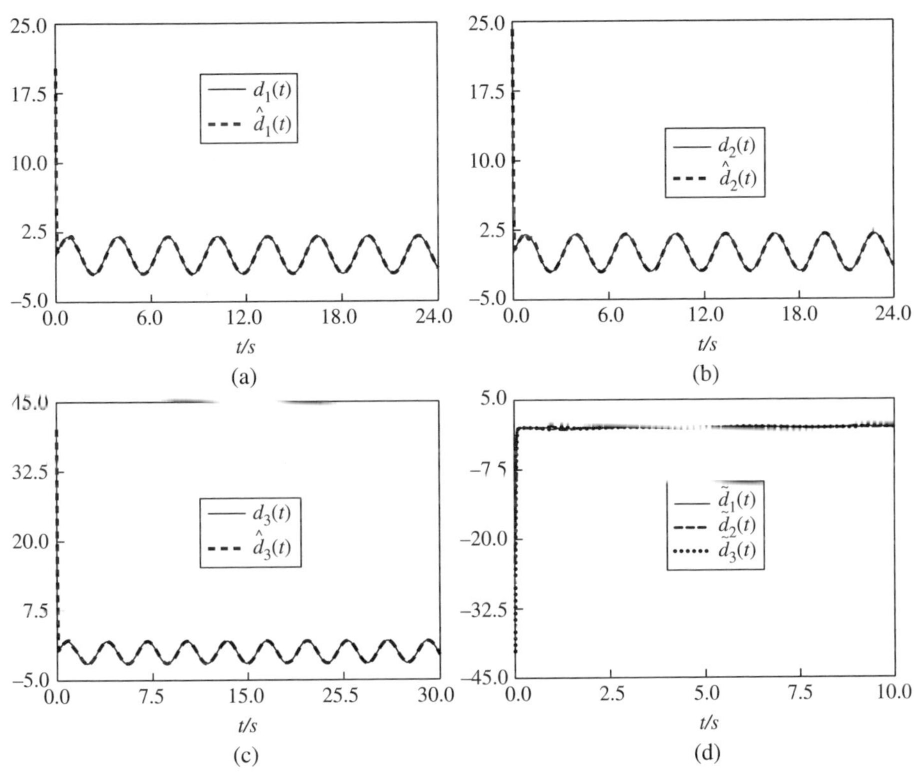

Figure 9.7 Disturbance observer results of the fractional-order Liu system: (a) $d_1(t)$ and $\hat{d}_1(t)$; (b) $d_2(t)$ and $\hat{d}_2(t)$; (c) $d_3(t)$ and $\hat{d}_3(t)$; (d) observation errors $\tilde{d}_1(t)$, $\tilde{d}_2(t)$, and $\tilde{d}_3(t)$.

According to these conditions and the proposed synchronization control scheme, numerical results are presented in Figure 9.6 and Figure 9.7. Good synchronization performance is shown in Figure 9.6a–c. Numerical results of the synchronization errors $e_1(t)$, $e_2(t)$, and $e_3(t)$ are given in Figure 9.6d. Furthermore, the observation performance of the proposed FODO (9.6) and (9.7) is presented in Figure 9.7. It shows that the disturbance observer is effective based on the estimation performance of the designed FODO. On the basis of the simulation results, the drive system (9.47) can synchronize the response system (9.48) well based on the designed sliding-mode controller (9.19) and the adaptive update law (9.20). Thus the proposed adaptive sliding-mode synchronization control method is effective for fractional-order chaotic systems with external disturbance.

9.5 Conclusion

In this chapter, a FODO-based adaptive sliding-mode synchronization control scheme has been studied for fractional-order chaotic systems in the presence of external disturbance. A FODO has been developed to approximate the unknown disturbances. A sliding-mode synchronization controller has been designed based on the FODO for synchronization of fractional-order chaotic systems. Furthermore, two examples are given, i.e., synchronization between two modified fractional-order jerk systems and synchronization between two fractional-order Liu systems. Numerical simulations show the effectiveness of the proposed FODO-based adaptive sliding-mode synchronization control scheme.

10

Anti-Synchronization Control for Fractional-Order Nonlinear Systems Using Disturbance Observer and Neural Networks

10.1 Problem Statement

According to the Caputo definition of the fractional derivative (2.17), the fractional-order nonlinear system will be described. Considering the following fractional-order nonlinear system as the master system:

$$D^{\alpha}x(t) = Ax(t) + g(x(t)), \tag{10.1}$$

where $A \in R^{n\times n}$ denotes the constant matrix, $x(t) = [x_1(t), x_2(t), \cdots, x_n(t)]^{\top} \in R^n$ is the state vector, $g(x(t)) = [g_1(x(t)), g_2(x(t)), \cdots, g_n(x(t))]^{\top} \in R^n$ is the nonlinear function vector, and the fractional order satisfies $0 < \alpha < 1$.

The slave system with system uncertainty and external disturbance is given as follows:

$$D^{\alpha}y(t) = Ay(t) + g(y(t)) + \Delta g(y(t)) + d(t) + u(t), \tag{10.2}$$

where $y(t) = [y_1(t), y_2(t), \cdots, y_n(t)]^{\top} \in R^n$ is the state vector, $g(y(t)) = [g_1(y(t)), g_2(y(t)), \cdots, g_n(y(t))]^{\top} \in R^n$ is the nonlinear function vector, $d(t) = [d_1(t), d_2(t), \cdots, d_n(t)]^{\top} \in R^n$ is the unknown bounded disturbance, $u(t) = [u_1(t), u_2(t), \cdots, u_n(t)]^{\top} \in R^n$ is the control input, and $\Delta g(y(t)) = [\Delta g_1(y(t)), \Delta g_2(y(t)), \cdots, \Delta g_n(y(t))]^{\top} \in R^n$ is the system uncertainty.

This chapter aims to develop a disturbance observer-based anti-synchronization control scheme, so that anti-synchronization is realized between two fractional-order nonlinear systems in the presence of external disturbances and system uncertainties. On the basis of the designed anti-synchronization controller, synchronization errors between the master system and the slave system are convergent under proper conditions. To obtain the main results, the following assumption is introduced.

Assumption 10.1 *The Caputo fractional derivative of external disturbances $d_i(t)(i = 1, 2, \dots, n)$ is bounded, that is $|D^{\alpha}d_i(t)| \leq \varpi_i$, where $\varpi_i > 0$ is an unknown positive constant.*

10.2 Design of Disturbance Observer

In this section, a neural-network-based disturbance observer will be designed to estimate the external disturbance $d(t)$ in the slave system (10.2). On the basis of the slave

Robust Adaptive Control for Fractional-Order Systems with Disturbance and Saturation, First Edition.
Mou Chen, Shuyi Shao, and Peng Shi.

system (10.2), the following equation of state is given:

$$D^\alpha y_i(t) = \vartheta_i + g_i(y(t)) + \Delta g_i(y(t)) + u_i(t) + d_i(t), \tag{10.3}$$

where ϑ_i is ith element of $Ay(t)$, $y_i(t)$ is the ith element of $y(t)$, $g_i(y(t))$ is the ith element of $g(y(t))$, $\Delta g_i(y(t))$ is the ith element of $\Delta g(y(t))$, $u_i(t)$ is the ith element of $u(t)$, $d_i(t)$ is the ith element of $d(t)$, and $i = 1, 2, \dots, n$.

On the basis of Lemma 2.4, a neural network is used to approximate the uncertainty $\rho_i \Delta g_i(y(t))$, with $\rho_i > 0$, and Equation (10.3) can be written as

$$D^\alpha y_i(t) = \vartheta_i + g_i(y(t)) + \frac{1}{\rho_i}\theta_i^{*\mathrm{T}}\phi_i(Z) + \frac{1}{\rho_i}\varepsilon_i^* + u_i(t) + d_i(t), \tag{10.4}$$

where $Z = y$.

To compensate for the effects of external disturbance $d(t)$ in Equation (10.2), a neural-network-based disturbance observer is designed.

To design the disturbance observer, an auxiliary variable is employed as follows [105]:

$$\xi_i(t) = d_i(t) - \kappa_i y_i(t), \tag{10.5}$$

where $\kappa_i > 0$ is a design constant.

According to Equations (10.4) and (10.5), the Caputo fractional derivative of $\xi_i(t)$ can be written as

$$\begin{aligned} D^\alpha \xi_i(t) &= D^\alpha d_i(t) - \kappa_i D^\alpha y_i(t) \\ &= -\kappa_i\left(\vartheta_i + g_i(y(t)) + \frac{1}{\rho_i}\theta_i^{*\mathrm{T}}\phi_i(Z) + \frac{1}{\rho_i}\varepsilon_i^* + d_i(t)\right) - \kappa_i u_i(t) + D^\alpha d_i(t) \\ &= -\kappa_i\left(\vartheta_i + g_i(y(t)) + \xi_i(t) + \kappa_i y_i(t) + \frac{1}{\rho_i}\theta_i^{*\mathrm{T}}\phi_i(Z) + \frac{1}{\rho_i}\varepsilon_i^*\right) \\ &\quad - \kappa_i u_i(t) + D^\alpha d_i(t). \end{aligned} \tag{10.6}$$

To estimate the auxiliary variable $\xi_i(t)$, we design the estimator as

$$D^\alpha \hat{\xi}_i = -\kappa_i(\vartheta_i + g_i(y(t)) + \kappa_i y_i(t) + \frac{1}{\rho_i}\hat{\theta}_i^{\mathrm{T}}\phi_i(Z)) - \kappa_i\hat{\xi}_i(t) - \kappa_i u_i(t), \tag{10.7}$$

where $\hat{\xi}_i$ is the estimate of ξ_i and $\hat{\theta}_i$ is the estimate of θ_i^*.

Combining Equations (10.5) and (10.7), the disturbance observer $\hat{d}_i(t)$ is given by

$$\hat{d}_i(t) = \hat{\xi}_i(t) + \kappa_i y_i(t). \tag{10.8}$$

Define the disturbance estimation error $\tilde{d}_i(t) = d_i(t) - \hat{d}_i(t)$. Considering Equations (10.5) and (10.8), we have

$$\tilde{\xi}_i(t) = \xi_i(t) - \hat{\xi}_i(t) = d_i(t) - \hat{d}_i(t) = \tilde{d}_i(t). \tag{10.9}$$

Considering Equations (10.6) and (10.7), the Caputo fractional derivative of $\tilde{\xi}_i(t)$ is described as

$$D^\alpha \tilde{\xi}_i(t) = -\kappa_i\tilde{\xi}_i(t) + \frac{\kappa_i}{\rho_i}\tilde{\theta}_i^{\mathrm{T}}\phi_i(Z) - \frac{\kappa_i}{\rho_i}\varepsilon_i^* + D^\alpha d_i(t), \tag{10.10}$$

where $\tilde{\theta}_i = \hat{\theta}_i - \theta_i^*$.

According to this design procedure of the neural-network-based disturbance observer, the following theorem is given.

Theorem 10.1 *Consider the slave system (10.2); the disturbance observer is designed as Equations (10.7) and (10.8). Then the disturbance estimation error is bounded.*

The proof of Theorem 10.1 and the adaptive law of $\hat{\theta}_i$ will be given in the next section.

10.3 Anti-Synchronization Control of Fractional-Order Nonlinear Systems

In this section, a disturbance-observer-based adaptive neural anti-synchronization control scheme will be proposed to guarantee that synchronization errors between the slave system (10.1) and the master system (10.2) are ultimately bounded. To design the anti-synchronization control scheme, the anti-synchronization error state $e(t) = y(t) + x(t)(e(t) = [e_1(t), e_2(t), \cdots, e_n(t)]^\top \in R^n)$ is defined. On the basis of Equations (10.1) and (10.2), the corresponding anti-synchronization error system is given as

$$D^\alpha e(t) = Ae(t) + g(y(t)) + g(x(t)) + \Delta f(y(t)) + d(t) + u(t). \tag{10.11}$$

The desired anti-synchronization control input is designed as

$$u(t) = -Ke(t) - Ae(t) - g(y(t)) - g(x(t)) - \hat{\Theta} - \hat{d}(t), \tag{10.12}$$

where $\hat{\Theta} = [\hat{\theta}_1^\top \phi_1(Z), \hat{\theta}_2^\top \phi_2(Z), \cdots, \hat{\theta}_n^\top \phi_n(Z)]^\top$ and K is a design diagonal positive definite matrix.

Furthermore, the adaptive law for $\hat{\theta}_i$ is given by

$$D^\alpha \hat{\theta}_i = \frac{\sigma_i}{\rho_i} e_i(t)\phi_i(Z) - \sigma_i h_i \hat{\theta}_i, \tag{10.13}$$

where $\sigma_i > 0$ is a design constant, $h_i > 0$ is a design constant, and $e_i(t)$ denotes the ith element of $e(t)$, with $i = 1, 2, \ldots, n$.

This design procedure can be summarized in the following theorem.

Theorem 10.2 *Consider the master system (10.1) and the slave system (10.2), with uncertain nonlinearities and bounded external disturbances. The neural network is used to approximate the uncertain nonlinearity. The external disturbance is handled using the deigned disturbance observer (10.7) and (10.8). The disturbance observer-based adaptive neural anti-synchronization controller $u(t)$ is designed as Equation (10.12) and the adaptive law $\dot{\theta}_i$ is given by Equation (10.13). Under the developed anti-synchronization controller (10.12), the anti-synchronization error $e(t)$ is ultimately bounded.*

Proof 10.1 *To prove the convergence of the anti-synchronization error $e(t)$, we first choose the following Lyapunov function candidate:*

$$V(t) = \frac{1}{2} e^\top(t)e(t) + \sum_{i=1}^{n} \frac{1}{2}\tilde{d}_i^2(t) + \sum_{i=1}^{n} \frac{1}{2\sigma_i}\tilde{\theta}_i^\top \tilde{\theta}_i \tag{10.14}$$

In particular, if the Caputo derivative of a constant is 0, we have

$$D^\alpha \tilde{\theta}_i = D^\alpha \hat{\theta}_i. \tag{10.15}$$

From Lemma 2.1, Lemma 2.5, and Equation (10.14), we obtain

$$D^{\alpha}V(t) \leq e^{\top}(t)D^{\alpha}e(t) + \sum_{i=1}^{n} \tilde{d}_i(t)D^{\alpha}\tilde{d}_i(t) + \sum_{i=1}^{n} \frac{1}{\sigma_i}\tilde{\theta}_i^{\top}D^{\alpha}\hat{\theta}_i. \tag{10.16}$$

Substituting Equation (10.11) into Equation (10.16), we have

$$\begin{aligned} D^{\alpha}V(t) \leq\, & e^{\top}(t)(Ae(t) + g(y(t)) + g(x(t)) + \Delta g(y(t)) + d(t) + u(t)) \\ & + \sum_{i=1}^{n} \tilde{d}_i(t)D^{\alpha}\tilde{d}_i(t) + \sum_{i=1}^{n} \frac{1}{\sigma_i}\tilde{\theta}_i^{\top}D^{\alpha}\hat{\theta}_i. \end{aligned} \tag{10.17}$$

On the basis of Lemma 2.4, the neural network is used to approximate the uncertain nonlinearity $\rho\Delta g(y(t))$ with $\rho = \text{diag}[\rho_1, \rho_2, \cdots, \rho_n]$. Equation (10.17) can be written as

$$\begin{aligned} D^{\alpha}V(t) \leq\, & e^{\top}(t)(Ae(t) + g(y(t)) + g(x(t)) + \Theta + \Xi + d(t) + u(t)) \\ & + \sum_{i=1}^{n} \tilde{d}_i(t)D^{\alpha}\tilde{d}_i(t) + \sum_{i=1}^{n} \frac{1}{\sigma_i}\tilde{\theta}_i^{\top}D^{\alpha}\hat{\theta}_i, \end{aligned} \tag{10.18}$$

where

$$\Xi = \left[\frac{1}{\rho_1}\varepsilon_1^*, \frac{1}{\rho_2}\varepsilon_2^*, \cdots, \frac{1}{\rho_n}\varepsilon_n^*\right]^{\top},$$

$$\Theta = \left[\frac{1}{\rho_1}\theta_1^{*\top}\phi_1(Z), \frac{1}{\rho_2}\theta_2^{*\top}\phi_2(Z), \cdots, \frac{1}{\rho_n}\theta_n^{*\top}\phi_n(Z)\right]^{\top}.$$

Combining Equation (10.12) with Equation (10.18), one has

$$\begin{aligned} D^{\alpha}V(t) \leq\, & -e^{\top}(t)Ke(t) - e^{\top}(t)\tilde{\Theta} + e^{\top}(t)\Xi + e^{\top}(t)\tilde{d}(t) \\ & + \sum_{i=1}^{n} \tilde{d}_i(t)D^{\alpha}\tilde{d}_i(t) + \sum_{i=1}^{n} \frac{1}{\sigma_i}\tilde{\theta}_i^{\top}D^{\alpha}\hat{\theta}_i, \end{aligned} \tag{10.19}$$

where

$$\tilde{\Theta} = \left[\frac{1}{\rho_1}\tilde{\theta}_1^{\top}\phi_1(Z), \frac{1}{\rho_2}\tilde{\theta}_2^{\top}\phi_2(Z), \cdots, \frac{1}{\rho_n}\tilde{\theta}_n^{\top}\phi_n(Z)\right]^{\top}$$

and $\tilde{\Theta} = \hat{\Theta} - \Theta$.

Referring to Equation (10.13), we obtain

$$\begin{aligned} \sum_{i=1}^{n} \frac{1}{\sigma_i}\tilde{\theta}_i^{\top}D^{\alpha}\hat{\theta}_i &= \sum_{i=1}^{n} \frac{1}{\rho_i}\tilde{\theta}_i^{\top}e_i(t)\phi_i(Z) - \sum_{i=1}^{n} \frac{h_i}{2}\|\tilde{\theta}_i\|^2 + \sum_{i=1}^{n} \frac{h_i}{2}\|\theta_i^*\|^2 - \sum_{i=1}^{n} \frac{h_i}{2}\|\hat{\theta}_i\|^2 \\ &\leq e^{\top}(t)\tilde{\Theta} - \sum_{i=1}^{n} \frac{h_i}{2}\|\tilde{\theta}_i\|^2 + \sum_{i=1}^{n} \frac{h_i}{2}\|\theta_i^*\|^2. \end{aligned} \tag{10.20}$$

According to Equations (10.19) and (10.20), we have

$$\begin{aligned} D^{\alpha}V(t) \leq\, & -e^{\top}(t)Ke(t) + e^{\top}(t)\Xi + e^{\top}(t)\tilde{d}(t) + \sum_{i=1}^{n} \tilde{d}_i(t)D^{\alpha}\tilde{d}_i(t) \\ & - \sum_{i=1}^{n} \frac{h_i}{2}\|\tilde{\theta}_i\|^2 + \sum_{i=1}^{n} \frac{h_i}{2}\|\theta_i^*\|^2. \end{aligned} \tag{10.21}$$

From Lemma 2.4, one has

$$e^{\top}(t)\Xi \leq \|e(t)\|\tau \leq \frac{1}{2}\|e(t)\|^2 + \frac{1}{2}\tau^2, \tag{10.22}$$

where $\|\Xi\| \leq \tau$, *and* $\tau > 0$ *is an unknown constant.*

Substituting Equation (10.22) into Equation (10.21) yields

$$D^{\alpha}V(t) \leq -e^{\top}(t)Ke(t) + \frac{1}{2}\|e(t)\|^2 + \frac{1}{2}\tau^2 + e^{\top}(t)\tilde{d}(t) + \sum_{i=1}^{n}\tilde{d}_i(t)D^{\alpha}\tilde{d}_i(t) - \sum_{i=1}^{n}\frac{h_i}{2}\|\tilde{\theta}_i\|^2 + \sum_{i=1}^{n}\frac{h_i}{2}\|\theta_i^*\|^2. \tag{10.23}$$

The following inequality can be obtained:

$$e^{\top}(t)\tilde{d}(t) \leq \|e(t)\|\|\tilde{d}(t)\| \leq \frac{1}{2}\|e(t)\|^2 + \frac{1}{2}\|\tilde{d}(t)\|^2. \tag{10.24}$$

Considering Equation (10.24), Equation (10.23) can be written as

$$D^{\alpha}V(t) \leq -(\lambda_{\min}(K) - 1)e^{\top}(t)e(t) + \frac{1}{2}\|\tilde{d}(t)\|^2 + \frac{1}{2}\tau^2 + \sum_{i=1}^{n}\tilde{d}_i(t)D^{\alpha}\tilde{d}_i(t) - \sum_{i-1}^{n}\frac{h_i}{2}\|\tilde{\theta}_i\|^2 + \sum_{i=1}^{n}\frac{h_i}{2}\|\theta_i^*\|^2, \tag{10.25}$$

where $\lambda_{\min}(K)$ *is the minimum eigenvalue of* K*. According to the designed disturbance observer (10.7) and (10.8), Lemma 2.4, and Assumption 10.1, we have*

$$\begin{aligned}
\sum_{i=1}^{n}\tilde{d}_i(t)D^{\alpha}\tilde{d}_i(t) &= -\sum_{i=1}^{n}\kappa_i\tilde{d}_i^2(t) + \sum_{i=1}^{n}\tilde{d}_i(t)\frac{\kappa_i}{\rho_i}\tilde{\theta}_i^{\top}\phi_i(Z) - \sum_{i=1}^{n}\tilde{d}_i(t)\frac{\kappa_i}{\rho_i}\varepsilon_i^* \\
&\quad + \sum_{i=1}^{n}\tilde{d}_i(t)D^{\alpha}d_i(t) \\
&\leq -\sum_{i=1}^{n}\kappa_i\tilde{d}_i^2(t) + \sum_{i=1}^{n}\frac{\kappa_i}{2\rho_i}\tilde{d}_i^2(t) + \sum_{i=1}^{n}\frac{\kappa_i}{2\rho_i}\|\tilde{\theta}_i\|^2 l_i + \sum_{i=1}^{n}\frac{1}{2\rho_i}\tilde{d}_i^2(t) \\
&\quad + \sum_{i=1}^{n}\frac{1}{2\rho_i}\kappa_i^2\varepsilon_i^{*2} + \sum_{i=1}^{n}\frac{1}{2}\tilde{d}_i^2(t) + \sum_{i=1}^{n}\frac{1}{2}\varpi_i^2 \\
&= -\sum_{i=1}^{n}\frac{2\kappa_i - \frac{\kappa_i}{\rho_i} - \frac{1}{\rho_i} - 1}{2}\tilde{d}_i^2(t) + \sum_{i=1}^{n}\frac{\kappa_i}{2\rho_i}\|\tilde{\theta}_i\|^2 l_i + \sum_{i=1}^{n}\frac{1}{2\rho_i}\kappa_i^2\varepsilon_i^{*2} \\
&\quad + \sum_{i=1}^{n}\frac{1}{2}\varpi_i^2,
\end{aligned} \tag{10.26}$$

where $\|\phi_i(Z)\|^2 \leq l_i$.

From Equation (10.26), we obtain

$$D^{\alpha}V(t) \leq -\lambda_{\min}(K)e^{\top}(t)e(t) - \sum_{i=1}^{n}\frac{2\kappa_i - \frac{\kappa_i}{\rho_i} - \frac{1}{\rho_i} - 2}{2}\tilde{d}_i^2(t) - \sum_{i=1}^{n}\frac{h_i - \frac{l_i\kappa_i}{\rho_i}}{2}\|\tilde{\theta}_i\|^2 + \sum_{i=1}^{n}\frac{h_i}{2}\|\theta_i^*\|^2 + \frac{1}{2}\tau^2 + \sum_{i=1}^{n}\frac{1}{2\rho_i}\kappa_i^2\varepsilon_i^{*2} + \sum_{i=1}^{n}\frac{1}{2}\varpi_i^2. \tag{10.27}$$

Furthermore, Equation (10.27) can be written as

$$D^{\alpha}V(t) \leq -\frac{2\lambda_{\min}(K)}{2}e^{\top}(t)e(t) - \sum_{i=1}^{n}\frac{2\kappa_i - \frac{\kappa_i}{\rho_i} - \frac{1}{\rho_i} - 2}{2}\tilde{d}_i^2(t) - \sum_{i=1}^{n}\frac{h_i - \frac{l_i\kappa_i}{\rho_i}}{2}\|\tilde{\theta}_i\|^2$$

$$+ \sum_{i=1}^{n}\frac{h_i}{2}\|\theta_i^*\|^2 + \frac{1}{2}\tau^2 + \sum_{i=1}^{n}\frac{1}{2\rho_i}\kappa_i^2\varepsilon_i^{*2} + \sum_{i=1}^{n}\frac{1}{2}\varpi_i^2$$

$$\leq -PV(t) + \overline{P}, \tag{10.28}$$

where

$$P = \min\left\{2\lambda_{\min}(K), 2\kappa_i - \frac{\kappa_i}{\rho_i} - \frac{1}{\rho_i} - 2, \sigma_i\left(h_i - \frac{l_i\kappa_i}{\rho_i}\right)\right\},$$

$$\overline{P} = \sum_{i=1}^{n}\frac{h_i}{2}\|\theta_i^*\|^2 + \frac{1}{2}\tau^2 + \sum_{i=1}^{n}\frac{1}{2\rho_i}\kappa_i^2\varepsilon_i^{*2} + \sum_{i=1}^{n}\frac{1}{2}\varpi_i^2.$$

To guarantee that the anti-synchronization error is bounded, the corresponding design parameters κ_i, h_i, ρ_i and l_i should be chosen to make

$$2\kappa_i - \frac{\kappa_i}{\rho_i} - \frac{1}{\rho_i} - 2 > 0,$$

$$h_i - \frac{l_i\kappa_i}{\rho_i} > 0.$$

According to Equation (10.28) and Lemma 2.3, this may show directly that the anti-synchronization error $e(t)$ and the disturbance estimation error $\tilde{d}_i(t)$ are ultimately bounded. On the basis of Lemma 2.3 and Equation (10.25), we obtain

$$|V(t)| \leq \frac{2\overline{P}}{P}, \tag{10.29}$$

which implies that

$$\|e(t)\| \leq \sqrt{\frac{4\overline{P}}{P\lambda_{\min}(M)}}, \tag{10.30}$$

$$\|\tilde{d}(t)\| \leq \sqrt{\frac{4\overline{P}}{P}}. \tag{10.31}$$

From Equations (10.30) and (10.31), we can obtain that the anti-synchronization error $e(t)$ and the disturbance estimation error $\tilde{d}(t)$ will be ultimately bounded as $t \to \infty$. Thus, the anti-synchronization error system (10.11) is bounded stable based on Lemma 2.3. The bounded anti-synchronization of the master system (10.1) and the slave system (10.2) is realized. This concludes the proof. □

10.4 Simulation Examples

In this section, the fractional-order Lorenz system [179] and the fractional-order Lü system [187] are used to illustrate the effectiveness of the proposed anti-synchronization control scheme.

10.4.1 Anti-Synchronization Control of Fractional-Order Lorenz System

To illustrate anti-synchronization control of the fractional-order Lorenz system, the model of fractional-order Lorenz system is described as follows [179]:

$$\begin{aligned} D^{\alpha}x_1(t) &= \hbar_1(x_2(t) - x_1(t)), \\ D^{\alpha}x_2(t) &= x_1(t)(\hbar_2 - x_3(t)) - x_2(t), \\ D^{\alpha}x_3(t) &= x_1(t)x_2(t) - \hbar_3 x_3(t), \end{aligned} \tag{10.32}$$

where α is the fractional order, $x_1(t)$, $x_2(t)$, and $x_3(t)$ are system state variables, and $\hbar_1$, $\hbar_2$, and $\hbar_3$ are system parameters. The fractional order is chosen as $\alpha = 0.995$, the system parameters are set as $\hbar_1 = 10$, $\hbar_2 = 28$, and $\hbar_3 = 8/3$, and the initial conditions are chosen as $x_1(0) = x_2(0) = x_3(0) = 0.1$; the chaotic dynamic behaviors of the fractional-order Lorenz system are shown in Figure 2.1.

To illustrate the effectiveness of the proposed anti-synchronization controller, the fractional-order Lorenz system (10.32) is regarded as the master system. From Equations (10.2) and (10.32), the slave system is described as follows:

$$\begin{aligned} D^{\alpha}y_1(t) &= \hbar_1(y_2(t) - y_1(t)) + \Delta f_1(y(t)) + d_1(t) + u_1(t), \\ D^{\alpha}y_2(t) &= y_1(t)(\hbar_2 - y_3(t)) - x_2(t) + \Delta f_2(y(t)) + d_2(t) + u_2(t), \\ D^{\alpha}y_3(t) &= y_1(t)y_2(t) - \hbar_3 y_3(t) + \Delta f_2(y(t)) + d_3(t) + u_3(t), \end{aligned} \tag{10.33}$$

where $y_1(t)$, $y_2(t)$, and $y_3(t)$ are system state variables, $\Delta f_1(y(t))$, $\Delta f_2(y(t))$, and $\Delta f_3(y(t))$ are system uncertainties, $d_1(t)$, $d_2(t)$, and $d_3(t)$ are unknown bounded disturbances, and $u_1(t)$, $u_2(t)$, and $u_3(t)$ are designed anti-synchronization control inputs.

According to Equations (10.32) and (10.33), the corresponding anti-synchronization error system is given as follows:

$$\begin{aligned} D^{\alpha}e_1(t) &= \hbar_1(e_2(t) - e_1(t)) + \Delta f_1(y(t)) + d_1(t) + u_1(t), \\ D^{\alpha}e_2(t) &= \hbar_2 e_1(t) - y_1(t)y_3(t) - x_1(t)x_3(t) - e_2(t) + \Delta f_2(y(t)) + d_2(t) + u_2(t), \\ D^{\alpha}e_3(t) &= y_1(t)y_2(t) + x_1(t)x_2(t) - \hbar_3 e_3(t) + \Delta f_2(y(t)) + d_3(t) + u_3(t), \end{aligned} \tag{10.34}$$

where $e_1(t)$, $e_2(t)$, and $e_3(t)$ are anti-synchronization errors.

To demonstrate the effectiveness of the proposed disturbance-observer-based adaptive neural anti-synchronization control scheme, numerical simulation results are shown for the fractional order Lorenz system (10.32) under the following conditions: the initial conditions are set as $x_0(t) = [0.1, 0.1, 0.1]^{\mathrm{T}}$, $y_0(t) = [0.7, 0.6, 0.5]^{\mathrm{T}}$, the designed parameters are chosen as $\alpha = 0.995$, $\kappa_1 = \kappa_2 = \kappa_3 = 20$, $h_1 = h_2 = h_3 = 40$, $\sigma_1 = \sigma_2 = \sigma_3 = 0.01$, $\rho_1 = \rho_2 = \rho_3 = 10$, $l_1 = l_2 = l_3 = 19$, and $K = \mathrm{diag}[300, 300, 300]$. The disturbances are assumed as $d_1(t) = -80\sin(0.1t) + 490\cos(0.8t)$, $d_2(t) = 200\sin(0.2t) - 210\cos(t)$, and $d_3(t) = 220\sin(0.3t) + 230\cos(0.5t)$. The system uncertainties are chosen as $\Delta f_1(y(t)) = 0.3\sin(y_1(t))$, $\Delta f_2(y(t)) = 0.4\sin(y_2(t))$, and $\Delta f_3(y(t)) = 0.5\sin(y_3(t))$.

Numerical results are presented under the proposed disturbance-observer-based adaptive neural anti-synchronization control scheme. The state synchronization results of the master system (10.32) and the slave system (10.33) are shown in Figure 10.1, Figure 10.2, and Figure 10.3. It is shown that good synchronization performance is

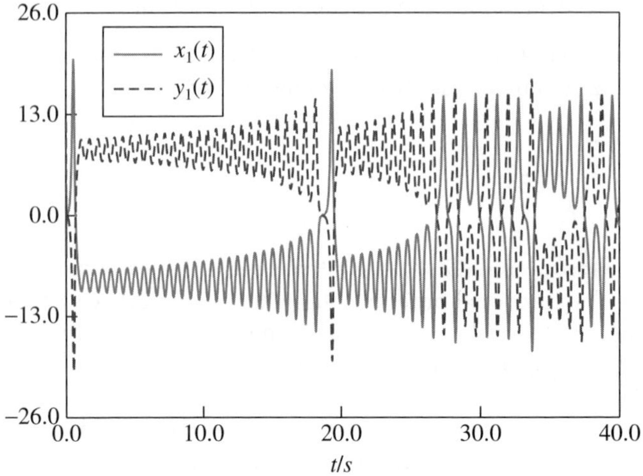

Figure 10.1 Anti-synchronization states of $x_1(t)$ and $y_1(t)$ of the master system (10.32) and the slave system (10.33).

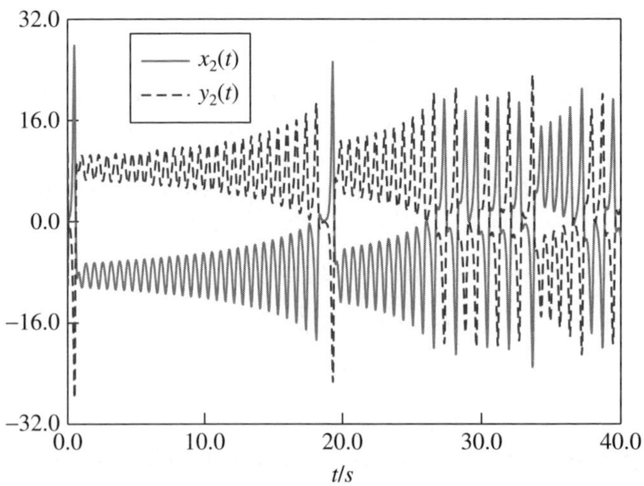

Figure 10.2 Anti-synchronization states of $x_2(t)$ and $y_2(t)$ of the master system (10.32) and the slave system (10.33).

obtained. Figure 10.4 shows that the synchronization errors $e_1(t)$, $e_2(t)$, and $e_3(t)$ are convergent. Furthermore, the estimation performance of the proposed disturbance observer (10.7 and 10.8) is shown in Figure 10.5, Figure 10.6, and Figure 10.7. It is evident from Figure 10.8 that the designed disturbance observer is effective and feasible. According to the simulation results, the master system (10.32) and the slave system (10.33) are bounded anti-synchronized under the designed adaptive neural anti-synchronization controller (10.12) and the adaptive update law (10.13). Thus, the proposed disturbance-observer-based adaptive neural anti-synchronization control scheme is valid for fractional-order nonlinear systems with external disturbances and system uncertainties.

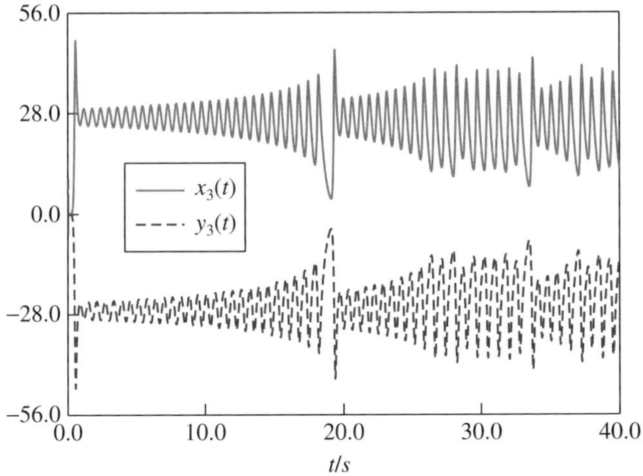

Figure 10.3 Anti-synchronization states of $x_3(t)$ and $y_3(t)$ of the master system (10.32) and the slave system (10.33).

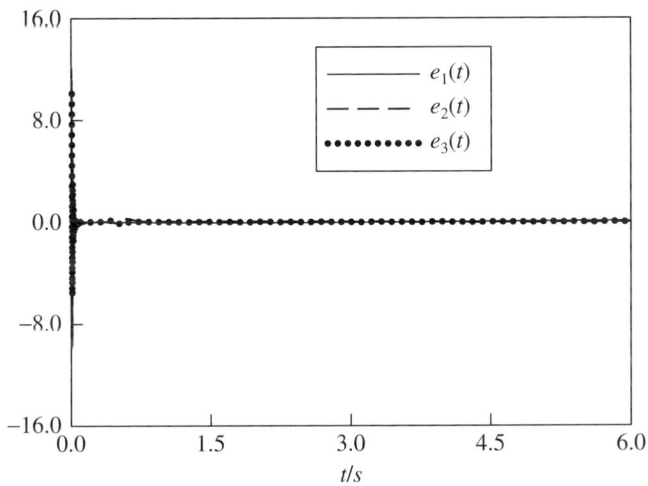

Figure 10.4 Anti-synchronization errors $e_1(t)$, $e_2(t)$, and $e_3(t)$ of the master system (10.32) and the slave system (10.33).

10.4.2 Anti-Synchronization Control of Fractional-Order Lü System

To further illustrate the effectiveness of the proposed anti-synchronization control scheme, the anti-synchronization of the fractional-order Lü system [187] is studied in this section. For convenience, the fractional-order Lü system is given as follows:

$$\begin{aligned} D^{\alpha}x_1(t) &= \eta_1(x_2(t) - x_1(t)), \\ D^{\alpha}x_2(t) &= -x_1(t)x_3(t) + \eta_2 x_2(t), \\ D^{\alpha}x_3(t) &= x_1(t)x_2(t) - \eta_3 x_3(t), \end{aligned} \tag{10.35}$$

where α is the fractional order, $x_1(t)$, $x_2(t)$, and $x_3(t)$ are system state variables, and η_1, η_2, and η_3 are system parameters. For the fractional order $\alpha = 0.95$, system parameters

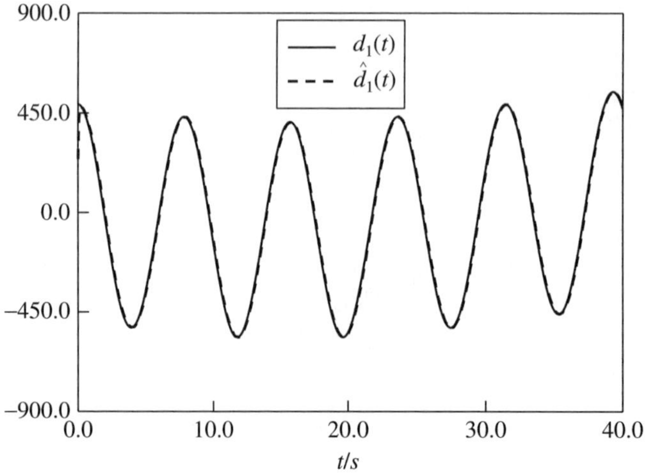

Figure 10.5 Disturbance estimation result of $d_1(t)$ and $\hat{d}_1(t)$ for fractional-order Lorenz system.

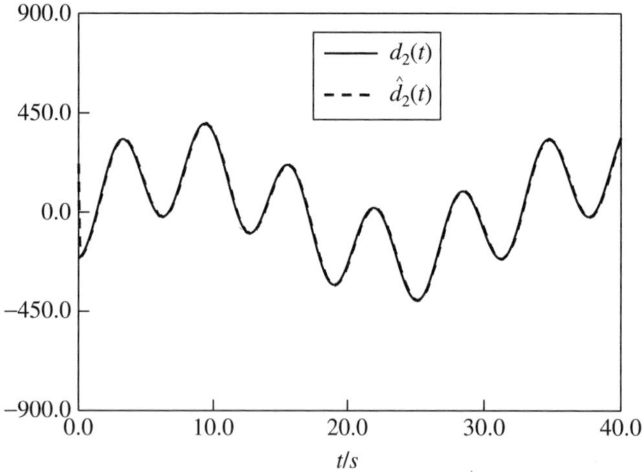

Figure 10.6 Disturbance estimation result of $d_2(t)$ and $\hat{d}_2(t)$ for fractional-order Lorenz system.

$\eta_1 = 36$, $\eta_2 = 20$, and $\eta_3 = 3$, and initial conditions chosen as $x_1(0) = 0.2$, $x_2(0) = 0.5$, and $x_3(0) = 0.3$, the chaotic dynamic behaviors of the fractional-order Lü system are presented in Figure 2.23.

To illustrate the effectiveness of the proposed anti-synchronization controller, the fractional-order Lorenz system (10.32) is regarded as the master system. From Equations (10.2) and (10.32), the slave system is described as follows:

$$
\begin{aligned}
D^\alpha y_1(t) &= \eta_1(y_2(t) - y_1(t)) + \Delta f_1(y(t)) + d_1(t) + u_1(t),\\
D^\alpha y_2(t) &= -y_1(t)y_3(t) + \eta_2 y_2(t) + \Delta f_2(y(t)) + d_2(t) + u_2(t),\\
D^\alpha y_3(t) &= y_1(t)y_2(t) - \eta_3 y_3(t) + \Delta f_2(y(t)) + d_3(t) + u_3(t),
\end{aligned}
\tag{10.36}
$$

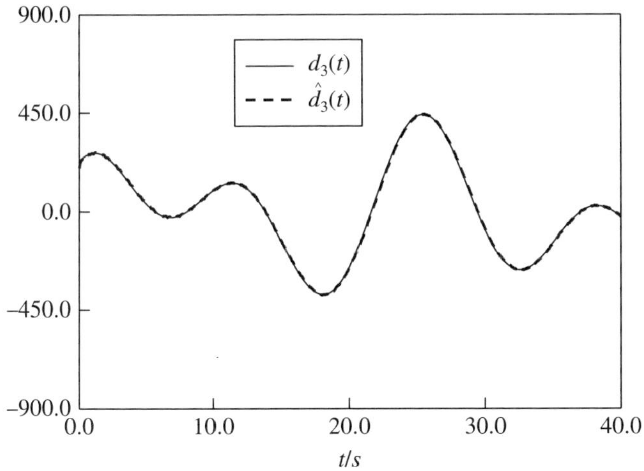

Figure 10.7 Disturbance estimation result of $d_3(t)$ and $\hat{d}_3(t)$ for fractional-order Lorenz system.

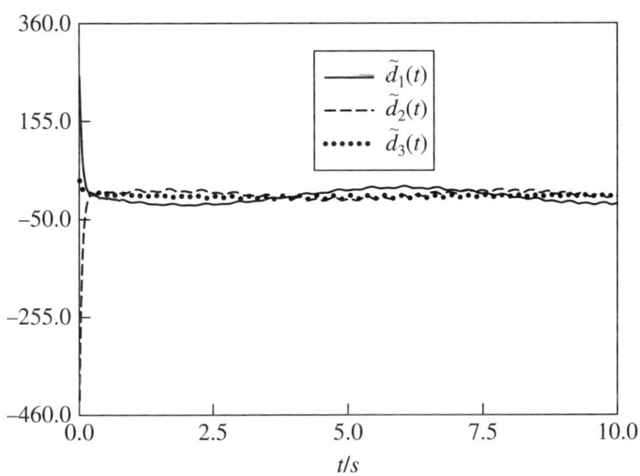

Figure 10.8 Disturbance observation errors $\tilde{d}_1(t)$, $\tilde{d}_2(t)$, and $\tilde{d}_3(t)$ of fractional-order Lorenz system.

where $y_1(t)$, $y_2(t)$, and $y_3(t)$ are system state variables, $\Delta f_1(y(t))$, $\Delta f_2(y(t))$, and $\Delta f_3(y(t))$ are system uncertainties, $d_1(t)$, $d_2(t)$, and $d_3(t)$ are unknown bounded disturbances, and $u_1(t)$, $u_2(t)$, and $u_3(t)$ are designed anti-synchronization control inputs.

According to Equations (10.32) and (10.33), the corresponding anti-synchronization error system is given as follows:

$$\begin{aligned} D^{\alpha}e_1(t) &= \eta_1(e_2(t) - e_1(t)) + \Delta f_1(y(t)) + d_1(t) + u_1(t),\\ D^{\alpha}e_2(t) &= -y_1(t)y_3(t) - x_1(t)x_3(t) + \eta_2 e_2(t) + \Delta f_2(y(t)) + d_2(t) + u_2(t),\\ D^{\alpha}e_3(t) &= y_1(t)y_2(t) + x_1(t)x_2(t) - \eta_3 e_3(t) + \Delta f_2(y(t)) + d_3(t) + u_3(t), \end{aligned} \tag{10.37}$$

where $e_1(t)$, $e_2(t)$, and $e_3(t)$ are anti-synchronization errors.

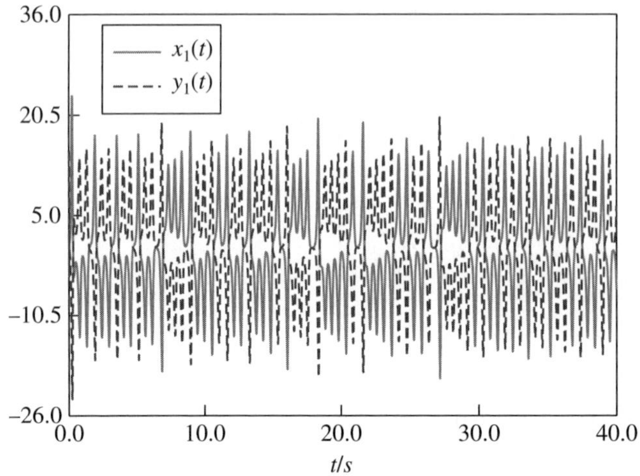

Figure 10.9 Anti-synchronization states of $x_1(t)$ and $y_1(t)$ of the master system (10.35) and the slave system (10.36).

For the numerical simulation, we choose the fractional order as $\alpha = 0.95$, the disturbances are assumed as $d_1(t) = -50\sin(0.1t) + 480\cos(0.8t)$, $d_2(t) = 210\sin(0.2t) - 220\cos(t)$, and $d_3(t) = 210\sin(0.3t) + 220\cos(0.5t)$, and the system uncertainties are chosen as $\Delta f_1(y(t)) = 0.5\sin(y_1(t))$, $\Delta f_2(y(t)) = 0.6\sin(y_2(t))$, and $\Delta f_3(y(t)) = 0.7\sin(y_3(t))$. The initial conditions are chosen as $x_0(t) = [0.2, 0.5, 0.3]^\top$ and $y_0(t) = [15, 12, 13]^\top$. The design parameters are chosen as $\kappa_1 = \kappa_2 = \kappa_3 = 20$, $h_1 = h_2 = h_3 = 40$, $\sigma_1 = \sigma_2 = \sigma_3 = 0.01$, $l_1 = l_2 = l_3 = 19$, $\rho_1 = \rho_2 = \rho_3 = 10$, and $K = \mathrm{diag}[200, 200, 200]$.

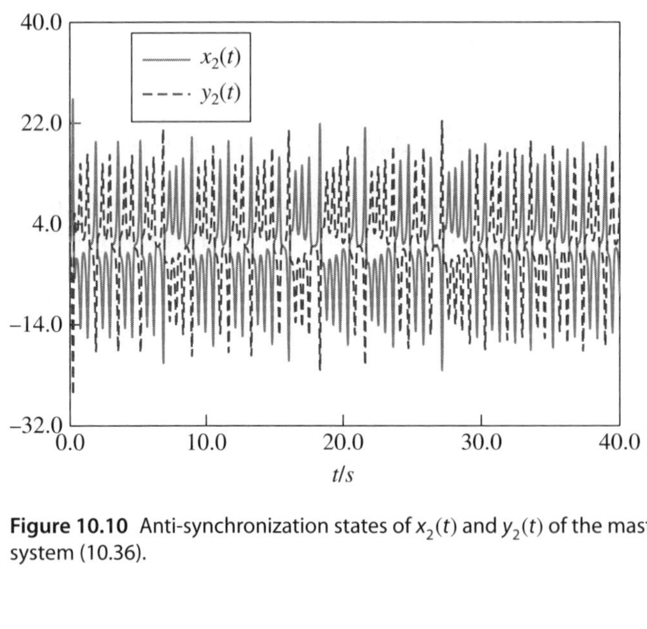

Figure 10.10 Anti-synchronization states of $x_2(t)$ and $y_2(t)$ of the master system (10.35) and the slave system (10.36).

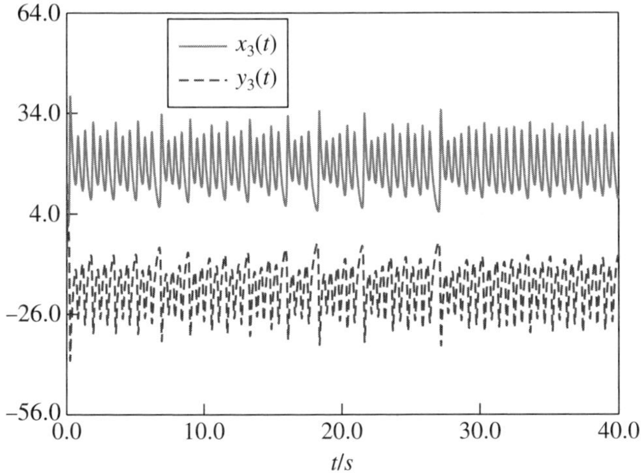

Figure 10.11 Anti-synchronization states of $x_3(t)$ and $y_3(t)$ of the master system (10.35) and the slave system (10.36).

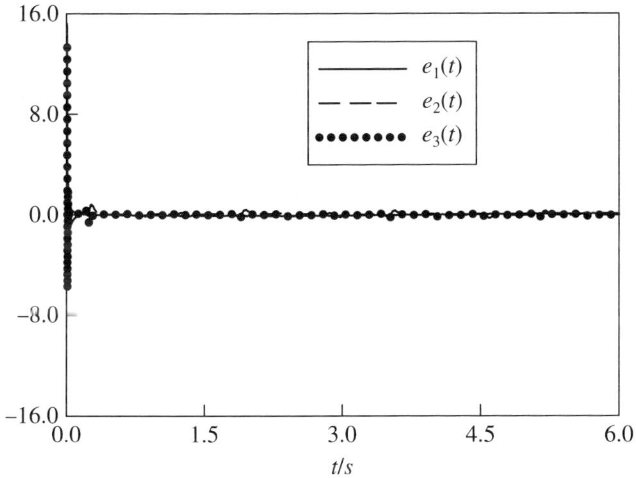

Figure 10.12 Anti-synchronization errors $e_1(t)$, $e_2(t)$, and $e_3(t)$ of the master system (10.35) and the slave system (10.36).

According to these conditions and the proposed synchronization control scheme, numerical results are presented. Good synchronization performance is shown in Figure 10.9, Figure 10.10, and Figure 10.11. Numerical results of the synchronization errors $e_1(t)$, $e_2(t)$, and $e_3(t)$ are given in Figure 10.12, and are bounded and convergent. Furthermore, the observation performance of the proposed disturbance observer (10.7 and 10.8) is presented in Figure 10.13, Figure 10.14, and Figure 10.15. From Figure 10.13, Figure 10.14, Figure 10.15, and Figure 10.16, the disturbance observer is valid. On the basis of the simulation results, the master system (10.35) can anti-synchronize the

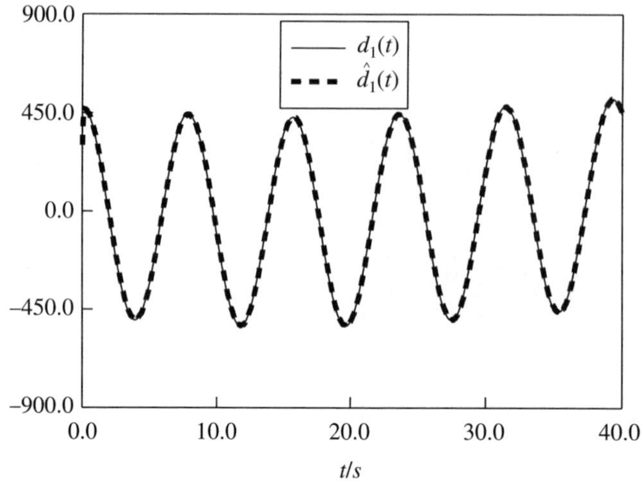

Figure 10.13 Disturbance estimation result of $d_1(t)$ and $\hat{d}_1(t)$ for the fractional-order Lü system.

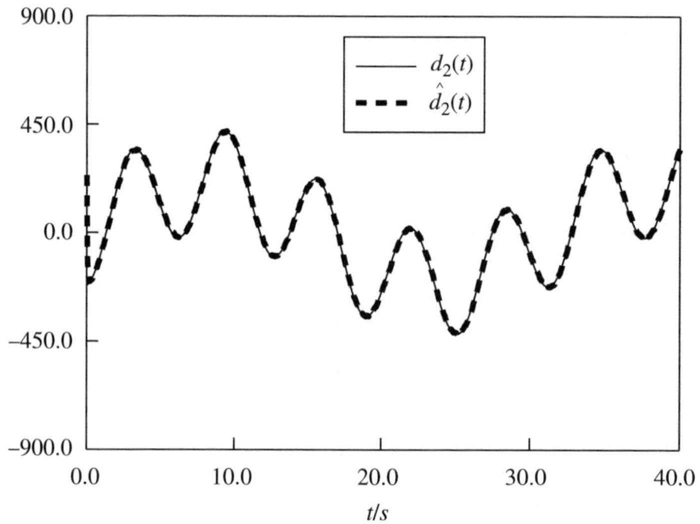

Figure 10.14 Disturbance estimation result of $d_2(t)$ and $\hat{d}_2(t)$ for the fractional-order Lü system.

slave system (10.36) well, based on the designed adaptive neural anti-synchronization controller (10.12) and the adaptive update law (10.13). Therefore, the proposed anti-synchronization control scheme is effective for fractional-order nonlinear systems with external disturbances and system uncertainties.

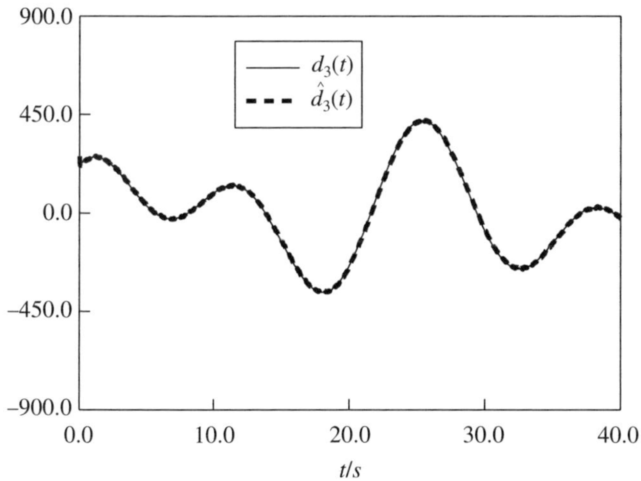

Figure 10.15 Disturbance estimation result of $d_3(t)$ and $\hat{d}_3(t)$ for the fractional-order Lü system.

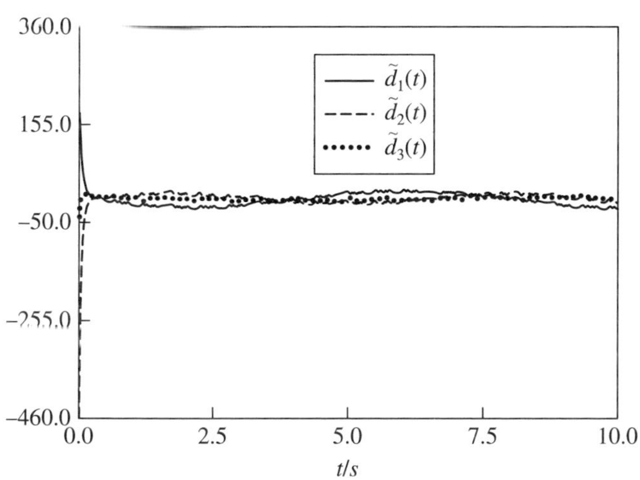

Figure 10.16 Observation errors $\tilde{d}_1(t)$, $\tilde{d}_2(t)$, and $\tilde{d}_3(t)$ of the fractional-order Lü system.

10.5 Conclusion

In this chapter, a disturbance-observer-based adaptive neural anti-synchronization control scheme has been studied for fractional-order nonlinear systems in the presence of external disturbances and system uncertainties. A disturbance observer has been developed to estimate the external disturbances. An anti-synchronization controller has been designed based on the disturbance observer and a neural network for the anti-synchronization of fractional-order nonlinear systems in the presence of external disturbances and system uncertainties. Furthermore, numerical simulations of two examples are shown in this chapter, i.e., the fractional-order Lorenz system and the fractional-order Lü system. Numerical simulations show the effectiveness of the proposed disturbance-observer-based adaptive neural anti-synchronization control scheme.

11

Synchronization Control for Fractional-Order Systems Subjected to Input Saturation

11.1 Problem Statement

According to the Caputo fractional derivative (2.17), the following nonlinear FOCS is considered:

$$D^\alpha x(t) = Ax(t) + h(x(t)), \tag{11.1}$$

where $A \in R^{n\times n}$ is a constant matrix, $x(t) = [x_1(t), x_2(t), \cdots, x_n(t)]^\top \in R^n$ is a state vector, $h(x(t)) = [h_1(x(t)), h_2(x(t)), \cdots, h_n(x(t))]^\top \in R^n$ is the nonlinear function vector, and the fractional order α satisfies $1 < \alpha < 2$.

To study the synchronization of nonlinear fractional-order systems, the FOCS (11.1) is taken as the master system, and the slave system is defined by

$$D^\alpha y(t) = Ay(t) + h(y(t)) + v(t), \tag{11.2}$$

where $y(t) = [y_1(t), y_2(t), \cdots, y_n(t)]^\top \in R^n$ is a state vector, $v(t) = [v_1(t), v_2(t), \cdots, v_n(t)]^\top \in R^n$ is the control input vector, and $h(y(t)) = [h_1(y(t)), h_2(y(t)), \cdots, h_n(y(t))]^\top \in R^n$ is the nonlinear function vector. Further, the nonlinear FOCS (11.2) subjected to input saturation with $1 < \alpha < 2$ is given as

$$D^\alpha y(t) = Ay(t) + h(y(t)) + \mathrm{sat}(u(t)), \tag{11.3}$$

where $v(t) = \mathrm{sat}(u(t))$ is the vector-valued input saturation function defined as $\mathrm{sat}(u(t)) = [\mathrm{sat}(u_1(t)), \mathrm{sat}(u_2(t)), \cdots, \mathrm{sat}(u_n(t))]^\top \in R^n$, with

$$\mathrm{sat}(u_i(t)) = \mathrm{sign}(u_i(t))\min(u_{0i}(t), |u_i(t)|), \tag{11.4}$$

where $i = 1, 2, \ldots, n$, $\min(\cdot)$ denotes the minimum value, and $u_{0i}(t)$ represents the symmetric saturation level relative to the ith control input and is known.

This chapter aims to develop the synchronization controller, so that the synchronization between two nonlinear fractional-order systems is realized based on the state-feedback method. On the basis of the designed synchronization controller, the synchronization can be realized between the master system and the slave system by giving the proper condition.

To facilitate the synchronization design of the master system (11.1) and the slave system (11.3), the following assumption is necessary in this study.

Robust Adaptive Control for Fractional-Order Systems with Disturbance and Saturation, First Edition.
Mou Chen, Shuyi Shao, and Peng Shi.

Assumption 11.1 *According to Lemma 2.9, the master system (11.1), and the slave system (11.3), nonlinear functions $h(x(t))$, $h(y(t))$, and $\varphi(t, w(t))$ satisfy $h(0) = 0$ and $\varphi(t, 0) = 0$, respectively. Furthermore, $\|h(y(t)) - h(x(t))\| \leq \kappa\|y(t) - x(t)\|(\kappa \geq 0)$ and $\|\varphi(t, w(t)\| \leq \vartheta\|w(t)\|(\vartheta \geq 0)$, where κ and ϑ are known constants and $w(t) = y(t) - x(t)$.*

11.2 Synchronization Control Design of Fractional-Order Systems with Input Saturation

In this section, a synchronization control scheme for two identical fractional-order systems with different initial conditions is investigated by considering the saturation function.

For the master system (11.1) and the slave system (11.3), we define the synchronization error state $w(t) = y(t) - x(t)(w(t) = [w_1(t), w_2(t), \cdots, w_n(t)]^\top \in R^n)$. The goal is to design the synchronization controller for the system. From Equations (11.1) and (11.3), the corresponding synchronization error system can be written as

$$D^\alpha w(t) = A(y(t) - x(t)) + h(y(t)) - h(x(t)) + \text{sat}(u(t)). \tag{11.5}$$

On the basis of the state-feedback control method, the control input is designed as

$$u(t) = Kw(t), \tag{11.6}$$

where $u(t) = [u_{(1)}(t), u_{(2)}(t), \cdots, u_{(n)}(t)] \in R^n$.

Then, the synchronization error system (11.5) is rewritten as

$$D^\alpha w(t) = A(y(t) - x(t)) + h(y(t)) - h(x(t)) + \text{sat}(Kw(t)). \tag{11.7}$$

From Lemma 2.9, the synchronization error system (11.7) has the following form:

$$\begin{aligned} D^\alpha w(t) &= A(y(t) - x(t)) + h(y(t)) - h(x(t)) + \text{sat}(Kw(t)) \\ &= Qw(t) + h(y(t)) - h(x(t)) + \varphi(t, w(t)), \end{aligned} \tag{11.8}$$

where $Q = A + LK$, based on Lemma 2.9.

This design procedure of synchronization control scheme can be summarized in the following theorem.

Theorem 11.1 *The synchronization error system (11.8) is asymptotically stable, if the design matrices L and K guarantee that Q is a stable matrix and $\eta = -\max\{\text{Re}\lambda(Q)\} > (M(\kappa + \vartheta)\Gamma(\alpha))^{1/\alpha}$, in which $M \geq 1$ and $\text{Re}\lambda(Q)$ denotes the real part of the eigenvalues of matrix Q.*

Proof 11.1 *By taking the Laplace transform of the synchronization error system (11.8), one gets*

$$s^\alpha W(s) = s^{\alpha-1}w_{00} + s^{\alpha-2}w_{01} + QW(s) + U_1, \tag{11.9}$$

where w_{00} and w_{01} are the initial conditions of Equation (11.8), $U_1 = \ell(h(y(t)) - h(x(t))) + \ell(\varphi(t, w(t)))$, and $W(s)$ is the Laplace transform of $w(t)$. Taking the Laplace inverse transform on Equation (11.9), one obtains

$$w(t) = E_{\alpha,1}(Qt^\alpha)w_{00} + tE_{\alpha,2}(Qt^\alpha)w_{01} + U_2, \tag{11.10}$$

where

$$U_2 = \int_0^t (t-\tau)^{\alpha-1}E_{\alpha,\alpha}(Q(t-\tau)^\alpha)(\varphi(\tau,w(\tau)) + (h(y(\tau)) - h(x(\tau))))\mathrm{d}\tau.$$

According to Lemma 2.11, since Q is a stable matrix, $M \geq 1$, $1 < \alpha < 2$, *and* $-\eta = \max(\mathrm{Re}\lambda(Q))(\eta > 0)$, *Equation (11.10) can be written as*

$$\|w(t)\| \leq M\mathrm{e}^{-\eta t}\|w_{00}\| + M\mathrm{e}^{-\eta t}\|w_{01}\|t + U_3, \tag{11.11}$$

where

$$U_3 = M\int_0^t (t-\tau)^{\alpha-1}\mathrm{e}^{-\eta(t-\tau)}(\|\varphi(\tau,w(\tau))\| + \|h(y(\tau)) - h(x(\tau))\|)\mathrm{d}\tau.$$

From Assumption 11.1, we have

$$\|w(t)\| \leq Me^{-\eta t}\|w_{00}\| + Me^{-\eta t}\|w_{01}\|t + U_4, \tag{11.12}$$

where

$$U_4 = M(\kappa+\vartheta)\int_0^t (t-\tau)^{\alpha-1}\mathrm{e}^{-\eta(t-\tau)}\|w(\tau)\|\mathrm{d}\tau.$$

Furthermore, Equation (11.12) can be rewritten as

$$\mathrm{e}^{\eta t}\|w(t)\| \leq M\|w_{00}\| + M\|w_{01}\|t + U_5, \tag{11.13}$$

where

$$U_5 = M(\kappa+\vartheta)\int_0^t (t-\tau)^{\alpha-1}\mathrm{e}^{\eta\tau}\|w(\tau)\|\mathrm{d}\tau.$$

Based on Lemma 2.12, $b(t) = M\|w_{00}\| + M\|w_{01}\|t$, $p(t) = M(\kappa+\vartheta)$, *and* $q(t) = \mathrm{e}^{\eta t}\|w(t)\|$, *one has*

$$\mathrm{e}^{\eta t}\|w(t)\| \leq M(\|w_{00}\| + \|w_{01}\|t)E_{\alpha,1}(M(\kappa+\vartheta)\Gamma(\alpha)t^\alpha). \tag{11.14}$$

Next, according to Lemma 2.10, there exist two constants $C_1 > 0$, $C_2 > 0$, *such that*

$$\mathrm{e}^{\eta t}\|w(t)\| \leq P + \frac{C_2 M(\|w_{00}\| + \|w_{01}\|t)}{1 + |M(\kappa+\vartheta)\Gamma(\alpha)t^\alpha|}, \tag{11.15}$$

where

$$P = C_1 M(\|w_{00}\| + \|w_{01}\|t)\mathrm{e}^{(M(\kappa+\vartheta)\Gamma(\alpha))^{1/\alpha}t}.$$

Moreover, Equation (11.15) is equivalent to

$$\|w(t)\| \leq P' + \frac{C_2 M(\|w_{00}\| + \|w_{01}\|t)}{(1 + |M(\kappa+\vartheta)\Gamma(\alpha)t^\alpha|)e^{\eta t}} \tag{11.16}$$

where

$$P' = C_1 M(\|w_{00}\| + \|w_{01}\|t)\mathrm{e}^{((M(\kappa+\vartheta)\Gamma(\alpha))^{1/\alpha}-\eta)t}.$$

When $t \to \infty, \eta > (M(\kappa+\vartheta)\Gamma(\alpha))^{1/\alpha}$, $\|w(t)\|$ *asymptotically tend to a stable value. This implies that the synchronization error system (11.8) is asymptotically stable. Therefore, synchronization of the master system (11.1) and the slave system (11.3) is achieved. This concludes the proof.* □

Remark 11.1 *Many chaotic systems can satisfy Assumption 11.1, including the classic Chua's system [234], the modified Chua's circuit with a sine function [234], the four-dimensional modified Chua's circuit [235], the hyperchaotic Chua's circuit [236], the five-dimensional modified Chua's circuit [237], and the Lur'e nonlinear system [238].*

11.3 Simulation Examples

In this section, to illustrate the effectiveness of the proposed synchronization method, two representative examples are analyzed; the fractional-order modified Chua's circuit with a sine function and the fractional-order four-dimensional modified Chua's circuit.

11.3.1 Fractional-Order Modified Chua's Circuit with Sine Function

Unlike the classic fractional-order Chua's circuit [239], the state equation of fractional-order modified Chua's circuit with a sine function is given as follows:

$$\begin{aligned}
D^{\alpha}x_1(t) &= \phi_1(x_2(t) - h_1(x(t))),\\
D^{\alpha}x_2(t) &= x_1(t) - x_2(t) + x_3(t),\\
D^{\alpha}x_3(t) &= -\phi_2 x_2(t),
\end{aligned} \tag{11.17}$$

where $x_1(t)$, $x_2(t)$, and $x_3(t)$ are system state variables, $\phi_1 > 0$, $\phi_2 > 0$, and $h_1(x(t))$ is a trigonometric function described by

$$h_1(x(t)) = \begin{cases} \frac{\kappa_2\pi}{2\kappa_1}(x_1(t) - 2\kappa_1\kappa_3), & x_1(t) \geq 2\kappa_1\kappa_3,\\ -\kappa_2 \sin\left(\frac{x_1(t)\pi}{2\kappa_1} + \kappa_4\right), & -2\kappa_1\kappa_3 < x_1(t) < 2\kappa_1\kappa_3,\\ \frac{\kappa_2\pi}{2\kappa_1}(x_1(t) + 2\kappa_1\kappa_3), & x_1(t) \leq -2\kappa_1\kappa_3. \end{cases} \tag{11.18}$$

Here, κ_1, κ_2, κ_3, κ_4 are constants, and $\kappa_1 > 0$, $\kappa_2 > 0$.

An N-scroll attractor is generated for the modified Chua's circuit (11.17) with the relationship

$$N = \kappa_3 + 1, \tag{11.19}$$

where

$$\kappa_4 = \begin{cases} \pi, & N \text{ is odd},\\ 0, & N \text{ is even}. \end{cases} \tag{11.20}$$

For parameters chosen as $\phi_1 = 10.814$, $\phi_2 = 14$, $\kappa_1 = 1.3$, $\kappa_2 = 0.11$, $\kappa_3 = 3$, and $\kappa_4 = 0$, initial values chosen as $[0, -0.1, 2]$, and the fractional order $\alpha : \alpha = 1.05$, the modified Chua's circuit (11.17) exhibits four-scroll attractor chaotic behaviors, as shown in Figure 11.1.

According to Equation (11.3), the slave system (11.17) is described as follows:

$$\begin{aligned}
D^{\alpha}y_1(t) &= \phi_1(y_2(t) - h_1(y(t))) + \text{sat}(u_1(t)),\\
D^{\alpha}y_2(t) &= y_1(t) - y_2(t) + y_3(t) + \text{sat}(u_2(t)),\\
D^{\alpha}y_3(t) &= -\phi_2 y_2(t) + \text{sat}(u_3(t)),
\end{aligned} \tag{11.21}$$

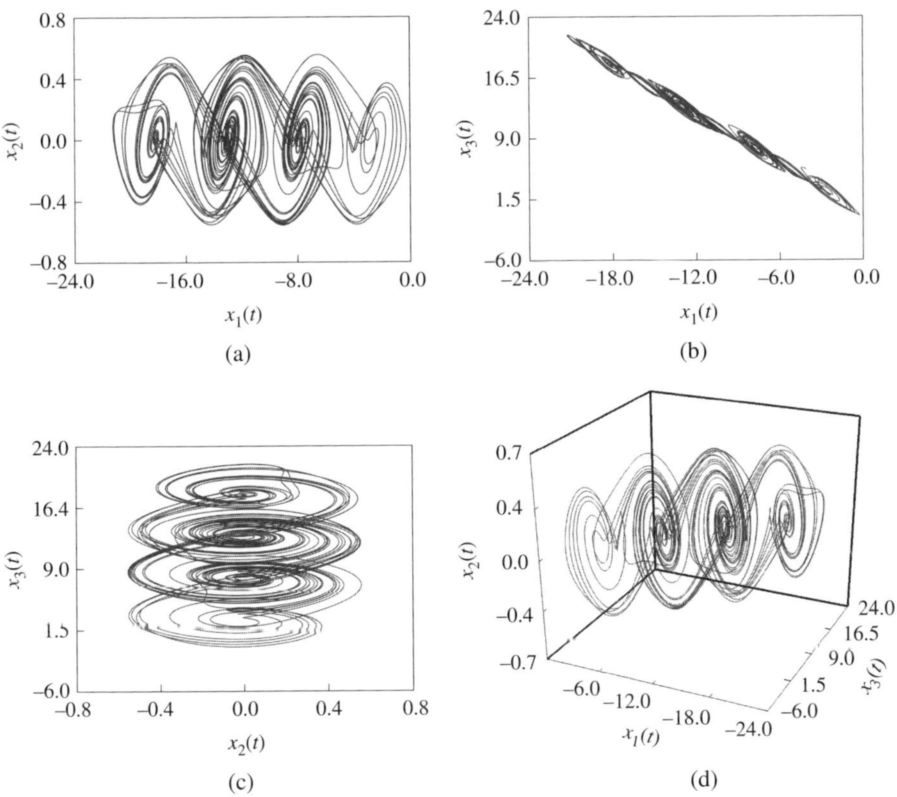

Figure 11.1 Chaotic behaviors of fractional-order modified Chua's circuit with sine function: (a) $x_1(t)$–$x_2(t)$ plane; (b) $x_1(t)$–$x_3(t)$ plane; (c) $x_2(t)$–$x_3(t)$ plane; (d) $x_3(t)$–$x_1(t)$–$x_2(t)$ space.

where $y_1(t)$, $y_2(t)$, and $y_3(t)$ are system state variables, and $u_1(t)$, $u_2(t)$, and $u_3(t)$ are synchronization control inputs.

From Equations (11.17) and (11.21), the synchronization error system can be written as the following:

$$\begin{aligned} w_1(t) &= \phi_1(w_2(t) - (h_1(y(t)) - h_1(x(t)))) + \text{sat}(u_1(t)), \\ w_2(t) &= w_1(t) - w_2(t) + w_3(t) + \text{sat}(u_2(t)), \\ w_3(t) &= -\phi_2 w_2(t) + \text{sat}(u_3(t)), \end{aligned} \tag{11.22}$$

where $w_1(t)$, $w_2(t)$, and $w_3(t)$ are synchronization errors.

Therefore, the constant matrix A can be obtained as

$$A = \begin{bmatrix} 0 & \phi_1 & 0 \\ 1 & -1 & 1 \\ 0 & -\phi_2 & 0 \end{bmatrix}. \tag{11.23}$$

Based on Equations (11.18) and (11.22), one has [234]

$$h_1(y(t)) - h_1(x(t)) = k_{y_1(t),x_1(t)}(y_1(t) - x_1(t)), \tag{11.24}$$

where $k_{y_1(t),x_1(t)}$ is dependent on $y_1(t)$ and $x_1(t)$, and satisfies the condition of $-\pi\kappa_2/2\kappa_1 \leq k_{y_1(t),x_1(t)} \leq \pi\kappa_2/2\kappa_1$. From Equation (11.24), one gets

$$\|h(y(t)) - h(x(t))\| \leq |10.814\pi\kappa_2/2\kappa_1| \|y(t) - x(t)\|. \tag{11.25}$$

Therefore, the nonlinear function $h(y(t)) - h(x(t))$ satisfies Assumption 11.1, with $\kappa = |10.814\pi\kappa_2/2\kappa_1| = 1.4373$. Furthermore, the matrices L and K are chosen as

$$\begin{cases} L = \mathrm{diag}[0.9, 0.9, 0.9], \\ K = \begin{bmatrix} k_1 & k_2 & 0 \\ k_3 & k_4 & k_5 \\ 0 & k_6 & k_7 \end{bmatrix}, \end{cases} \tag{11.26}$$

where $k_1 = -11.1111$, $k_2 = -12.0156$, $k_3 = -1.1111$, $k_4 = -10$, $k_5 = -1.1111$, $k_6 = -15.5556$, and $k_7 = -11.1111$.

On the basis of Equations (11.23) and (11.26), we have

$$Q = \begin{bmatrix} -10 & 0 & 0 \\ 0 & -10 & 0 \\ 0 & 0 & -10 \end{bmatrix}. \tag{11.27}$$

According to this discussion, we have

$$\|e^{Qt^\alpha}\| \leq e^{-10t}(M = 1)$$

and

$$\begin{aligned} \eta &= -\max\{\mathrm{Re}\lambda(Q)\} = 10 \\ &> (M(\kappa + \vartheta)\Gamma(\alpha))^{1/\alpha} = 3.5296(\vartheta = \|K - LK\| = 2.4243); \end{aligned}$$

Theorem 11.1 is satisfied. Let $u_{01}(t) = u_{02}(t) = u_{03}(t) = 15$ and choose the initial values of the master system and the slave system as $[0, -0.1, 2]^\mathrm{T}$ and $[-3, 1.3, 6]^\mathrm{T}$, respectively. The state synchronization results of the master system (11.17) and the slave system (11.21) are given in Figure 11.2a–c. It is shown that good synchronization performance is achieved. The responses of the synchronization errors $w_1(t)$, $w_2(t)$, and $w_3(t)$, which are convergent, are presented in Figure 11.3. Figure 11.4 shows the responses of saturation functions $\mathrm{sat}(u_1(t))$, $\mathrm{sat}(u_2(t))$, and $\mathrm{sat}(u_3(t))$. Obviously, two fractional-order modified Chua's circuits with a sine function under different initial conditions are synchronized.

11.3.2 Fractional-Order Four-Dimensional Modified Chua's Circuit

According to the integral-order four-dimensional modified Chua's model [235], the fractional-order four-dimensional modified Chua's circuit can be described by the following:

$$\begin{aligned} D^\alpha x_1(t) &= \sigma_1(x_2(t) - x_1(t) - h_1(x(t))), \\ D^\alpha x_2(t) &= x_1(t) - x_2(t) + x_3(t), \\ D^\alpha x_3(t) &= -\sigma_2(x_2(t) - x_4(t)), \\ D^\alpha x_4(t) &= -\sigma_3(x_2(t) + \sigma_4 x_4(t)), \end{aligned} \tag{11.28}$$

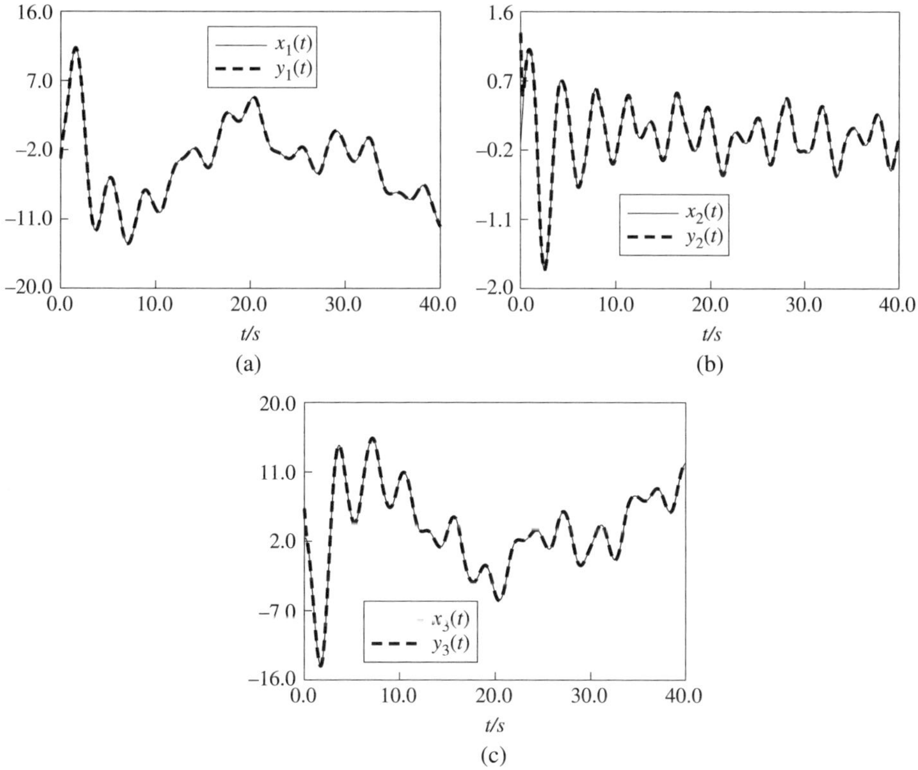

Figure 11.2 Synchronization control results of the master system (11.17) and the slave system (11.21): (a) $x_1(t)$ and $y_1(t)$; (b) $x_2(t)$ and $y_2(t)$; (c) $x_3(t)$ and $y_3(t)$.

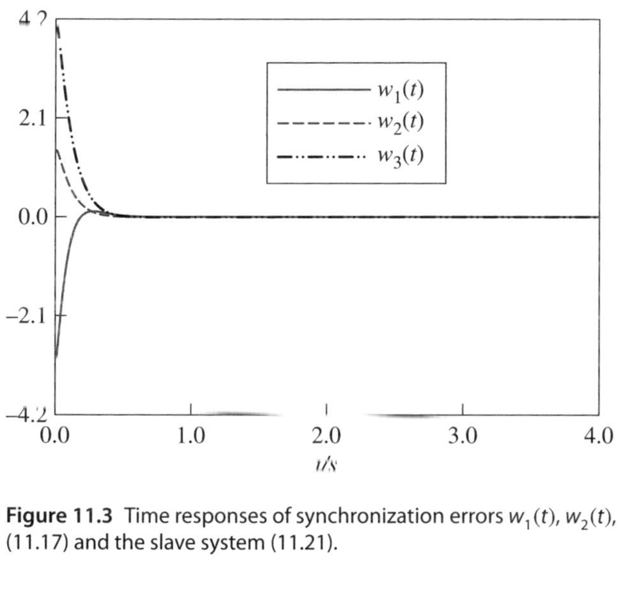

Figure 11.3 Time responses of synchronization errors $w_1(t)$, $w_2(t)$, and $w_2(t)$ of the master system (11.17) and the slave system (11.21).

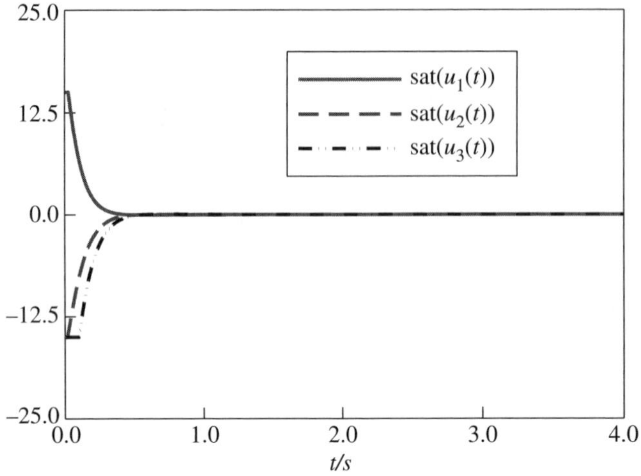

Figure 11.4 Time responses of synchronization control inputs sat($u_1(t)$), sat($u_2(t)$), and sat($u_3(t)$) of the master system (11.17) and the slave system (11.21).

where $x_1(t)$, $x_2(t)$, $x_3(t)$, and $x_4(t)$ are system state variables, $\sigma_1 > 0$, $\sigma_2 > 0$, $\sigma_3 > 0$, $\sigma_4 > 0$, and $h_1(x(t))$ is a piecewise linear function described by

$$h_1(x(t)) = m_1x_1(t) + \frac{1}{2}(m_0 - m_1)(|x_1(t) + 1| - |x_1(t) - 1|), \tag{11.29}$$

with $m_0 < -1 < m_1 < 0$. For parameters chosen as $\sigma_1 = 9.934$, $\sigma_2 = 14.334$, $\sigma_3 = 0.0497$, $\sigma_4 = 27.333$, $m_0 = -1.246$, $m_1 = -0.6724$, and $\alpha = 1.01$, initial values chosen as $[0.1, 0.2, 0.3, 0.4]^{\mathrm{T}}$, the fractional-order four-dimensional modified Chua's system (11.28) exhibits chaotic behaviors, as shown in Figure 11.5.

From Equation (11.3) and the master system (11.28), the slave system is defined as follows:

$$\begin{aligned}
D^\alpha y_1(t) &= \sigma_1(y_2(t) - y_1(t) - h_1(y(t))) + \mathrm{sat}(u_1(t)),\\
D^\alpha y_2(t) &= y_1(t) - y_2(t) + y_3(t) + \mathrm{sat}(u_2(t)),\\
D^\alpha y_3(t) &= -\sigma_2(y_2(t) - y_4(t)) + \mathrm{sat}(u_3(t)),\\
D^\alpha y_4(t) &= -\sigma_3(y_2(t) + \sigma_4 y_4(t)) + \mathrm{sat}(u_4(t)),
\end{aligned} \tag{11.30}$$

where $y_1(t)$, $y_2(t)$, $y_3(t)$, and $y_4(t)$ are system state variables, and $u_1(t)$, $u_2(t)$, $u_3(t)$, and $u_4(t)$ are synchronization control inputs.

According to the master system (11.28) and the slave system (11.30), the synchronization error system is written as follows:

$$\begin{aligned}
D^\alpha w_1(t) &= \sigma_1(w_2(t) - w_1(t) - (h_1(y(t)) - h_1(x(t)))) + \mathrm{sat}(u_1(t)),\\
D^\alpha w_2(t) &= w_1(t) - w_2(t) + w_3(t) + \mathrm{sat}(u_2(t)),\\
D^\alpha w_3(t) &= -\sigma_2(w_2(t) - w_4(t)) + \mathrm{sat}(u_3(t)),\\
D^\alpha w_4(t) &= -\sigma_3(w_2(t) + \sigma_4 w_4(t)) + \mathrm{sat}(u_4(t)).
\end{aligned} \tag{11.31}$$

Based on the properties of the piecewise linear function [234], we have

$$h_1(y(t)) - h_1(x(t)) = m_{y_1(t),x_1(t)}(y_1(t) - x_1(t)), \tag{11.32}$$

where $m_{y_1(t),x_1(t)}$ is dependent on $y_1(t)$ and $x_1(t)$, and varies within the interval $[m_0, m_1]$ for $t \geq 0$. Therefore, $m_{y_1(t),x_1(t)}$ is bounded as $m_0 \leq m_{y_1(t),x_1(t)} \leq m_1 < 0$.

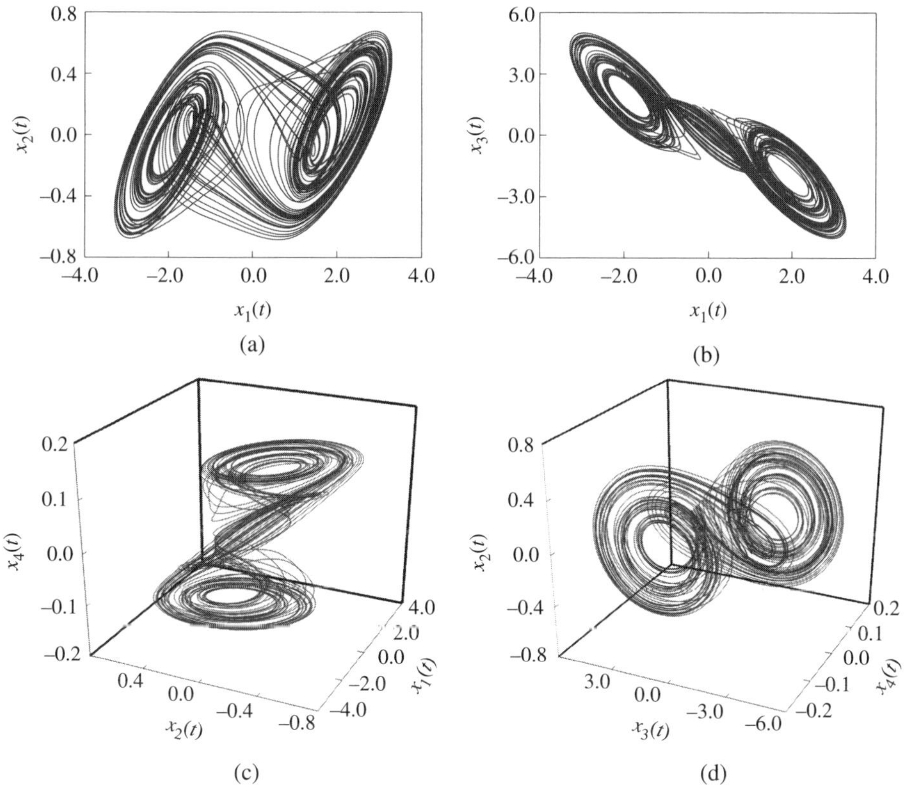

Figure 11.5 Chaotic behaviors of fractional-order four-dimensional modified Chua's system: (a) $x_1(t)$–$x_2(t)$ plane; (b) $x_1(t)$–$x_3(t)$ plane; (c) $x_1(t)$–$x_2(t)$–$x_4(t)$ space; (d) $x_4(t)$–$x_3(t)$–$x_2(t)$ space.

Invoking Equations (11.31) and (11.32), one obtains

$$\|h(y(t)) - h(x(t))\| \leq |9.943m_0|\|(y(t) - x(t))\|. \tag{11.33}$$

Therefore, the nonlinear function $h(y(t)) - h(x(t))$ satisfies Assumption 11.1, with $\kappa = 9.943\,|m_0| = 12.3778$.

Furthermore, from Equation (11.31), the constant matrix A is obtained as

$$A = \begin{bmatrix} -\sigma_1 & \sigma_1 & 0 & 0 \\ 1 & -1 & 1 & 0 \\ 0 & -\sigma_2 & 0 & \sigma_2 \\ 0 & 0 & -\sigma_3 & -\sigma_3\sigma_4 \end{bmatrix}. \tag{11.34}$$

The control matrices L and K are chosen as

$$\begin{cases} L = \mathrm{diag}[0.9, 0.9, 0.9, 0.9], \\ K = \begin{bmatrix} k_1 & k_2 & 0 & 0 \\ k_3 & k_4 & k_5 & 0 \\ 0 & k_6 & k_7 & k_8 \\ 0 & 0 & k_9 & k_{10} \end{bmatrix}, \end{cases} \tag{11.35}$$

where $k_1 = -6.6667$, $k_2 = -11.0378$, $k_3 = -1.1111$, $k_4 = -16.6667$, $k_5 = -1.1111$, $k_6 = 15.9267$, $k_7 = -17.7778$, $k_8 = -15.9267$, $k_9 = 0.0552$, and $k_{10} = -17.7778$.

On the basis of Equations (11.34) and (11.35), one obtains

$$Q = \begin{bmatrix} -15.934 & 0 & 0 & 0 \\ 0 & -16 & 0 & 0 \\ 0 & 0 & -16 & 0 \\ 0 & 0 & 0 & -17.3583 \end{bmatrix}. \tag{11.36}$$

Based on this discussion, we have

$$\|e^{Qt^{\alpha}}\| \leq e^{-15.934t}(M = 1)$$

and

$$\eta = -\max\{\mathrm{Re}\lambda(Q)\} = 15.934$$
$$> (M(\kappa + \vartheta)\Gamma(\alpha))^{1/\alpha} = 15.1429(\vartheta = \|K - LK\| = 3.2711).$$

According to Theorem 11.1, the synchronization error system (11.31) can be stabilized, i.e., the master system (11.28) and the slave system (11.30) are synchronized. For the simulation, we choose $u_{01}(t) = u_{02}(t) = u_{03}(t) = u_{04}(t) = 20$ and the initial conditions $x_0(t) = [0.1, 0.2, 0.3, 0.4]^{\top}$ and $y_0(t) = [-1, -0.6, -1, -0.2]^{\top}$. The state synchronization results of the master system (11.28) and the slave system (11.30) are given in Figure 11.6a–d. It is shown that good synchronization performance is achieved.

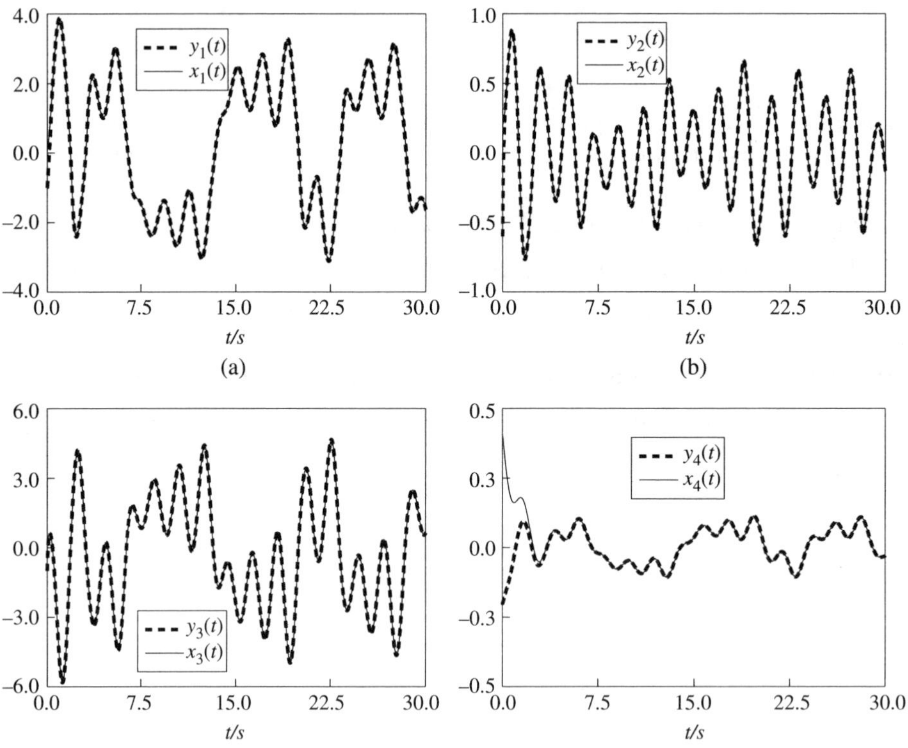

Figure 11.6 Synchronization control results of the master system (11.28) and the slave system (11.30): (a) $x_1(t)$ and $y_1(t)$; (b) $x_2(t)$ and $y_2(t)$; (c) $x_3(t)$ and $y_3(t)$; (d) $x_4(t)$ and $y_4(t)$.

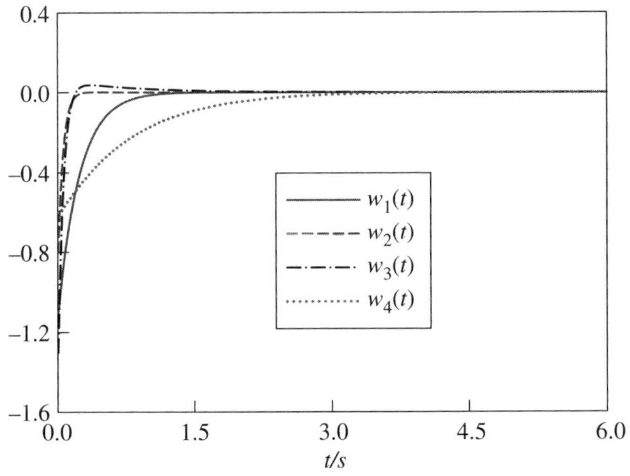

Figure 11.7 Time responses of synchronization errors $w_1(t)$, $w_2(t)$, $w_3(t)$, and $w_4(t)$ of the master system (11.28) and the slave system (11.30).

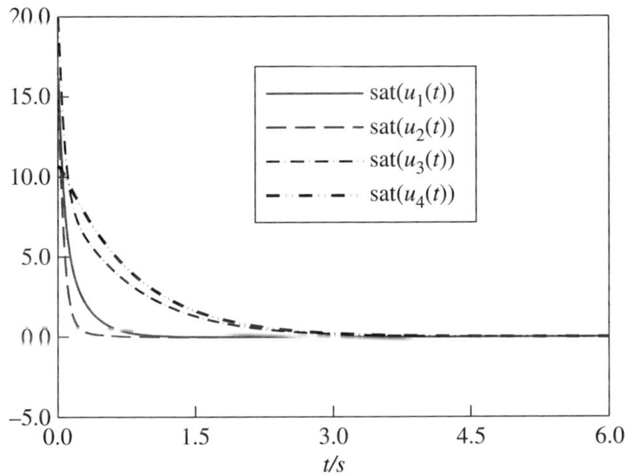

Figure 11.8 Time responses of synchronization control inputs $\mathrm{sat}(u_1(t))$, $\mathrm{sat}(u_2(t))$, $\mathrm{sat}(u_3(t))$, and $\mathrm{sat}(u_4(t))$ of the master system (11.28) and the slave system (11.30).

Figure 11.7 shows that the error functions $w_1(t), w_2(t), w_3(t)$, and $w_4(t)$ are stabilized to origin asymptotically. The saturation functions $\mathrm{sat}(u_1(t))$, $\mathrm{sat}(u_2(t))$, $\mathrm{sat}(u_3(t))$, and $\mathrm{sat}(u_4(t))$ are presented in Figure 11.8.

From these simulation results for the synchronization of two fractional-order modified Chua's circuits, we obtain that the proposed synchronization control scheme based on the stability theory of fractional-order systems is valid for a class of nonlinear fractional-order systems subjected to input saturation.

11.4 Conclusion

In this chapter, the synchronization control problem has been investigated for a class of nonlinear fractional-order systems subjected to input saturation. According to the generalized Gronwall inequality, the ML function, and the properties of fractional calculus, a synchronous control theorem has been given for the synchronization of a class of nonlinear fractional-order systems with fractional order $1 < \alpha < 2$. It has been shown that the synchronization can be guaranteed for such systems by using the designed linear state-feedback controller. Two examples are given in this chapter, i.e., the synchronization between two modified Chua's circuits with a sine function, and the synchronization between two fractional-order four-dimensional modified Chua's systems. Numerical simulations have confirmed the effectiveness of the developed synchronization control method.

12

Synchronization Control for Fractional-Order Chaotic Systems with Input Saturation and Disturbance

12.1 Problem Statement

Based on the Caputo definition of the fractional derivative (2.17), we consider a class of fractional-order chaotic systems with unknown constant disturbance in the form of

$$D^{\alpha}x(t) = Ax(t) + g(x(t)) + Bd, \tag{12.1}$$

where $A \in R^{n\times n}$ is a constant matrix, $x(t) = [x_1(t), x_2(t), \cdots, x_n(t)]^{\top} \in R^n$ is a state vector, $g(x(t)) = [g_1(x(t)), g_2(x(t)), \cdots, g_n(x(t))]^{\top} \in R^n$ is a nonlinear function vector, $B = [0, 0, \cdots, 1]^{\top} \in R^n$, $d \in R$ is an unknown bounded disturbance, and $0 < \alpha < 1$.

To design the synchronization control scheme, the FOCS (12.1) is taken as the drive system, and the response system is constructed as

$$D^{\alpha}y(t) = Ay(t) + g(y(t)) + B\hat{d} + G\text{sat}(u(t)), \tag{12.2}$$

where $G = B$, $y(t) = [y_1(t), y_2(t), \cdots, y_n(t)]^{\top} \in R^n$ is a state vector, $g(y(t)) = [g_1(y(t)), g_2(y(t)), \cdots, g_n(y(t))]^{\top} \in R^n$ is a nonlinear function vector, $0 < \alpha < 1$, $\hat{d}$ denotes the estimated value of d, and the control input $\text{sat}(u(t))$ is the vector-valued input saturation function, which is defined as

$$\text{sat}(u(t)) = \text{sign}(u(t))\min(u_0(t), |u(t)|), \tag{12.3}$$

where $u_0(t)$ represents the symmetric saturation level of the control input $u(t)$, which is assumed to be known.

This chapter aims to develop a synchronization control scheme for fractional-order chaotic systems subjected to input saturation and unknown disturbance. In the synchronization control design, a FODO is developed to estimate the unknown disturbance.

12.2 Design of Fractional-Order Disturbance Observer

In this section, a FODO is designed to estimate the external unknown bounded disturbance of the FOCS. This kind of FODO was introduced in Chapter 8. To further illustrate the design method of the FODO, the design process of this kind of FODO will be redescribed. On the basis of the integer-order disturbance observer design technique presented by Chen [117], the FODO will be designed for the FOCS (12.1) in this

Robust Adaptive Control for Fractional-Order Systems with Disturbance and Saturation, First Edition.
Mou Chen, Shuyi Shao, and Peng Shi.

section. In general, there is no prior information about the fractional derivative of the disturbance d. Thus, it is reasonable to assume that the disturbance d is an unknown bounded constant in this chapter and

$$D^\alpha d = 0. \tag{12.4}$$

From the system (12.1), d can be written as

$$d = D^\alpha x_n(t) - Hx(t) - g_n(x(t)), \tag{12.5}$$

where H is the nth line of matrix A. An initial FODO is proposed as

$$D^\alpha \hat{d} = -a\hat{d} + aD^\alpha x_n(t) - a(Hx(t) + g_n(x(t))), \tag{12.6}$$

where $a > 0$ is a design parameter of the FODO. The disturbance estimation error is defined as

$$\tilde{d} = d - \hat{d}. \tag{12.7}$$

According to Equations (12.4) to (12.7), one obtains

$$D^\alpha \tilde{d} = a(\hat{d} - d) = -a\tilde{d}. \tag{12.8}$$

If the design parameter a is chosen to be such that

$$D^\alpha \tilde{d} + a\tilde{d} = 0, \tag{12.9}$$

the disturbance estimation error system tends to be stable. Then the disturbance estimate output $\hat{d}$ can approximate d, as $t \to \infty$.

However, $D^\alpha x(t)$ is not available; hence, the FODO (12.6) cannot be realized. To solve this problem, an auxiliary design variable is defined as

$$\phi(t) = \hat{d} - ax_n(t). \tag{12.10}$$

Considering Equations (12.6) and (12.10) yields

$$\begin{aligned} D^\alpha \phi(t) &= D^\alpha \hat{d} - aD^\alpha x_n(t) \\ &= -a(\phi(t) + ax_n(t)) - a(Hx(t) + g_n(x(t))) \\ &= -a\phi(t) - f(x, u), \end{aligned} \tag{12.11}$$

where $f(x, u) = a(ax_n(t) + Hx(t) + g_n(x(t)))$.

From Equation (12.10), the disturbance estimate output $\hat{d}$ can be written as

$$\hat{d} = \phi(t) + ax_n(t). \tag{12.12}$$

The designed FODO consists of Equations (12.11) and (12.12). Invoking Equations (12.7), (12.11), and (12.12), the disturbance estimation error is given by

$$\begin{aligned} D^\alpha \tilde{d} &= D^\alpha d - D^\alpha \hat{d} \\ &= -D^\alpha \phi(t) - aD^\alpha x_n(t) \\ &= a\phi(t) + a(ax_n(t) - d + D^\alpha x_n(t)) - aD^\alpha x_n(t) \\ &= a(\hat{d} - d) = -a\tilde{d}. \end{aligned} \tag{12.13}$$

From Equation (12.13), we know that the disturbance estimation output $\hat{d}$ can approximate the disturbance d well if a is chosen to be such that $\tilde{d}$ is asymptotically stable, based on Equation (12.13).

Remark 12.1 *The disturbance estimation error systems (12.9) and (12.13) are asymptotically stable if the design parameter a satisfies* $|\arg(-a)| > \alpha\pi/2$, *where* $\arg(-a)$ *denotes the argument of the parameter* $-a$ *[240].*

12.3 Design of Synchronization Control

In this section, the synchronization control scheme will be investigated for two fractional-order chaotic systems with different initial conditions. To design the synchronization control scheme, the following assumption is required.

Assumption 12.1 *According to Lemma 2.9, the drive system (12.1), and the response system (12.2), nonlinear functions* $g(x(t))$ *and* $g(y(t))$ *satisfy* $g(0) = 0$, *and* $\varphi(t, w(t))$ *satisfy* $\varphi(t, 0) = 0$. *At the same time,* $\|g(y(t)) - g(x(t))\| \leq \kappa\|w(t)\|(\kappa \geq 0)$, *and* $\varphi(t, w(t)) \leq \vartheta\|w(t)\|(\vartheta \geq 0)$, *where* κ *and* ϑ *are known constants and* $w(t) = y(t) - x(t)$ *is the synchronization error.*

From Equations (12.1) and (12.2), the synchronization error system can be written as

$$D^{\alpha}w(t) = A(y(t) - x(t)) + g(y(t)) - g(x(t)) + G\text{sat}(u(t)) - G\tilde{d}. \tag{12.14}$$

Using the state-feedback control method, the desired synchronization control input is designed as

$$u(t) = Kw(t), \tag{12.15}$$

where K is a design vector. Then the synchronization error system (12.14) is

$$D^{\alpha}w(t) = A(y(t) - x(t)) + g(y(t)) - g(x(t)) + G\text{sat}(Kw(t)) - G\tilde{d}. \tag{12.16}$$

From Lemma 2.9, we have $\varphi(t, w(t)) = \text{sat}(Kw(t)) - LKw(t)$, where L is a design constant. Considering Equation (12.16), we obtain

$$\begin{aligned} D^{\alpha}w(t) &= A(y(t) - x(t)) + g(y(t)) - g(x(t)) + G\text{sat}(Kw(t)) - G\tilde{d} \\ &= Aw(t) + g(y(t)) - g(x(t)) + G(LKw(t)) - G\tilde{d} + G\varphi(t, w(t)). \end{aligned} \tag{12.17}$$

To analyze the stability of synchronization error system (12.17), an augmented system is given based on Equations (12.13) and (12.17) as follows:

$$\begin{aligned} &D^{\alpha}\tilde{d} = -a\tilde{d}, \\ &D^{\alpha}w(t) = Aw(t) + g(y(t)) - g(x(t)) + G(LKw(t)) - G\tilde{d} + G\varphi(t, w(t)). \end{aligned} \tag{12.18}$$

On the basis of Equation (12.18), we have

$$D^{\alpha}e(t) = Qe(t) + g_{x(t),y(t)} + G_1\varphi(t, w(t)), \tag{12.19}$$

where

$$Q = \begin{bmatrix} -a & O_1 \\ O_2 & A \end{bmatrix} + \begin{bmatrix} O_2 & O_3 \\ 0 & LK \end{bmatrix} - \begin{bmatrix} O_2 & O_3 \\ 1 & O_1 \end{bmatrix},$$

$O_1 \in R^{1\times n}$ denotes a zero vector, the nonlinear function

$$g_{x(t),y(t)} = \begin{bmatrix} 0 \\ g(y(t)) - g(x(t)) \end{bmatrix},$$

$O_2 \in R^{n\times 1}$ is a zero vector, $O_3 \in R^{n\times n}$ is a zero matrix, and $G_1 = [0, 0, \cdots, 1]^{\top} \in R^{(n+1)\times 1}$.

The synchronization control scheme for two fractional-order chaotic systems can be summarized in the following theorem.

Theorem 12.1 *For the synchronization error system (12.14), the disturbance observer is designed according to Equations (12.11) and (12.12) with $a > 0$. Then, the synchronization control scheme is proposed as Equation (12.15). Then the synchronization error $w(t)$ is asymptotically stable under the proposed synchronization control scheme for $0 < \alpha < 1$, if the design parameters L and K render $|\arg(\mathrm{eig}(Q))| > \alpha\pi/2$, $-\max(\mathrm{Re}(\mathrm{eig}(Q^{1/\alpha}))) > 0$.*

Proof 12.1 *To demonstrate that the synchronization error is asymptotically convergent, it only remains to prove that the compound system (12.19) is asymptotically stable. Assume that e_0 is the initial condition of Equation (12.19). By taking the Laplace transform for the system (12.19), we have*

$$s^\alpha E(s) = s^{\alpha-1}e_0 + QE(s) + \mathscr{L}(g_{x(t),y(t)}) + G_1\mathscr{L}(\varphi(t, w(t))), \tag{12.20}$$

where $E(s)$ is the Laplace transform of $e(t)$.

Taking the Laplace inverse transform of Equation (12.20), one obtains

$$e(t) = E_{\alpha,1}(Qt^\alpha)e_0 + \bar{e}, \tag{12.21}$$

where

$$\bar{e} = \int_0^t (t-\tau)^{\alpha-1}E_{\alpha,\alpha}(Q(t-\tau)^\alpha)g_{x(\tau),y(\tau)}\mathrm{d}\tau + G_1\int_0^t (t-\tau)^{\alpha-1}E_{\alpha,\alpha}(Q(t-\tau)^\alpha)\varphi(\tau, w(\tau))\mathrm{d}\tau. \tag{12.22}$$

Then Equation (12.21) satisfies

$$\|e(t)\| \le \|E_{\alpha,1}(Qt^\alpha)e_0\| + \|\bar{e}\|, \tag{12.23}$$

where

$$\|\bar{e}\| = \int_0^t (t-\tau)^{\alpha-1}\|E_{\alpha,\alpha}(Q(t-\tau)^\alpha)\|\|g_{x(\tau),y(\tau)}\|\mathrm{d}\tau + \int_0^t (t-\tau)^{\alpha-1}\|E_{\alpha,\alpha}(Q(t-\tau)^\alpha)\|\|\varphi(\tau, w(\tau))\|\mathrm{d}\tau. \tag{12.24}$$

Based on Assumption 12.1, Equation (12.23) can be written as

$$\|e(t)\| \le \|E_{\alpha,1}(Qt^\alpha)e_0\| + \bar{e}_0, \tag{12.25}$$

where

$$\bar{e}_0 = (\kappa + \vartheta)\int_0^t (t-\tau)^{\alpha-1}\|E_{\alpha,\alpha}(Q(t-\tau)^\alpha)\|\|e(\tau)\|\mathrm{d}\tau.$$

According to Lemma 2.14 and considering the stable matrix Q, $|\arg(\mathrm{eig}(Q))| > \alpha\pi/2$, $-\rho = \max(\mathrm{Re}(\mathrm{eig}(Q^{1/\alpha})))(\rho > 0)$, $M_1 > 0$, $M_2 > 0$, and $0 < \alpha < 1$, Equation (12.25) can be written as

$$\begin{aligned}\|e(t)\| &\le M_1\|e_0\|\left\|e^{Q^{1/\alpha}t}\right\| + (\kappa+\vartheta)\tilde{e}\\ &\le C_1e^{-\rho t} + C_2\tilde{e},\end{aligned} \tag{12.26}$$

where $C_1 = M_1 M_2 \|w_0\|$, $C_2 = M_2(\kappa + \vartheta)$,

$$C_3 = \sum_{k=0}^{\infty} \frac{\|Q^{(\alpha-1)k/\alpha}\| \Gamma(k+1)}{\Gamma(k\alpha + \alpha)},$$

and

$$\tilde{e} = \int_0^t C_3(t-\tau)^{(\alpha-1)(k+1)} \mathrm{e}^{-\rho(t-\tau)} \|e(\tau)\| \mathrm{d}\tau.$$

Multiplying both sides of Equation (12.26) by $\mathrm{e}^{\rho t}$, *we obtain*

$$\|e(t)\|\mathrm{e}^{\rho t} \leq C_1 + C_2 C_3 \overline{C}, \tag{12.27}$$

where

$$\overline{C} = \int_0^t (t-\tau)^{(\alpha-1)(k+1)} \mathrm{e}^{\rho\tau} \|e(\tau)\| \mathrm{d}\tau.$$

From Lemma 2.8, considering $b(t) = C_1$, $h(t) = \mathrm{e}^{\rho t}\|e(t)\|$, *and*

$$p(\tau) = C_2 \sum_{k=0}^{\infty} \frac{\|Q^{(\alpha-1)k/\alpha}\| \Gamma(k+1)}{\Gamma(k\alpha + \alpha)} (t-\tau)^{(\alpha-1)(k+1)},$$

one has

$$\begin{aligned} \|e(t)\|\mathrm{e}^{\rho t} &\leq C_1 \exp\left\{ C_2 C_3 \int_0^t (t-\tau)^{(\alpha-1)(k+1)} \mathrm{d}\tau \right\} \\ &= C_1 \exp\left\{ C_2 C_3 \frac{t^{(\alpha-1)(k+1)+1}}{(\alpha-1)(k+1)+1} \right\}. \end{aligned} \tag{12.28}$$

Multiplying both sides of Equation (12.28) by $\mathrm{e}^{-\rho t}$, *we have*

$$\|e(t)\| \leq C_1 \exp\left\{ -\rho t + C_2 C_3 \frac{t^{(\alpha-1)(k+1)+1}}{(\alpha-1)(k+1)+1} \right\}. \tag{12.29}$$

Referring to Lemma 2.13, we have

$$\lim_{t\to\infty} \|e(t)\| = 0. \tag{12.30}$$

Thus, $\|w(t)\|$ *asymptotically tends to zero as* $t \to \infty$. *Namely, the synchronization error system (12.17) is asymptotically stable. Thus, the synchronization of the drive system (12.1) and the response system (12.2) is achieved under the effect of the input saturation and the unknown disturbance. This concludes the proof.* □

12.4 Simulation Examples

To illustrate the effectiveness of the proposed synchronization control scheme, two simulation examples are studied in this section.

12.4.1 Fractional-Order Chua's Circuit

Synchronization between two identical fractional-order Chua's circuits [241] with fractional-order $\alpha = 0.98$ under different initial conditions is investigated. In this simulation, phase portraits of fractional-order chaotic systems are obtained by the method described by Petráš [164]. The fractional-order Chua's circuit [241] with external disturbance is described as follows:

$$\begin{aligned} D^{\alpha}x_1(t) &= \varepsilon_1(x_2(t) - x_1(t) - g_1(x(t))), \\ D^{\alpha}x_2(t) &= x_1(t) - x_2(t) + x_3(t), \\ D^{\alpha}x_3(t) &= -\varepsilon_2 x_2(t) + d, \end{aligned} \tag{12.31}$$

where $x_i, i = 1, 2, 3$ are the system states, $\varepsilon_1 > 0$, $\varepsilon_2 > 0$, d is the constant disturbance, and $g_1(x(t))$ is a piecewise linear function described by

$$g_1(x(t)) = m_1 x_1(t) + g_{10}(x(t)), \tag{12.32}$$

where

$$g_{10}(x(t)) = \frac{1}{2}(m_0 - m_1)(|x_1(t) + 1| - |x_1(t) - 1|),$$

with $m_0 < -1 < m_1 < 0$. The parameters are chosen as $\varepsilon_1 = 10, \varepsilon_2 = 14.97, m_0 = -1.31$, $m_1 = -0.75$, the constant disturbance is assumed as $d = 1$, and the initial values are chosen as $x_0 = [-0.1, -0.2, 0.3]^{\top}$.

Figure 12.1 shows the fractional-order system (12.31) that exhibits chaotic behaviors, which is different from the chaotic behaviors of a fractional-order Chua's circuit [241].

Consider the fractional-order Chua's circuit [241] with the saturation function and the disturbance observer. The system can be written as follows:

$$\begin{aligned} D^{\alpha}y_1(t) &= \varepsilon_1(y_2(t) - y_1(t) - g_1(y(t))), \\ D^{\alpha}y_2(t) &= y_1(t) - y_2(t) + y_3(t), \\ D^{\alpha}y_3(t) &= -\varepsilon_2 y_2(t) + \hat{d} + \mathrm{sat}(u(t)), \end{aligned} \tag{12.33}$$

where $y_i, i = 1, 2, 3$ are the system states, $\hat{d}$ is the disturbance observer, and $u(t)$ is the synchronization control input.

Based on the drive system (12.31) and the response system (12.33), the synchronization error system and the disturbance estimation error system can be written as follows:

$$\begin{aligned} D^{\alpha}w_1(t) &= \varepsilon_1(w_2(t) - w_1(t) - (g_1(y(t)) - g_1(x(t)))), \\ D^{\alpha}w_2(t) &= w_1(t) - w_2(t) + w_3(t), \\ D^{\alpha}w_3(t) &= -\varepsilon_2 w_2(t) - \tilde{d} + \mathrm{sat}(u(t)), \end{aligned} \tag{12.34}$$

where w_i, with $i = 1, 2, 3$, are the synchronization errors.

From Equation (12.34), all the parameter matrices in the synchronization error system (12.14) are given by

$$A = \begin{bmatrix} -\varepsilon_1 & \varepsilon_1 & 0 \\ 1 & -1 & 1 \\ 0 & -\varepsilon_2 & 0 \end{bmatrix}, \qquad G = [0, 0, 1]^{\top}. \tag{12.35}$$

According to Equations (12.32) and (12.34), one has [242]

$$g_1(y(t)) - g_1(x(t)) = k_{y(t),x(t)}(y_1(t) - x_1(t)), \tag{12.36}$$

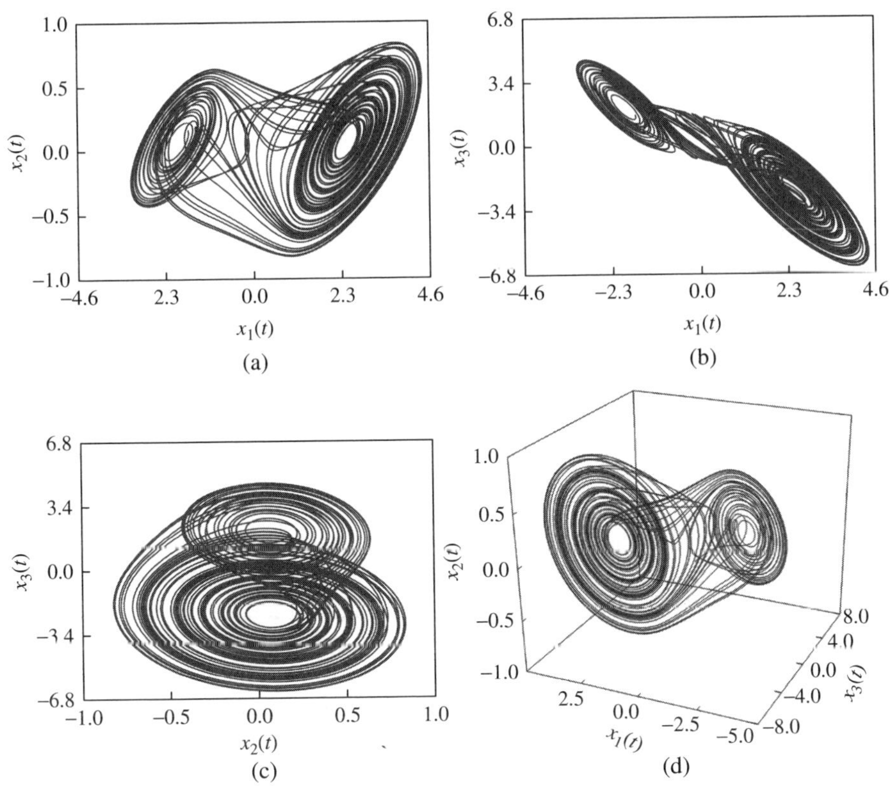

Figure 12.1 Chaotic behaviors of fractional-order Chua's circuit (12.31) with disturbance: (a) $x_1(t)$–$x_2(t)$ plane; (b) $x_1(t)$–$x_3(t)$ plane; (c) $x_2(t)$–$x_3(t)$ plane, (d) $x_3(t)$–$x_1(t)$–$x_2(t)$ space.

where $k_{y(t),x(t)}$ depends on $y_1(t)$ and $x_1(t)$, and changes within the interval $[m_0, m_1]$ for $t \geq 0$. Namely, $k_{y(t),x(t)}$ is bounded and satisfies $m_0 < k_{y(t),x(t)} \leq m_1 < 0$.

From Equations (12.31) and (12.36), one obtains

$$\|g(y(t)) - g(x(t))\| \leq |10m_0|\|(y(t) - x(t))\|. \tag{12.37}$$

Therefore, the nonlinear function $g(y(t)) - g(x(t))$ satisfies Assumption 12.1 and $\kappa = 10\,|m_0| = 13.1$.

In this simulation study, the parameter of the disturbance observer is chosen as $a = 0.2$. It follows that $|\arg(-a)| > 1.5394 = \alpha\pi/2$. The initial values are chosen as $x_1(0) = -0.1$, $x_2(0) = -0.2$, $x_3(0) = 0.3$, $y_1(0) = 0.4$, $y_2(0) = 0.1$, $y_3(0) = 0.2$, and $\phi(0) = -1$. The saturation level value is chosen as $u_0(t) = 10$, and the unknown disturbance is taken as $d = 1$.

To ensure that $-u_0(t) \leq LKw(t) \leq u_0(t)$ and Q is a stable matrix, the design parameters K and L are chosen as

$$K = \begin{bmatrix} 0 & -40 & 0 \end{bmatrix}, \qquad L = [0.8]. \tag{12.38}$$

From Equations (12.34) and (12.38), we obtain $\max(LKw(t)) = 6.4184$ and $\min(LKw(t)) = -9.6$ by numerical computation; thus, the memoryless nonlinearity $\varphi(t, w(t))$ satisfies Assumption 12.1.

Then, we obtain the following control matrix:

$$Q = \begin{bmatrix} -0.2 & 0 & 0 & 0 \\ 0 & -10 & 10 & 0 \\ 0 & 1 & -1 & 1 \\ -1 & 0 & -46.97 & 0 \end{bmatrix}. \tag{12.39}$$

From Equation (12.39), the eigenvalues of Q are $\lambda_{01} = -10.7095$, $\lambda_{02} = -0.2$, $\lambda_{03} = -0.1453 + 6.6210i$, $\lambda_{04} = -0.1453 - 6.6210i$, and all eigenvalues of $Q^{1/0.98}$ are $\lambda_{11} = -0.3745 + 6.8729i$, $\lambda_{12} = -0.3745 - 6.8729i$, $\lambda_{13} = -11.2173 - 0.7202i$, $\lambda_{14} = -0.1931 - 0.0124i$. Thus, $|\arg(\mathrm{eig}(Q))| > \alpha\pi/2$ and $-\max(\mathrm{Re}(\mathrm{eig}(Q^{1/\alpha}))) > 0$ satisfy the given conditions of Theorem 12.1. The disturbance observer is designed based on Equations (12.11) and (12.12). The synchronization control scheme is designed according to Equation (12.15).

Synchronization control results are obtained under the developed chaotic synchronization control scheme. The state synchronization results of the drive system (12.31) and the response system (12.33), given in Figure 12.2, Figure 12.3, and Figure 12.4, show that good synchronization performance is achieved. Figure 12.5 shows that the synchronization errors $w_1(t)$, $w_2(t)$, and $w_3(t)$ are convergent. The disturbance estimation result is presented in Figure 12.6 and the estimation error of the disturbance observer is shown in Figure 12.7. From Figure 12.6 and Figure 12.7, we note that satisfactory disturbance estimation is obtained. The synchronization control input is given in Figure 12.8; a bounded synchronization control input is observed. According to these simulation results, the synchronization error and the disturbance estimation error are asymptotically stable. This implies that the developed chaotic synchronization control scheme is valid.

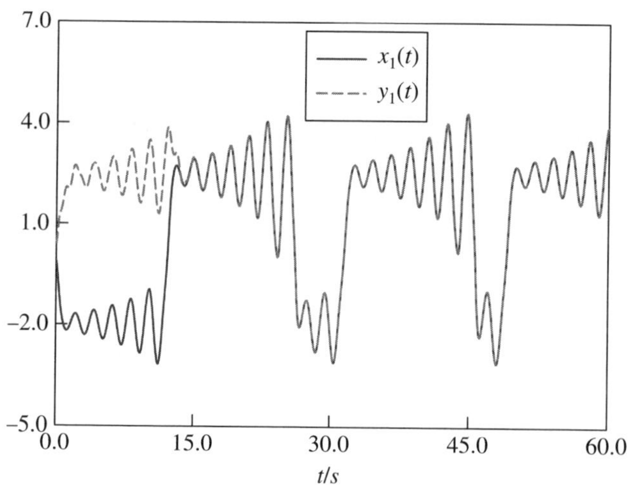

Figure 12.2 Synchronization states of $x_1(t)$ and $y_1(t)$ for the systems (12.31 and 12.33).

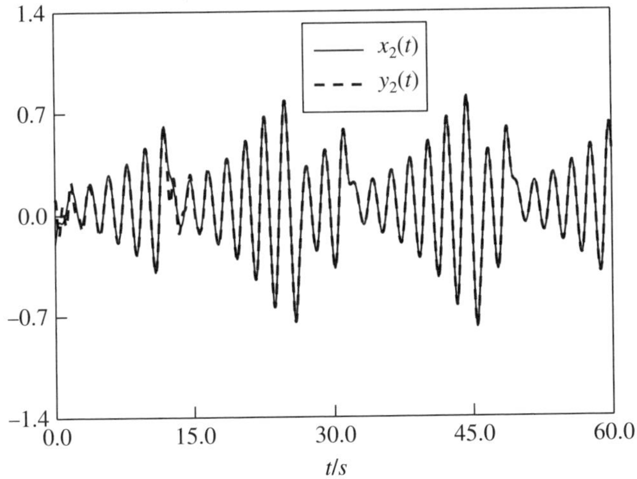

Figure 12.3 Synchronization states of $x_2(t)$ and $y_2(t)$ for the systems (12.31 and 12.33).

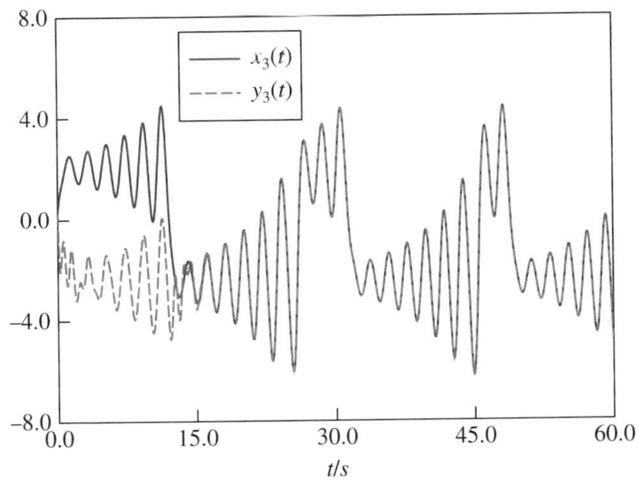

Figure 12.4 Synchronization states of $x_3(t)$ and $y_3(t)$ for the systems (12.31 and 12.33).

12.4.2 Fractional-Order Hyperchaos Chua's Circuit

According to the model of the integer-order hyperchaos Chua's circuit [236], synchronization between two fractional-order hyperchaos Chua's circuits with fractional-order $\alpha = 0.98$ under different initial conditions is investigated. First, the fractional-order hyperchaos Chua's circuit is given by the following:

$$\begin{aligned} D^{\alpha}x_1(t) &= \hbar_1(x_3(t) - 2.895x_1(t) - g_1(x(t))), \\ D^{\alpha}x_2(t) &= \hbar_2 x_2(t) - x_3(t) - x_4(t), \\ D^{\alpha}x_3(t) &= \hbar_3(x_2(t) - x_1(t) - x_3(t)), \\ D^{\alpha}x_4(t) &= \hbar_4 x_2(t), \end{aligned} \tag{12.40}$$

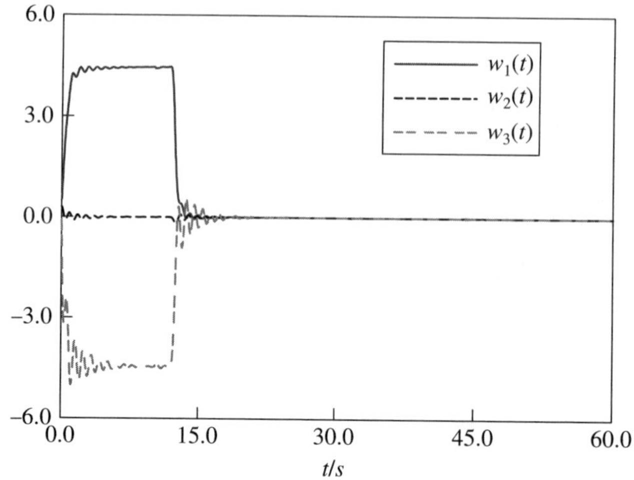

Figure 12.5 Synchronization errors $w_1(t)$, $w_2(t)$, and $w_3(t)$ for the systems (12.31 and 12.33).

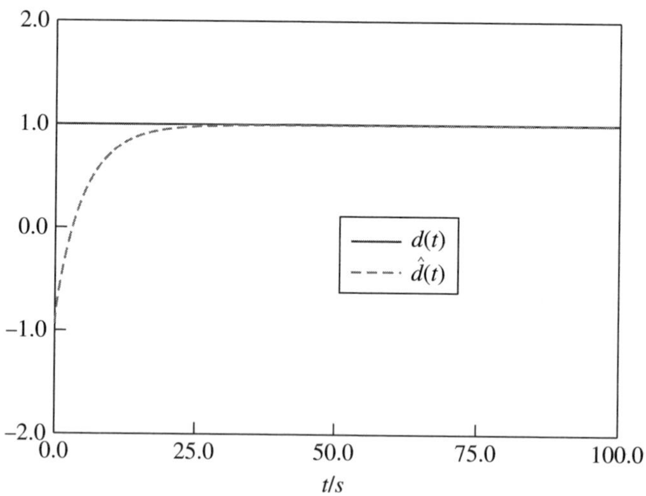

Figure 12.6 Disturbance estimation $\hat{d}$ in system (12.33) versus actual d in system (12.31).

where x_i, with $i = 1, 2, 3, 4$, are the system states, $\hbar_1$, $\hbar_2$, $\hbar_3$, and $\hbar_4$ are system parameters, and $g_1(x(t))$ is a piecewise linear function described by

$$g_1(x(t)) = \hbar_5 x_1(t) + g_{10}(x(t)), \tag{12.41}$$

where $g_{10}(x(t)) = \frac{1}{2}(\hbar_6 - \hbar_5)(|x_1(t) + 1| - |x_1(t) - 1|)$ with $\hbar_6 < -1 < \hbar_5 < 0$. The parameters are set as $\hbar_1 = 2.1428$, $\hbar_2 = 0.1285$, $\hbar_3 = 0.0393$, $\hbar_4 = 0.001532$, $\hbar_5 = -0.9$, and $\hbar_6 = -2.9249$, and the initial values are chosen as $x_0 = [0.1, 0.1, 0.1, 0.1]^{\mathrm{T}}$. Figure 12.9 shows the fractional-order hyperchaos Chua's circuit (12.40).

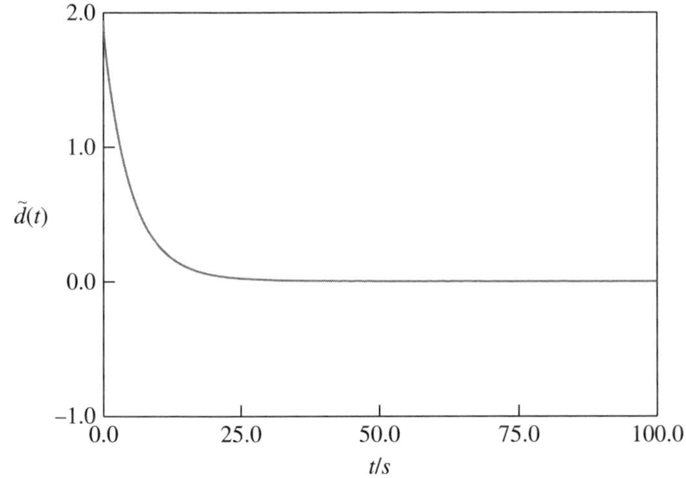

Figure 12.7 Disturbance estimation error $\tilde{d}$ for $\hat{d}$ in system (12.33) and d in system (12.31).

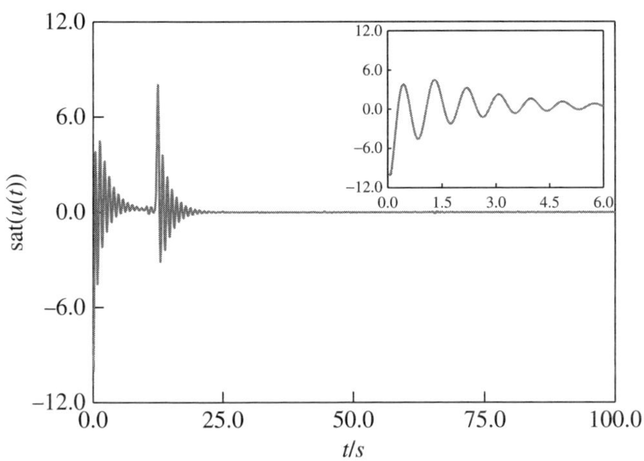

Figure 12.8 Synchronization control input for the systems (12.31 and 12.33).

Consider the external disturbance in the system (12.40), we have the following:

$$
\begin{aligned}
D^{\alpha}x_1(t) &= \hbar_1(x_3(t) - 2.895x_1(t) - g_1(x(t))),\\
D^{\alpha}x_2(t) &= \hbar_2 x_2(t) - x_3(t) - x_4(t),\\
D^{\alpha}x_3(t) &= \hbar_3(x_2(t) - x_1(t) - x_3(t)),\\
D^{\alpha}x_4(t) &= \hbar_4 x_2(t) + d,
\end{aligned}
\tag{12.42}
$$

where d is the constant disturbance.

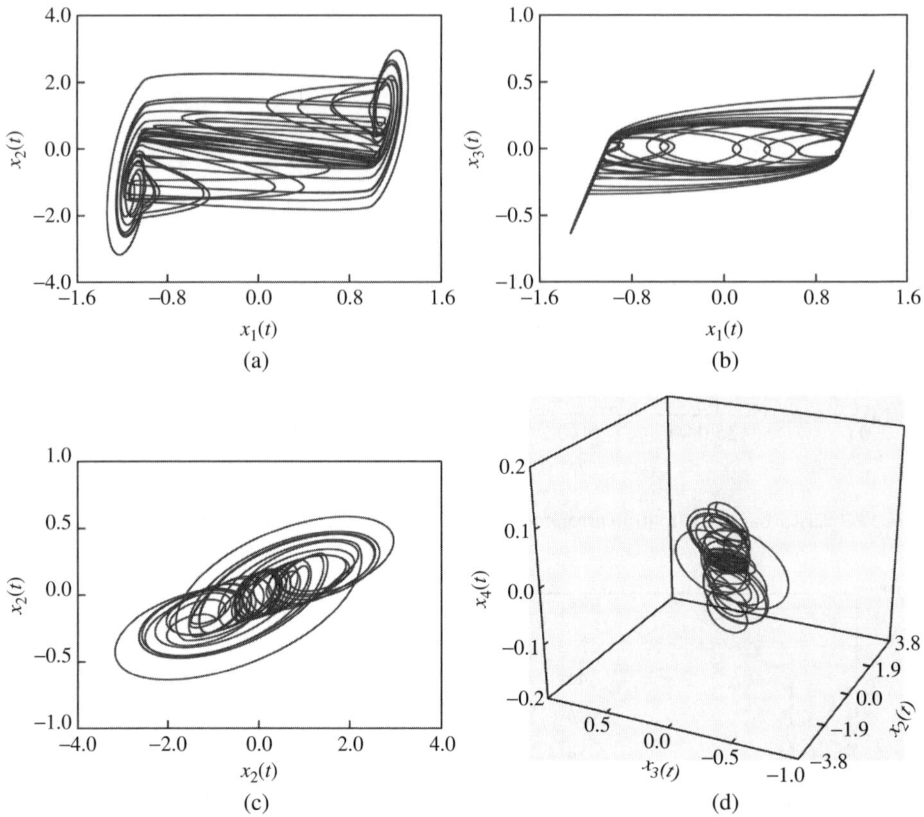

Figure 12.9 Chaotic behaviors of fractional-order hyperchaos Chua's circuit: (a) $x_1(t)$–$x_2(t)$ plane; (b) $x_1(t)$–$x_3(t)$ plane; (c) $x_2(t)$–$x_3(t)$ plane; (d) $x_2(t)$–$x_3(t)$–$x_4(t)$ space.

Furthermore, the fractional-order hyperchaos Chua's circuit with the saturation function and the disturbance observer can be written as follows:

$$
\begin{aligned}
D^\alpha y_1(t) &= \hbar_1(y_3(t) - 2.895y_1(t) - g_1(y(t))),\\
D^\alpha y_2(t) &= \hbar_2 y_2(t) - y_3(t) - y_4(t),\\
D^\alpha y_3(t) &= \hbar_3(y_2(t) - y_1(t) - x_3(t)),\\
D^\alpha y_4(t) &= \hbar_4 y_2(t) + \hat{d} + \mathrm{sat}(u(t)),
\end{aligned}
\tag{12.43}
$$

where y_i, with $i = 1, 2, 3, 4$, are the system states, $\hat{d}$ is the disturbance observer, and $u(t)$ is the synchronization control input.

Based on the drive system (12.42) and the response system (12.43), the synchronization error of the system and the disturbance estimation error of the system can be written as follows:

$$
D^\alpha w_1(t) = \hbar_1(w_3(t) - 2.895w_1(t) - g_1(y(t)) + g_1(x(t))),
$$

$$D^{\alpha} w_2(t) = \hbar_2 w_2(t) - w_3(t) - w_4(t),$$
$$D^{\alpha} w_3(t) = \hbar_3 (w_2(t) - w_1(t) - w_3(t)),$$
$$D^{\alpha} w_4(t) = \hbar_4 w_2(t) - \tilde{d} + \mathrm{sat}(u(t)), \tag{12.44}$$

where w_i, with $i = 1, 2, 3, 4$, are the synchronization errors.

From Equation (12.44), the parameter matrices in the synchronization error system (12.14) are given by

$$A = \begin{bmatrix} -2.895\hbar_1 & 0 & \hbar_1 & 0 \\ 0 & \hbar_2 & -1 & -1 \\ -\hbar_3 & \hbar_3 & -\hbar_3 & 0 \\ 0 & \hbar_4 & 0 & 0 \end{bmatrix}, \quad G = [0, 0, 0, 1]^{\mathsf{T}}. \tag{12.45}$$

According to Equations (12.41) and (12.44), one has [242]

$$g_1(y(t)) - g_1(x(t)) = \Upsilon_{y(t),x(t)} (y_1(t) - x_1(t)), \tag{12.46}$$

where $\Upsilon_{y(t),x(t)}$ depends on $y_1(t)$ and $x_1(t)$, and changes within the interval $[\hbar_6, \hbar_5]$ for $t \geq 0$. Namely, $\Upsilon_{y(t),x(t)}$ is bounded and satisfies $\hbar_6 \leq \kappa_{y(t),x(t)} \leq \hbar_5 < 0$.

From Equations (12.42) and (12.46), one obtains

$$\|g(y(t)) - g(x(t))\| \leq |\hbar_1 \hbar_6| \|(y(t) - x(t))\|. \tag{12.47}$$

Therefore, the nonlinear function $g(y(t)) - g(x(t))$ satisfies Assumption 12.1 and $\kappa = \hbar_1 |\hbar_6| = 6.27$.

In this simulation study, the parameter of the disturbance observer is chosen as $a = 0.5$. It follows that $|\arg(-a)| > 1.5394 = \alpha\ \pi/2$. The initial values are chosen as $x_1(0) = x_2(0) = x_3(0) = x_4(0) = 0.1$, $y_1(0) = y_2(0) = y_3(0) = y_4(0) = -0.1$, and $\phi(0) = -1$. The saturation level value is chosen as $u_0(t) = 8$, and the unknown disturbance is taken as $d = 0.01$.

To ensure that $-u_0(t) \leq LKw(t) \leq u_0(t)$ and Q is a stable matrix, the design parameters K and L are chosen as

$$K = \begin{bmatrix} 0 & -2 & 50 & -2 & -1 \end{bmatrix}, \qquad L = [0.8]. \tag{12.48}$$

From Equations (12.44) and (12.48), we obtain $\max(LKw(t)) = 7.3159$ and $\min(LKw(t)) = -7.2$ by numerical computation; thus, the memoryless nonlinearity $\varphi(t, w(t))$ satisfies Assumption 12.1.

Then, we obtain the following matrix:

$$Q = \begin{bmatrix} -0.2 & 0 & 0 & 0 & 0 \\ 0 & -6.2034 & 0 & 2.1428 & 0 \\ 0 & 0 & 0.1285 & -1 & -1 \\ 0 & -0.0393 & 0.0393 & -0.0393 & 0 \\ -1 & -1.6 & 40.0015 & -1.6 & -0.8 \end{bmatrix} \tag{12.49}$$

From Equation (12.49), the eigenvalues of Q are $\lambda_{01} = -0.3363 + 6.3109i$, $\lambda_{02} = -0.3363 - 6.3109i$, $\lambda_{03} = -0.0516$, $\lambda_{04} = -6.19$, $\lambda_{05} = -0.2$, and all eigenvalues of $Q^{1/0.99}$ are $\lambda_{11} = -0.448 + 6.423i$, $\lambda_{12} = -0.448 - 6.423i$, $\lambda_{13} = -0.0501 - 0.0016i$,

$\lambda_{14} = -6.3019 - 0.2i$, $\lambda_{15} = -0.1967 - 0.0062i$. Thus, $|\arg(\text{eig}(Q))| > \alpha\pi/2$, and $-\max(\text{Re}(\text{eig}(Q^{1/\alpha}))) > 0$ satisfy the given conditions of Theorem 12.1. The disturbance observer is designed based on Equations (12.11) and (12.12). The synchronization control scheme is developed according to Equation (12.15).

On the basis of the proposed chaotic synchronization control scheme, synchronization control results are presented. The state synchronization results of the drive system (12.42) and the response system (12.43) given in Figure 12.10, Figure 12.11, Figure 12.12, and Figure 12.13 show that good synchronization performance is achieved. Figure 12.14

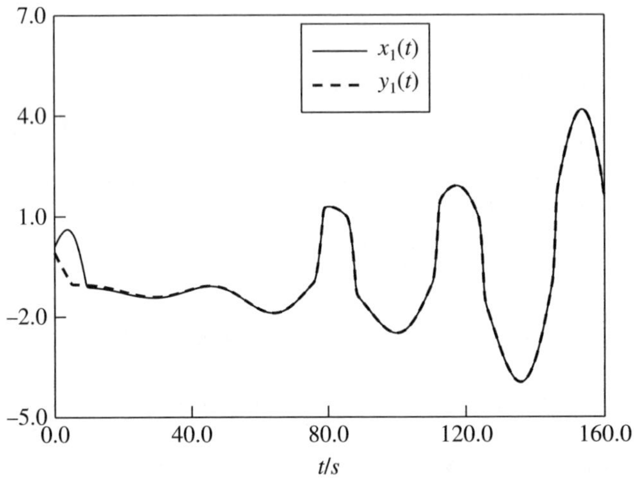

Figure 12.10 Synchronization states of $x_1(t)$ and $y_1(t)$ for the systems (12.42 and 12.43).

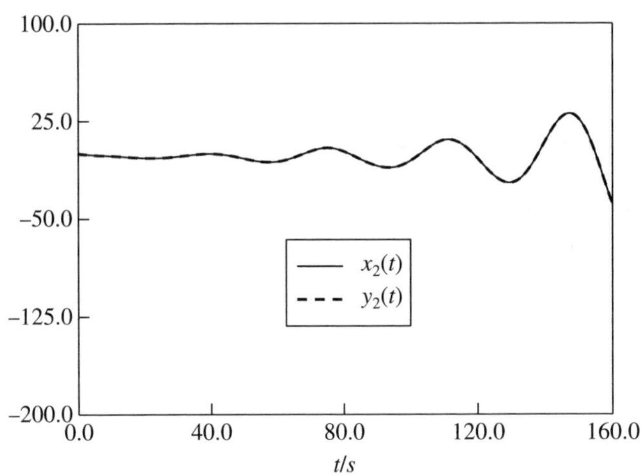

Figure 12.11 Synchronization states of $x_2(t)$ and $y_2(t)$ for the systems (12.42 and 12.43).

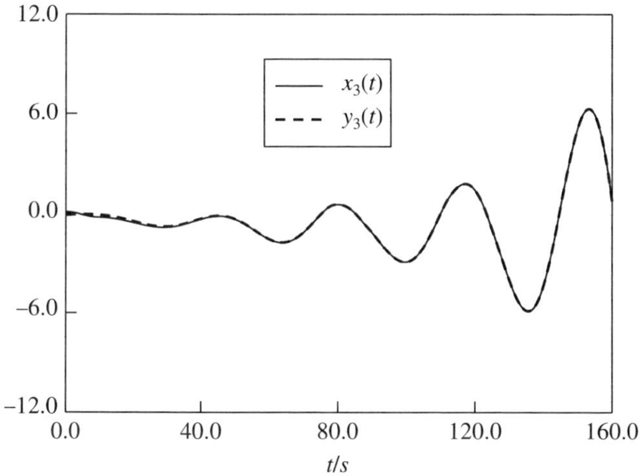

Figure 12.12 Synchronization states of $x_3(t)$ and $y_3(t)$ for the systems (12.42 and 12.43).

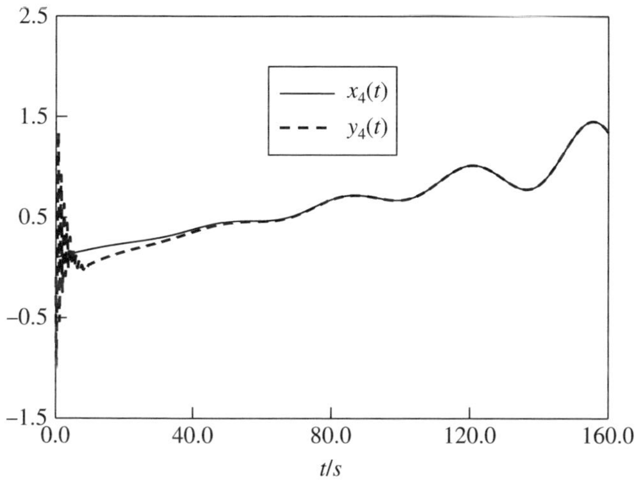

Figure 12.13 Synchronization states of $x_4(t)$ and $y_4(t)$ for the systems (12.42 and 12.43).

shows that the synchronization errors $w_1(t)$, $w_2(t)$, and $w_3(t)$ are convergent. The disturbance estimation result is presented in Figure 12.15 and the estimation error of the disturbance observer is shown in Figure 12.16. From Figure 12.15 and Figure 12.16, we note that satisfactory disturbance estimation performance is obtained. The synchronization control input is given in Figure 12.17; bounded synchronization control input is observed. Based on these simulation results, the synchronization error and the disturbance estimation error are asymptotically stable. This implies that the developed chaotic synchronization control scheme is valid.

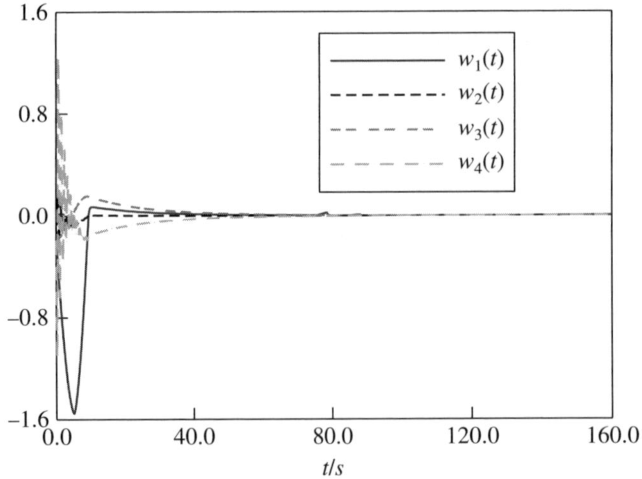

Figure 12.14 Synchronization errors $w_1(t)$, $w_2(t)$, $w_3(t)$, and $w_4(t)$ for the systems (12.42 and 12.43).

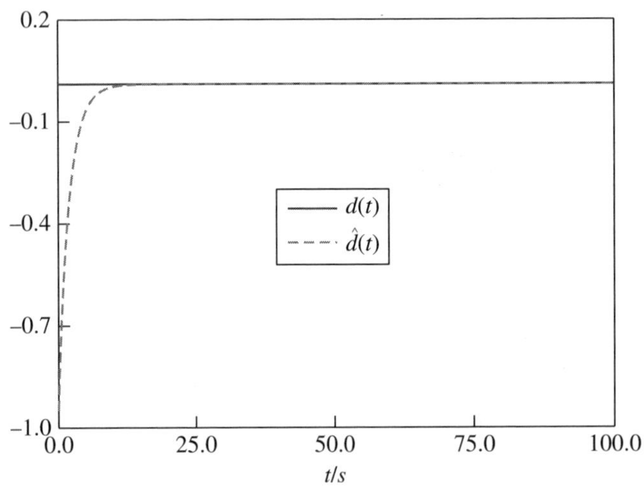

Figure 12.15 Disturbance estimate $\hat{d}$ in system (12.43) versus actual d in system (12.42).

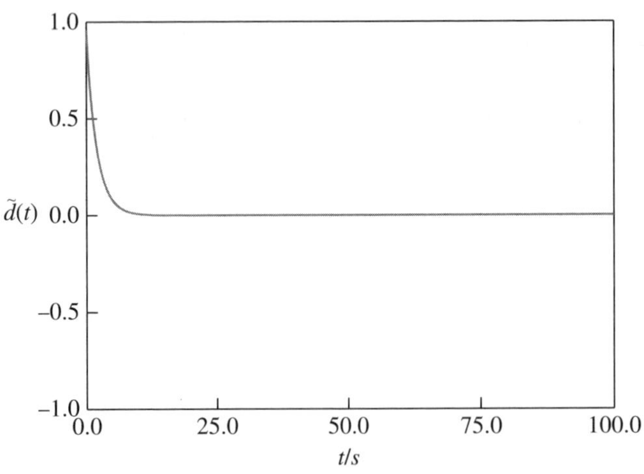

Figure 12.16 Disturbance estimation error $\tilde{d}$ for $\hat{d}$ in system (12.43) and d in system (12.42).

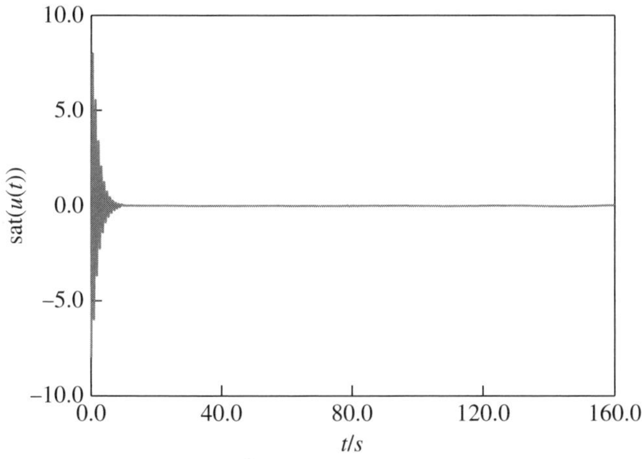

Figure 12.17 Synchronization control input for the systems (12.42 and 12.43).

12.5 Conclusion

In this chapter, a FODO based on the integer-order design method has been proposed, and incorporated in a synchronization control scheme for a class of fractional-order nonlinear systems. Synchronization has been realized between two fractional-order chaotic systems. Through numerical simulation, we have demonstrated the effectiveness of the given synchronization control method operating under the fractional-order Chua's circuit and the fractional-order hyperchaos Chua's circuit.

Appendix A

Fractional Derivatives of Some Functions

Fractional derivatives of some functions are discussed, e.g., the constant, the power and the exponential function, as well as the sine and cosine functions. The Caputo fractional derivatives of these functions are given and a comparison with the RL fractional derivative is shown [225].

A.1 Fractional Derivative of Constant

From a physical point of view, it is reasonable to have the fractional derivative of a constant equal to zero. For the Riemann–Liouville fractional derivative, it holds [225] that

$$ {}_{\mathrm{RL}}D^{\alpha}_{0,t}c = \frac{c}{\Gamma(1-\alpha)}t^{-\alpha} \neq 0, \tag{A.1}$$

where c is a constant. The following property is one of the advantages of the Caputo fractional derivative over the RL fractional derivative.

Lemma A.1 *For the Caputo fractional derivative, it holds that*

$$D^{\alpha}c = 0. \tag{A.2}$$

Proof A.1 *According the definition of the Caputo fractional derivative, we have*

$$D^{\alpha}c = \frac{1}{\Gamma(m-\alpha)}\int_0^t \frac{c^{(m)}}{(t-\tau)^{\alpha-m+1}}\mathrm{d}\tau = 0, \tag{A.3}$$

where $m-1 \leq \alpha < m$ *and* $m \in Z^{+}$. □

A.2 Fractional Derivative of the Power Function

To facilitate the analysis of the Caputo fractional derivative for the power function, we first give the Taylor expansion, as follows [225]:

$$f(t) = f(0) + f'(0)t + \frac{f''(0)}{2!}t^2 + \frac{f'''(0)}{3!}t^3 + \cdots \tag{A.4}$$

Robust Adaptive Control for Fractional-Order Systems with Disturbance and Saturation, First Edition.
Mou Chen, Shuyi Shao, and Peng Shi.

If $D^{\alpha} t^{p}$ is known, the Caputo fractional derivative for arbitrary function can be described as

$$D^{\alpha} f(t) = D^{\alpha} \sum_{k=0}^{\infty} \frac{f^{(k)}(0)}{k!} t^{k} = \sum_{k=0}^{\infty} \frac{f^{(k)}(0)}{k!} D^{\alpha} t^{k}. \tag{A.5}$$

According to the literature [154], the RL fractional derivative of the power function satisfies

$${}_{\mathrm{RL}}D_{0,t}^{\alpha} t^{p} = \frac{\Gamma(p+1)}{\Gamma(p-\alpha+1)} t^{p-\alpha}, \tag{A.6}$$

where $m-1 < \alpha < m, p > -1$, and $p \in \Re$.

Furthermore, the Caputo fractional derivative of the power function satisfies

$$D^{\alpha} t^{p} = \begin{cases} \dfrac{\Gamma(p+1)}{\Gamma(p-\alpha+1)} t^{p-\alpha} = {}_{\mathrm{RL}}D_{0,t}^{\alpha} t^{p}, & m-1 < \alpha < m, \quad p > n-1, \quad p \in \Re, \\ 0, & m-1 < \alpha < m, \quad p \leq n-1, \quad p \in Z^{+}. \end{cases} \tag{A.7}$$

For $p > m-1$, the Caputo fractional derivative of the power function (A.7) is a generalization of the integer-order derivative of the power function. We have

$$\begin{aligned} (t^{p})^{(n)} &= (pt^{p-1})^{(n-1)} = (p(p-1)t^{p-2})^{(n-2)} \\ &= \cdots = p(p-1)\cdots(p-n+1)t^{p-n} \\ &= \frac{\Gamma(p+1)}{\Gamma(p-n+1)} t^{p-n}, \quad n \in Z^{+}, p \in \Re. \end{aligned} \tag{A.8}$$

On the basis of this analysis, the Caputo fractional derivative for an arbitrary function $f(t)$ can be computed by the following equation:

$$\begin{aligned} D^{\alpha} f(t) &= \sum_{k=0}^{\infty} \frac{f^{(k)}(0)}{k!} D^{\alpha} t^{k} = \sum_{k=m}^{\infty} \frac{f^{(k)}(0)\Gamma(k+1)}{k!\Gamma(k-\alpha+1)} t^{k-\alpha} \\ &= \sum_{k=m}^{\infty} \frac{f^{(k)}(0)}{\Gamma(k-\alpha+1)} t^{k-\alpha}. \end{aligned} \tag{A.9}$$

A.3 Fractional Derivative of the Exponential Function

This section considers the exponential function $e^{\lambda t}$. According to Ishteva [225], the Caputo fractional derivative of the exponential function $e^{\lambda t}$ has the following form [225]:

$$D^{\alpha} e^{\lambda t} = \sum_{k=0}^{\infty} \frac{\lambda^{k+m} t^{k+m-\alpha}}{\Gamma(k+1+m-\alpha)} = \lambda^{m} t^{m-\alpha} E_{1,m-\alpha+1}(\lambda t), \tag{A.10}$$

where λ is a positive constant.

A.4 Fractional Derivatives of Sine and Cosine Functions

The behavior of the Caputo fractional derivative for sine and cosine functions is discussed in this section.

For the sine function [225], we have

$$\begin{aligned} D^{\alpha} \sin \lambda t &= D^{\alpha} \frac{e^{i\lambda t} - e^{-i\lambda t}}{2i} \\ &= \frac{1}{2i}(D^{\alpha} e^{i\lambda t} - D^{\alpha} e^{-i\lambda t}) \\ &= \frac{1}{2i}((i\lambda)^m t^{m-\alpha} E_{1,m-\alpha+1}(i\lambda t) - (-i\lambda)^m t^{m-\alpha} E_{1,m-\alpha+1}(-i\lambda t)) \\ &= \frac{1}{2i}(i\lambda)^m t^{m-\alpha} (E_{1,m-\alpha+1}(i\lambda t) - (-1)^m t^{m-\alpha} E_{1,m-\alpha+1}(-i\lambda t)). \end{aligned} \tag{A.11}$$

For the cosine function, we obtain

$$D^{\alpha} \cos \lambda t = \frac{1}{2}(i\lambda)^m t^{m-\alpha} (E_{1,m-\alpha+1}(i\lambda t) + (-1)^m t^{m-\alpha} E_{1,m-\alpha+1}(-i\lambda t)). \tag{A.12}$$

Appendix B

Table of Caputo Derivatives

In this appendix, Caputo derivatives of particular functions are given in Tables B.1 and B.2 [225].

Table B.1 Caputo derivatives of particular functions for the fractional orders α, 1/2, and 1/3.

$f(t)$	$D^{\alpha}f(t)$	$D^{1/3}f(t)$	$D^{1/2}f(t)$
Constant	0	0	0
t	$\frac{1}{\Gamma(2-\alpha)}t^{1-\alpha}$	$1.1077t^{2/3}$	$1.128t^{1/2}$
t^2	$\frac{2}{\Gamma(3-\alpha)}t^{2-\alpha}$	$1.3293t^{5/3}$	$1.5045t^{3/2}$
t^3	$\frac{6}{\Gamma(4-\alpha)}t^{3-\alpha}$	$1.4954t^{8/3}$	$1.8054t^{5/2}$
t^4	$\frac{24}{\Gamma(5-\alpha)}t^{4-\alpha}$	$1.6314t^{11/3}$	$2.0633t^{7/2}$
t^5	$\frac{120}{\Gamma(6-\alpha)}t^{5-\alpha}$	$1.7479t^{14/3}$	$2.2926t^{9/2}$
$t^{1/2}$	$\frac{\sqrt{\pi}}{2\Gamma(3/2-\alpha)}t^{1/2-\alpha}$	$0.9553t^{1/6}$	0.8862
$t^{3/2}$	$\frac{3\sqrt{\pi}}{4\Gamma(5/2-\alpha)}t^{3/2-\alpha}$	$1.2282t^{7/6}$	$1.3292t$
e^t	$t^{m-\alpha}E_{1,m-\alpha+1}(t)$	$t^{2/3}E_{1,5/3}(t)$	$t^{1/2}E_{1,3/2}(t)$

Robust Adaptive Control for Fractional-Order Systems with Disturbance and Saturation, First Edition.
Mou Chen, Shuyi Shao, and Peng Shi.

Table B.2 List of $D^{1/2}D^{1/2}f(t)$, $D^{1/2}D^{1/2}D^{1/2}f(t)$, and $D^{1/2}D^{1/2}D^{1/2}D^{1/2}f(t)$.

$f(t)$	$D^{1/2}D^{1/2}f(t)$	$D^{1/2}D^{1/2}D^{1/2}f(t)$	$D^{1/2}D^{1/2}D^{1/2}D^{1/2}f(t)$
Constant	0	0	0
t	1	0	0
t^2	$2t$	$2.2568t^{1/2}$	2
t^3	$3t^2$	$4.5135t^{3/2}$	$6t$
t^4	$4t^3$	$7.2216t^{5/2}$	$12t^2$
t^5	$5t^4$	$10.3166t^{7/2}$	$20t^3$
$t^{1/2}$	0	0	0
$t^{3/2}$	$1.5t^{1/2}$	1.3293	0
e^t	e^t	$t^{1/2}E_{1,3/2}(t)$	e^t

Appendix C

Laplace Transforms Involving Fractional Operations

Like integer-order systems, the Laplace transform and its inversion are very important. In this appendix, the Laplace transform definition is given first. Then some essential special functions are described. Finally, tables of inversions of Laplace transforms involving fractional and irrational-order operators are given [243].

C.1 Laplace Transforms

For a time-domain function $f(t)$, its Laplace transform in the s-domain is defined as

$$\mathscr{L}(f(t)) = \int_0^\infty f(t)\mathrm{e}^{-st}\mathrm{d}t = F(s), \tag{C.1}$$

where $\mathscr{L}(\cdot)$ is the notation of Laplace transform.

If the Laplace transform of a signal $f(t)$ is $F(s)$, the inverse Laplace transform of $F(s)$ is defined as

$$f(t) = \mathscr{L}^{-1}(F(s)) = \frac{1}{\mathrm{i}2\pi}\int_{\sigma-\mathrm{i}\infty}^{\sigma+\mathrm{i}\infty} F(s)\mathrm{e}^{st}\mathrm{d}s, \tag{C.2}$$

where σ is greater than the real part of all the poles of function $F(s)$.

C.2 Special Functions for Laplace Transforms

Some special functions for Laplace transforms are listed in Table C.1.

C.3 Laplace Transform Tables

Inversions of Laplace transforms involving fractional and irrational operators are collected in Table C.2.

Robust Adaptive Control for Fractional-Order Systems with Disturbance and Saturation, First Edition.
Mou Chen, Shuyi Shao, and Peng Shi.

Table C.1 Some special functions for Laplace transforms.

Special functions	Definition
Mittag–Leffler function	$\mathscr{E}^{\gamma}_{\alpha,\beta}(z) = \sum_{k=0}^{\infty} \frac{(\gamma)k}{\Gamma(\alpha k+\beta)} \frac{z^k}{k!}$, $\mathscr{E}_{\alpha,\beta}(z) = \mathscr{E}^{1}_{\alpha,\beta}(z)$, $\mathscr{E}_{\alpha}(z) = \mathscr{E}_{\alpha,1}(z)$
Dawson function	$\mathrm{daw}(t) = \mathrm{e}^{-t^2} \int_0^t \mathrm{e}^{\tau^2}\,\mathrm{d}\tau$
erf function	$\mathrm{erf}(t) = \frac{2}{\sqrt{\pi}} \int_0^t \mathrm{e}^{-\tau^2}\,\mathrm{d}\tau$
erfc function	$\mathrm{erfc}(t) = \frac{2}{\sqrt{\pi}} \int_t^{\infty} \mathrm{e}^{-\tau^2}\,\mathrm{d}\tau = 1 - \mathrm{erf}(t)$
Hermite polynomial	$\mathscr{H}_n(t) = \mathrm{e}^{t^2} \frac{\mathrm{d}^n}{\mathrm{d}t^n} \mathrm{e}^{-t^2}$
Bessel function	$\mathscr{J}_{\nu}(t)$ is the solution to $t^2\ddot{y} + t\dot{y} + (t^2 - \nu^2)y = 0$
Extended Bessel function	$\mathscr{G}_{\nu}(t) = j^{-\nu}\mathscr{G}_{\nu}(jt)$

Table C.2 Inversions of Laplace transforms with fractional and irrational operators.

$F(s)$	$f(t) = \mathscr{L}^{-1}(F(s))$
$\frac{s^{\alpha\gamma-\beta}}{(s^\alpha + a)^\gamma}$	$t^{\beta-1}\mathscr{E}^{\gamma}_{\alpha,\beta}(-at^\alpha)$
$\frac{1}{s + \sqrt{s^2 + a^2}}$	$\frac{\mathscr{J}_1(at)}{at}$
$\frac{k}{s^2 + k^2} \coth\left(\frac{\pi s}{2k}\right)$	$\lvert\sin(kt)\rvert$
$\frac{1}{s^n\sqrt{n}}$, $n = 1, 2, \ldots$	$\frac{2^n t^{n-0.5}}{1 \cdot 3 \cdot 5 \cdots (2n-1)\sqrt{\pi}}$
$\log \frac{s^2 - a^2}{s^2}$	$\frac{2}{t}(1 - \cosh at)$
$\arctan \frac{k}{s}$	$\frac{1}{t} \sin kt$
$\log \frac{s^2 + a^2}{s^2}$	$\frac{2}{t}(1 - \cos at)$
$\frac{1}{s\sqrt{s}} \mathrm{e}^{-k\sqrt{s}}$	$2\sqrt{\frac{t}{\pi}} \mathrm{e}^{-\frac{1}{4t}k^2} - k\mathrm{erfc}\left(\frac{k}{2\sqrt{t}}\right)$
$\frac{(1-s)^n}{s^{n+0.5}}$	$\frac{n!}{(2n)\sqrt{\pi t}} \mathscr{H}_{2n}(\sqrt{t})$

Table C.2 (Continued)

$F(s)$	$f(t) = \mathscr{L}^{-1}(F(s))$
$\dfrac{e^{-k\sqrt{s}}}{\sqrt{s}(a+\sqrt{s})}$	$e^{ak+a^2t}\text{erfc}\left(a\sqrt{t}+\dfrac{k}{2\sqrt{t}}\right)$
$\dfrac{1}{\sqrt{s^2+a^2}}$	$\mathscr{J}_0(at)$
$\dfrac{1}{\sqrt{s+b}(s+a)}$	$\dfrac{1}{\sqrt{b-a}}e^{-at}\text{erf}(\sqrt{(b-a)t})$
$\dfrac{1}{\sqrt{s^2-a^2}}$	$\mathscr{G}_0(at)$
$\dfrac{(1-s)^n}{s^{n+1.5}}$	$-\dfrac{n!}{(2n+1)!\sqrt{\pi}}\mathscr{H}_{2n+1}(\sqrt{t})$
$\dfrac{\sqrt{s+2a}-\sqrt{s}}{\sqrt{s+2a}+\sqrt{s}}$	$\dfrac{1}{t}e^{-at}\mathscr{G}_1(at)$
$\dfrac{(a-b)^k}{(\sqrt{s+a}+\sqrt{s+b})^{2k}}$	$\dfrac{k}{t}e^{-0.5(a+b)t}\mathscr{G}_k\left(\dfrac{a-b}{2}t\right), \quad k>0$
$\dfrac{(\sqrt{s^2+a^2}-s)^\nu}{\sqrt{s^2+a^2}}$	$a^\nu\mathscr{J}_\nu(at), \quad \nu>-1$
$\dfrac{1}{(s^2-a^2)^k}$	$\dfrac{\sqrt{\pi}}{\Gamma(k)}\left(\dfrac{t}{2a}\right)^{k-0.5}G_{k-0.5}(at)$
$\dfrac{(\sqrt{s^2-a^2}+s)^\nu}{\sqrt{s^2-a^2}}$	$a^\nu\mathscr{G}_\nu(at), \quad \nu>-1$
$\dfrac{1}{(\sqrt{s^2+a^2})^k}$	$\dfrac{\sqrt{\pi}}{\Gamma(k)}\left(\dfrac{t}{2a}\right)^{k-0.5}\mathscr{J}_{k-0.5}(at)$
$(\sqrt{s^2+a^2}-s)^k$	$\dfrac{ka^k}{t}\mathscr{J}_k(at), \quad k>0$
$\log\dfrac{s-a}{s-b}$	$\dfrac{1}{t}(e^{bt}-e^{at})$
$\dfrac{1}{\sqrt{s+a}\sqrt{s+b}}$	$e^{-0.5(a+b)t}\mathscr{G}_0\left(\dfrac{a-b}{2}t\right)$
$\dfrac{1}{(s+\sqrt{s^2+a^2})^N}$	$\dfrac{N\mathscr{J}_N(at)}{at}, \quad N>0$
$\sqrt{s-a}-\sqrt{s-b}$	$\dfrac{1}{2\sqrt{\pi t^3}}(e^{bt}-e^{at})$
$\dfrac{1}{s}e^{-k/s}$	$\mathscr{J}_0(2\sqrt{kt})$
$\dfrac{1}{\sqrt{s}}e^{k/s}$	$\dfrac{1}{\sqrt{\pi t}}\cosh 2\sqrt{kt}$

(Continued)

Table C.2 (Continued)

$F(s)$	$f(t) = \mathcal{L}^{-1}(F(s))$
$\frac{1}{s\sqrt{s}}\mathrm{e}^{k/s}$	$\frac{1}{\sqrt{\pi k}}\sinh 2\sqrt{kt}$
$\mathrm{e}^{-k\sqrt{s}}$	$\frac{k}{2\sqrt{\pi t^3}}\mathrm{e}^{-\frac{1}{4k}k^2}$
$\frac{1}{s}\mathrm{e}^{-k\sqrt{s}}$	$\mathrm{erfc}\left(\frac{k}{2\sqrt{t}}\right)$
$\frac{1}{\sqrt{s}}\mathrm{e}^{-k\sqrt{s}}$	$\frac{1}{\sqrt{\pi t}}\mathrm{e}^{-\frac{1}{4k}k^2}$
$\frac{1}{(s+a)^{\alpha}}$	$\frac{t^{\alpha-1}}{\Gamma(\alpha)}\mathrm{e}^{-at}$
$\frac{a}{s(s^{\alpha}+a)}$	$1-\mathscr{E}_{\alpha}(-at^{\alpha})$
$\frac{1}{s^{\alpha}(s-a)}$	$t^{\alpha}\mathscr{E}_{1,1+\alpha}(at)$
$\frac{1}{\sqrt{s}}$	$\frac{1}{\sqrt{\pi t}}$
$\frac{1}{\sqrt{s}(s+1)}$	$\frac{2}{\sqrt{\pi}}\mathrm{daw}(\sqrt{t})$
$\frac{1}{\sqrt{s}(s+a^2)}$	$\sqrt{t}\mathscr{E}_{1,1.5}(-a^2t)$
$\frac{\sqrt{s}}{s+a^2}$	$\frac{1}{\sqrt{t}}\mathscr{E}_{1,0.5}(-a^2t)$
$\frac{1}{s\sqrt{s+1}}$	$\mathrm{erf}(\sqrt{t})$
$\frac{1}{\sqrt{s}(s-a^2)}$	$\frac{1}{a}\mathrm{e}^{a^2t}\mathrm{erf}(a\sqrt{t})$
$\frac{1}{\sqrt{s}(\sqrt{s}+a)}$	$\mathrm{e}^{a^2t}\mathrm{erfc}(a\sqrt{t})$
$\frac{1}{\sqrt{s+1}}$	$\frac{\mathrm{e}^{-t}}{\sqrt{\pi t}}$
$\frac{b^2-a^2}{(s-a^2)(\sqrt{s}+b)}$	$\mathrm{e}^{a^2t}[b-a\mathrm{erf}(a\sqrt{t})]-b\mathrm{e}^{b^2t}\mathrm{erfc}(b\sqrt{t})$
$\frac{1}{s^{\alpha}}$	$\frac{t^{\alpha-1}}{\Gamma(\alpha)}$

Table C.2 (Continued)

$F(s)$	$f(t) = \mathscr{L}^{-1}(F(s))$
$\dfrac{\sqrt{s}}{s-1}$	$\dfrac{1}{\sqrt{\pi t}} + \mathrm{e}^{t}\mathrm{erf}(\sqrt{t})$
$\dfrac{1}{\sqrt{s}}\mathrm{e}^{-k/s}$	$\dfrac{1}{\sqrt{\pi t}}\cos 2\sqrt{kt}$
$\dfrac{\sqrt{s+2a}-\sqrt{s}}{\sqrt{s}}$	$a\mathrm{e}^{-at}[\mathscr{G}_1(at)+\mathscr{G}_0(at)]$
$\dfrac{1}{s\sqrt{s}}\mathrm{e}^{-k/s}$	$\dfrac{1}{\sqrt{\pi k}}\sin 2\sqrt{kt}$
$\dfrac{1}{s^{\nu}}\mathrm{e}^{-k/s}$	$\left(\dfrac{t}{k}\right)^{0.5(\nu-1)}\mathscr{J}_{\nu-1}(2\sqrt{kt}), \quad \nu > 0$
$\dfrac{s^{\alpha}}{s(s^{\alpha}+a)}$	$\mathscr{E}_{\alpha}(-at^{\alpha})$
$\dfrac{1}{s\sqrt{s}}\mathrm{e}^{-\sqrt{s}}$	$2\sqrt{\dfrac{t}{\pi}}\mathrm{e}^{-\dfrac{1}{4t}} - \mathrm{erfc}\left(\dfrac{1}{2\sqrt{t}}\right)$
$\dfrac{1}{s^{\nu}}\mathrm{e}^{k/s}$	$\left(\dfrac{t}{k}\right)^{0.5(\nu-1)}\mathscr{G}_{\nu-1}(2\sqrt{kt})$
$\dfrac{\mathrm{e}^{-\sqrt{s}}}{\sqrt{s}(\sqrt{s}+1)}$	$\mathrm{e}^{t+1}\mathrm{erfc}\left(\sqrt{t}+\dfrac{1}{2\sqrt{t}}\right)$
$\dfrac{1}{s^{\alpha}+a}$	$t^{\alpha-1}\mathscr{E}_{\alpha,\alpha}(-at^{\alpha})$
$\dfrac{s^{\alpha}}{s-a}$	$-t^{\alpha}\mathscr{E}_{1,1-\alpha}(at), \quad 0<\alpha<1$
$\dfrac{1}{s\sqrt{s}}$	$2\sqrt{\dfrac{t}{\pi}}$
$\dfrac{\sqrt{s}}{s+1}$	$\dfrac{1}{\sqrt{\pi t}} - \dfrac{2}{\sqrt{t}}\mathrm{daw}(\sqrt{t})$
$\dfrac{s}{(s-a)\sqrt{s-a}}$	$\dfrac{1}{\sqrt{\pi t}}\mathrm{e}^{at}(1+2at)$
$\dfrac{1}{\sqrt{s}+a}$	$\dfrac{1}{\sqrt{\pi t}} - a\mathrm{e}^{a^2 t}\mathrm{erfc}(a\sqrt{t})$
$\dfrac{\sqrt{s}}{s-a^2}$	$\dfrac{1}{\sqrt{\pi t}} + a\mathrm{e}^{a^2 t}\mathrm{erf}(a\sqrt{t})$
$\dfrac{1}{\sqrt{s}(s+a^2)}$	$\dfrac{2}{a\sqrt{\pi}}\mathrm{e}^{-a^2 t}\displaystyle\int_0^{a\sqrt{t}}\mathrm{e}^{\tau^2}\,\mathrm{d}\tau$

(Continued)

Table C.2 (Continued)

$F(s)$	$f(t) = \mathscr{L}^{-1}(F(s))$
$\frac{s\sqrt{s}}{s+1}$	$2\sqrt{\frac{t}{\pi}} - \frac{2}{\sqrt{\pi}}\mathrm{daw}(\sqrt{t})$
$\frac{1}{\sqrt{s}(s-1)}$	$\mathrm{e}^{t}\mathrm{erf}(\sqrt{t})$
$\frac{1}{\sqrt{s^2+a^2}(s+\sqrt{s^2+a^2})^N}$	$\frac{J_N(at)}{a^N}$
$\frac{s^{\alpha-1}}{s^\alpha \pm \lambda}$	$\mathscr{E}_\alpha(\mp\lambda t^\alpha), \quad \Re(s) > \|\lambda\|^{1/\alpha}$
$\frac{1}{\sqrt{s^2+a^2}(s+\sqrt{s^2+a^2})}$	$\frac{\mathscr{J}_1(at)}{a}$
$\frac{1}{\sqrt{s(s+a)}(\sqrt{s+a}+\sqrt{s})^{2\nu}}$	$\frac{1}{a^\nu}\mathrm{e}^{-at/2}G_\nu\left(\frac{a}{2}t\right), \quad k>0$
$\frac{\Gamma(k)}{(s+a)^k(s+b)^k}$	$\sqrt{\pi}\left(\frac{t}{a-b}\right)^{k-0.5}\mathrm{e}^{-0.5(a+b)t}G_{k-0.5}\left(\frac{a-b}{2}t\right)$
$\frac{k!}{\sqrt{s}\pm\lambda}$	$t^{(k-1)/2}\mathscr{E}^{(k)}_{0.5,0.5}(\mp\lambda\sqrt{t}), \quad \Re(s) > \lambda^2$
$\frac{b^2-a^2}{\sqrt{s}(s-a^2)(\sqrt{s}+b)}$	$\mathrm{e}^{a^2t}\left[\frac{b}{a}\mathrm{erf}(a\sqrt{t})-1\right]+\mathrm{e}^{b^2t}\mathrm{erfc}(b\sqrt{t})$
$\frac{a\mathrm{e}^{-k/\sqrt{s}}}{s(\sqrt{s}+a)}$	$-\mathrm{e}^{ak}\mathrm{e}^{a^2t}\mathrm{erfc}\left[a\sqrt{t}+\frac{k}{2\sqrt{t}}\right]+\mathrm{erfc}\left(\frac{k}{2\sqrt{t}}\right)$
$\frac{1}{\sqrt{s+a}(s+b)\sqrt{s+b}}$	$t\mathrm{e}^{-0.5(a+b)t}\left[G_0\left(\frac{a-b}{2}t\right)+G_1\left(\frac{a-b}{2}t\right)\right]$
$\frac{\mathrm{e}^{-\sqrt{s}}}{\sqrt{s}+1}$	$\frac{\mathrm{e}^{-\frac{1}{4k}}}{\sqrt{\pi t}} - \mathrm{e}^{t+1}\mathrm{erfc}\left[\sqrt{t}+\frac{1}{2\sqrt{t}}\right]$
$\frac{\mathrm{e}^{-\sqrt{s}}}{s(\sqrt{s}+1)}$	$\mathrm{erfc}\left[\frac{1}{2\sqrt{t}}\right] - \mathrm{e}^{t+1}\mathrm{erfc}\left[\sqrt{t}+\frac{1}{2\sqrt{t}}\right]$

References

1 Podlubny, I. (1999) Fractional-order systems and $PI^{\lambda}D^{\mu}$-controllers. *IEEE Transactions on Automatic Control*, **44** (1), 208–214.
2 Moghadasianx, M., Betin, F., Yazidi, A., *et al.* (2012) Position control of six-phase induction machine using fractional-order controller. International Conference on Electrical Machines (ICEM), Marseilles, France.
3 Das, S. and Pan, I. (2012) *Fractional Order Signal Processing: Introductory Concepts and Applications*, Springer, New York.
4 Shokooh, A. and Suáez, L. (1999) A comparison of numerical methods applied to a fractional model of damping materials. *Journal of Vibration and Control*, **5** (3), 331–354.
5 West, B., Bologna, M., and Grigolini, P. (2003) *Physics of Fractal Operators*, Springer, New York.
6 Dadras, S. and Momeni, H.R. (2010) Control of a fractional-order economical system via sliding mode. *Physica A*, **389** (12), 2434–2442.
7 El-Sayed, A.M.A., Rida, S.Z., and Arafa, A.A.M. (2009) Exact solutions of fractional-order biological population model. *Communications in Theoretical Physics*, **52** (6), 992–996.
8 Chen, W.C. (2008) Nonlinear dynamics and chaos in a fractional-order financial system. *Chaos, Solitons & Fractals*, **36** (5), 1305–1314.
9 Lu, J.G. (2006) Chaotic dynamics of the fractional-order Lü system and its synchronization. *Physics Letters A*, **354** (4), 305–311.
10 Wang, X.Y. and Wang, M.J. (2007) Dynamic analysis of the fractional-order Liu system and its synchronization. *Chaos*, **17** (3), 033106.
11 Liu, L., Liu, C.X., and Zhang, Y.B. (2009) Experimental verification of a four-dimensional Chua's system and its fractional order chaotic attractors. *International Journal of Bifurcation and Chaos*, **19** (8), 2473–2486.
12 Wang, X.Y. and Song, J.M. (2009) Synchronization of the fractional order hyperchaos Lorenz systems with activation feedback control. *Communications in Nonlinear Science and Numerical Simulation*, **14** (8), 3351–3357.
13 Min, F.H., Yu, Y., and Ge, C.J. (2009) Circuit implementation and tracking control of the fractional-order hyper chaotic Lü system. *Acta Physica Sinica*, **58** (3), 1456–1461.
14 Liu, L., Liang, D.L., and Liu, C.X. (2012) Nonlinear state-observer control for projective synchronization of a fractional-order hyperchaotic system. *Nonlinear Dynamics*, **69** (4), 1929–1939.

Robust Adaptive Control for Fractional-Order Systems with Disturbance and Saturation, First Edition.
Mou Chen, Shuyi Shao, and Peng Shi.

15 Liu, L. and Liu, C.X. (2014) Theoretical analysis and circuit verification for fractional-order chaotic behavior in a new hyperchaotic system. *Mathematical Problems in Engineering*, **2014**, 682408.

16 Šabanovic, A. (2011) Variable structure systems with sliding modes in motion control—a survey. *IEEE Transactions on Industrial Informatics*, **7** (2), 212–223.

17 Shi, P., Xia, Y.Q., Liu, G.P., and Rees, D. (2006) On designing of sliding-mode control for stochastic jump systems. *IEEE Transactions on Automatic Control*, **51** (1), 97–103.

18 Zhang, J.H., Shi, P., and Xia, Y.Q. (2010) Robust adaptive sliding-mode control for fuzzy systems with mismatched uncertainties. *IEEE Transactions on Fuzzy Systems*, **18** (4), 700–711.

19 Gao, Z., Jiang, B., Shi, P., *et al.* (2012) Active fault tolerant control design for reusable launch vehicle using adaptive sliding mode technique. *Journal of the Franklin Institute*, **349** (4), 1543–1560.

20 Jiang, B., Shi, P., and Mao, Z.H. (2011) Sliding mode observer-based fault estimation for nonlinear networked control systems. *Circuits, Systems, and Signal Processing*, **30** (1), 1–16.

21 Utkin, V., Guldner, J., and Shi, J. (2009) *Sliding Mode Control in Electro-Mechanical Systems*, CRC Press, New York.

22 Feng, Y., Yu, X.H., and Man, Z.H. (2002) Non-singular terminal sliding mode control of rigid manipulators. *Automatica*, **38** (12), 2159–2167.

23 Yu, X.H., Man, Z.H., and Wu, B.L. (1998) Design of fuzzy sliding-mode control systems. *Fuzzy Sets and Systems*, **95** (3), 295–306.

24 Yu, X.H. and Man, Z.H. (2002) Fast terminal sliding-mode control design for nonlinear dynamical systems. *IEEE Transaction on Circuits and Systems I: Fundamental Theory and Applications*, **49** (2), 261–264.

25 Byungkook, Y. and Ham, W. (1998) Adaptive fuzzy sliding mode control of nonlinear systems. *IEEE Transactions on Fuzzy Systems*, **6**(2), 315–321.

26 Chen, D.Y., Liu, Y.X., Ma, X.Y., and Zhang, R.F. (2012) Control of a class of fractional-order chaotic systems via sliding mode. *Nonlinear Dynamics*, **67** (1), 893–901.

27 Yin, C., Zhong, S.M., and Chen, W.F. (2012) Design of sliding mode controller for a class of fractional-order chaotic systems. *Communications in Nonlinear Science and Numerical Simulation*, **17** (1), 356–366.

28 Chen, D.Y., Liu, Y.X., Ma, X.Y., and Zhang, R.F. (2011) No-chattering sliding mode control in a class of fractional-order chaotic systems. *Chinese Physics B*, **20** (12), 120506.

29 Yin, C., Dadras, S., Zhong, S.M., and Chen, Y.Q. (2013) Control of a novel class of fractional-order chaotic systems via adaptive sliding mode control approach. *Applied Mathematical Modelling*, **37** (4): 2469–2483.

30 Tavazoei, M.S. and Haeri, M. (2008) Synchronization of chaotic fractional-order systems via active sliding mode controller. *Physica A*, **387** (1), 57–70.

31 Wang, B., Zhou, Y.G., Xue, J.Y., and Zhu, D.L. (2014) Active sliding mode for synchronization of a wide class of four-dimensional fractional-order chaotic systems. *ISRN Applied Mathematics*, **2014**, 472371.

32 Hosseinnia, S.H., Ghaderi, R., Ranjbar, A.N., *et al.* (2010) Sliding mode synchronization of an uncertain fractional order chaotic system. *Computers & Mathematics with Applications*, **59** (5), 1637–1643.

33 Aghababa, M.P. (2012) Robust stabilization and synchronization of a class of fractional-order chaotic systems via a novel fractional sliding mode controller. *Communications in Nonlinear Science and Numerical Simulation*, **17** (6), 2670–2681.
34 Zhang, L., and Yan, Y. (2014) Robust synchronization of two different uncertain fractional-order chaotic systems via adaptive sliding mode control. *Nonlinear Dynamics*, **76** (3), 1761–1767.
35 Li, C., Su, K., and Wu, L. (2013) Adaptive sliding mode control for synchronization of a fractional-order chaotic system. *Journal of Computational and Nonlinear Dynamics*, **8** (3), 031005.
36 Liu, L., Ding, W., Liu, C., *et al.* (2014) Hyperchaos synchronization of fractional-order arbitrary dimensional dynamical systems via modified sliding mode control. *Nonlinear Dynamics*, **76** (4), 2059–2071.
37 Tao, G. (2003) *Adaptive Control Design and Analysis*, John Wiley & Sons, Inc., Hoboken.
38 Wang, L.X. (1993) Stable adaptive fuzzy control of nonlinear systems. *IEEE Transactions on Fuzzy Systems*, **1** (2), 146–155.
39 Polycarpou, M.M. (1996) Stable adaptive neural control scheme for nonlinear systems. *IEEE Transactions on Automatic Control*, **41** (3), 447–451.
40 Fukao, T., Nakagawa, H., and Adachi, N. (2000) Adaptive tracking control of a nonholonomic mobile robot. *IEEE Transactions on Robotics and Automation*, **16** (5), 609–615.
41 Wang, H., Chen, B., Liu, X., *et al.* (2014) Adaptive neural tracking control for stochastic nonlinear strict-feedback systems with unknown input saturation. *Information Sciences*, **269**, 300–315.
42 Li, Z., Ge, S.S., Adams, M., and Wijesoma, W.S. (2008) Robust adaptive control of uncertain force/motion constrained nonholonomic mobile manipulators. *Automatica*, **44** (3), 776–784.
43 Li, Y., Tong, S., Li, T., and Jing, X. (2014) Adaptive fuzzy control of uncertain stochastic nonlinear systems with unknown dead zone using small-gain approach. *Fuzzy Sets and Systems*, **235**, 1–24.
44 Li, Y., Tong, S., Liu, Y., and Li, T. (2014) Adaptive fuzzy robust output feedback control of nonlinear systems with unknown dead zones based on a small-gain approach. *IEEE Transactions on Fuzzy Systems*, **22** (1), 164–176.
45 Zhou, Q., Shi, P., Lu, J., and Xu, S. (2011) Adaptive output-feedback fuzzy tracking control for a class of nonlinear systems. *IEEE Transactions on Fuzzy Systems*, **19** (5), 972–982.
46 Li, Z., Cao, X., and Ding, N. (2011) Adaptive fuzzy control for synchronization of nonlinear teleoperators with stochastic time-varying communication delays. *IEEE Transactions on Fuzzy Systems*, **19** (4), 745–757.
47 Li, T., Wang, D., and Chen, N. (2011) Adaptive fuzzy control of uncertain MIMO non-linear systems in block-triangular forms. *Nonlinear Dynamics*, **63** (1), 105–123.
48 Vinagre, B.M., Petráš, I., Podlubny, I., and Chen, Y.Q. (2002) Using fractional order adjustment rules and fractional order reference models in model-reference adaptive control. *Nonlinear Dynamics*, **29** (1), 269–279.
49 Odibat, Z.M. (2010) Adaptive feedback control and synchronization of non-identical chaotic fractional order systems. *Nonlinear Dynamics*, **60** (4), 479–487.

50 Yuan, J., Shi, B., and Yu, Z. (2015) Adaptive sliding control for a class of fractional commensurate order chaotic systems. *Mathematical Problems in Engineering*, **2015**, 972914.

51 Zhou, P., and Ding, R. (2012) Adaptive function projective synchronization between different fractional-order chaotic systems. *Indian Journal of Physics*, **86** (6), 497–501.

52 Yin, C., Cheng, Y., Chen, Y.Q., *et al.* (2015) Adaptive fractional-order switching-type control method design for 3D fractional-order nonlinear systems. *Nonlinear Dynamics*, **82** (1), 39–55.

53 Lin, T.C. and Lee, T.Y. (2011) Chaos synchronization of uncertain fractional-order chaotic systems with time delay based on adaptive fuzzy sliding mode control. *IEEE Transactions on Fuzzy Systems*, **19** (4), 623–635.

54 Lin, T.C., Lee, T.Y., and Balas, V.E. (2011) Adaptive fuzzy sliding mode control for synchronization of uncertain fractional order chaotic systems. *Chaos, Solitons & Fractals*, **44** (10), 791–801.

55 Li, C. and Tong, Y. (2013) Adaptive control and synchronization of a fractional-order chaotic system. *Pramana*, **80** (4), 583–592.

56 Lin, T.C. and Kuo, C.H. (2011) H^{∞} synchronization of uncertain fractional order chaotic systems: Adaptive fuzzy approach. *ISA Transactions*, **50** (4), 548–556.

57 Bhalekar, S. and Daftardar-Gejji, V. (2010) Synchronization of different fractional order chaotic systems using active control. *Communications in Nonlinear Science and Numerical Simulation*, **15** (11), 3536–3546.

58 Taghvafard, H. and Erjaee, G.H. (2011) Phase and anti-phase synchronization of fractional order chaotic systems via active control. *Communications in Nonlinear Science and Numerical Simulation*, **16** (10), 4079–4088.

59 Agrawal, S.K., Srivastava, M., and Das, S. (2012) Synchronization of fractional order chaotic systems using active control method. *Chaos, Solitons & Fractals*, **45** (6), 737–752.

60 Razminia, A., Majd, V.J., and Baleanu, D. (2011) Chaotic incommensurate fractional order Rössler system: Active control and synchronization. *Advances in Difference Equations*, **15**. doi: 10.1186/1687-1847-2011-15

61 Radwan, A.G., Moaddy, K., Salama, K.N., *et al.* (2014) Control and switching synchronization of fractional order chaotic systems using active control technique. *Journal of Advanced Research*, **5** (1), 125–132.

62 Li, M., Li, D., Wang, J., and Zhao, C. (2013) Active disturbance rejection control for fractional-order system. *ISA Transactions*, **52** (3), 365–374.

63 Senejohnny, D.M. and Delavari, H. (2012) Active sliding observer scheme based fractional chaos synchronization. *Communications in Nonlinear Science and Numerical Simulation*, **17** (11), 4373–4383.

64 Srivastava, M., Ansari, S.P., Agrawal, S.K., *et al.* (2014) Anti-synchronization between identical and non-identical fractional-order chaotic systems using active control method. *Nonlinear Dynamics*, **76** (2), 905–914.

65 Bhalekar, S. (2014) Synchronization of non-identical fractional order hyperchaotic systems using active control. *World Journal of Modelling and Simulation*, **10** (1), 60–68.

66 Baskonus, H.M., Mekkaoui, T., Hammouch, Z., and Hammouch, B. (2015) Active control of a chaotic fractional order economic system. *Entropy*, **17** (8), 5771–5783.

67 Li, X., Wang, X., and Chen, G. (2004) Pinning a complex dynamical network to its equilibrium. *IEEE Transactions on Circuits and Systems I: Regular Papers*, **51** (10), 2074–2087.

68 Wang, X.F. and Chen, G. (2002) Pinning control of scale-free dynamical networks. *Physica A*, **310** (3), 521–531.

69 Liu, Z.X., Chen, Z.Q., and Yuan, Z.Z. (2007) Pinning control of weighted general complex dynamical networks with time delay. *Physica A*, **375** (1), 345–354.

70 Xiang, L., Chen, Z., Liu, Z., *et al.* (2008) Pinning control of complex dynamical networks with heterogeneous delays. *Computers & Mathematics with Applications*, **56** (5), 1423–1433.

71 Zhou, J., Lu, J., and Lü, J. (2008) Pinning adaptive synchronization of a general complex dynamical network. *Automatica*, **44** (4), 996–1003.

72 Yu, W., Chen, G., and Lü, J. (2009) On pinning synchronization of complex dynamical networks. *Automatica*, **45** (2), 429–435.

73 Grigoriev, R.O., Cross, M.C., and Schuster, H.G. (1997) Pinning control of spatiotemporal chaos. *Physical Review Letters*, **79** (15), 2795–2798.

74 Chai, Y., Chen, L., Wu, R., and Sun, J. (2012) Adaptive pinning synchronization in fractional-order complex dynamical networks. *Physica A*, **391** (22), 5746–5758.

75 Sun, W., Chen, Y.Q., and Li, C. (2011) Multi-group consensus of heterogeneous fractional-order nonlinear agents via pinning control. ASME 2011 International Design Engineering Technical Conferences and Computers and Information in Engineering Conference, Washington, DC.

76 Tang, Y., Wang, Z., and Fang, J. (2009) Pinning control of fractional-order weighted complex networks. *Chaos*, **19** (1), 013112.

77 Yu, Z., Jiang, H., Hu, C., and Yu, J. (2015) Leader-following consensus of fractional-order multi-agent systems via adaptive pinning control. *International Journal of Control*, **88** (9), 1746–1756.

78 Liang, S., Wu, R., and Chen, L. (2016) Adaptive pinning synchronization in fractional-order uncertain complex dynamical networks with delay. *Physica A*, **444**, 49–62.

79 Liu, J., Huang, L., and Meng, Y. (2013) Adaptive synchronization of fractional-order complex networks via pinning control. *International Journal of Adaptive Control and Signal Processing*, **27** (12), 1086–1096.

80 Yang, L., He, W., Zhang, F., and Jia, J. (2014) Cluster projective synchronization of fractional-order complex network via pinning control. *Abstract and Applied Analysis*, **2014**, 314742.

81 Wang, F., Yang, Y., Hu, A., and Xu, X. (2015) Exponential synchronization of fractional-order complex networks via pinning impulsive control. *Nonlinear Dynamics*, **82** (4), 1979–1987.

82 Wang, J., Ma, Q., Chen, A., and Liang, Z. (2015) Pinning synchronization of fractional-order complex networks with Lipschitz-type nonlinear dynamics. *ISA Transactions*, **57**, 111–116.

83 Wang, F., Yang, Y., Hu, M., and Xu, X. (2015) Projective cluster synchronization of fractional-order coupled-delay complex network via adaptive pinning control, *Physica A*, **434**, 134–143.

84 Pan, L., Zhou, W., Fang, J., and Li, D. (2010) Synchronization and anti-synchronization of new uncertain fractional-order modified unified chaotic

systems via novel active pinning control. *Communications in Nonlinear Science and Numerical Simulation*, **15** (12), 3754–3762.

85 Zhang, T.P. and Ge, S.S. (2007) Adaptive neural control of MIMO nonlinear state time-varying delay systems with unknown dead-zones and gain signs. *Automatica*, **43** (6), 1021–1033.

86 Chen, M., Ge, S.S., and How, B.V.E. (2010) Robust adaptive neural network control for a class of uncertain MIMO nonlinear systems with input nonlinearities. *IEEE Transactions on Neural Networks*, **21** (5), 796–812.

87 Liu, Y.J., Chen, C.L.P., Wen, G.X., and Tong, S.C. (2011) Adaptive neural output feedback tracking control for a class of uncertain discrete-time nonlinear systems. *IEEE Transactions on Neural Networks*, **22** (7), 1162–1167.

88 Wang, H., Chen, B., Liu, K., *et al.* (2014) Adaptive neural tracking control for a class of nonstrict-feedback stochastic nonlinear systems with unknown backlash-like hysteresis. *IEEE Transactions on Neural Networks and Learning Systems*, **25** (5), 947–958.

89 Liu, Y.J., Tong, S.C., Wang, D. *et al.* (2011) Adaptive neural output feedback controller design with reduced-order observer for a class of uncertain nonlinear SISO systems. *IEEE Transactions on Neural Networks*, **22** (8), 1328–1334.

90 Wang, D. and Huang, J. (2002) Adaptive neural network control for a class of uncertain nonlinear systems in pure-feedback form. *Automatica*, **38** (8), 1365–1372.

91 Hovakimyan, N., Nardi, F., Calise, A., and Kim, N. (2002) Adaptive output feedback control of uncertain nonlinear systems using single-hidden-layer neural networks. *IEEE Transactions on Neural Networks*, **13** (6), 1420–1431.

92 Ge, S.S. and Wang, J. (2002) Robust adaptive neural control for a class of perturbed strict feedback nonlinear systems. *IEEE Transactions on Neural Networks*, **13** (6), 1409–1419.

93 Wang, D. and Huang, J. (2005) Neural network-based adaptive dynamic surface control for a class of uncertain nonlinear systems in strict-feedback form. *IEEE Transactions on Neural Networks*, **16** (1), 195–202.

94 Calise, A.J., Hovakimyan, N., and Idan, M. (2001) Adaptive output feedback control of nonlinear systems using neural networks. *Automatica*, **37**(8), 1201–1211.

95 Ge, S.S., Hong, F., and Lee, T. H. (2003) Adaptive neural network control of nonlinear systems with unknown time delays. *IEEE Transactions on Automatic Control*, **48** (11), 2004–2010.

96 Ge, S.S., Hong, F., and Lee, T.H. (2004) Adaptive neural control of nonlinear time-delay systems with unknown virtual control coefficients. *IEEE Transactions on Systems, Man, and Cybernetics, Part B: Cybernetics*, **34** (1), 499–516.

97 Yu, J., Shi, P., Dong, W., *et al.* (2015) Neural network-based adaptive dynamic surface control for permanent magnet synchronous motors. *IEEE Transactions on Neural Networks and Learning Systems*, **26** (3), 640–645.

98 Kiumarsi, B., Lewis, F.L., and Levine, D.S. (2015) Optimal control of nonlinear discrete time-varying systems using a new neural network approximation structure. *Neurocomputing*, **156**, 157–165.

99 Chen, M., Jiang, C., and Wu, Q. (2007) Backstepping control for a class of uncertain nonlinear systems with neural network. *International Journal of Nonlinear Science*, **3** (2), 137–143.

100 Chen, M. and Chen, W. (2009) Robust adaptive neural network synchronization controller design for a class of time delay uncertain chaotic systems. *Chaos, Solitons & Fractals*, **41** (5), 2716–2724.

101 Chen, M., Jiang, C., and Wu, Q. (2007) Synchronization scheme for uncertain chaotic systems via RBF neural network. *Chinese Physics Letters*, **24** (4), 890–893.

102 Li, S.H., Yang, J., Chen, W.H., and Chen, X.S. (2014) *Disturbance Observer-Based Control: Methods and Applications*, CRC Press, London.

103 Chen, W.H., Ballance, D.J., Gawthrop, P.J., and O'Reilly, J. (2000) A nonlinear disturbance observer for robotic manipulators. *IEEE Transactions on Industrial Electronics*, **47** (4), 932–938.

104 Chen, W.H. (2004) Disturbance observer based control for nonlinear systems. *IEEE/ASME Transactions on Mechatronics*, **9** (4), 706–710.

105 Chen, M., Chen, W., and Wu, Q. (2014) Adaptive fuzzy tracking control for a class of uncertain MIMO nonlinear systems using disturbance observer. *Science China Information Sciences*, **57** (1), 1–13.

106 Chen, W.H., Yang, J., Guo, L., and Li, S.H. (2015) Disturbance observer-based control and related methods—an overview. *IEEE Transactions on Industrial Electronics*, **63** (2), 1083–1095.

107 Yang, J., Li, S., and Yu, X. (2013) Sliding-mode control for systems with mismatched uncertainties via a disturbance observer. *IEEE Transactions on Industrial Electronics*, **60** (1), 160–169.

108 Chen, M. and Yu, J. (2015) Adaptive dynamic surface control of NSVs with input saturation using a disturbance observer. *Chinese Journal of Aeronautics*, **28** (3), 853–864.

109 Chen, M. and Yu, J. (2015) Disturbance observer-based adaptive sliding mode control for near-space vehicles. *Nonlinear Dynamics*, **82** (4), 1671–1682.

110 Chen, M. and Chen, W.H. (2010) Disturbance-observer-based robust control for time delay uncertain systems. *International Journal of Control, Automation and Systems*, **8** (2), 445–453.

111 Chen, M., Jiang, B., Zou, J., and Feng, X. (2010) Robust adaptive tracking control of the underwater robot with input nonlinearity using neural networks. *International Journal of Computational Intelligence Systems*, **3** (5), 646–655.

112 Chen, M., Wu, Q., and Jiang, C. (2012) Disturbance-observer-based robust synchronization control of uncertain chaotic systems. *Nonlinear Dynamics*, **70** (4), 2421–2432.

113 Chen, M. and Ge, S.S. (2015) Adaptive neural output feedback control of uncertain nonlinear systems with unknown hysteresis using disturbance observer. *IEEE Transactions on Industrial Electronics*, **62** (12), 7706–7716.

114 Chen, M. and Jiang, B. (2013) Robust attitude control of near space vehicles with time-varying disturbances. *International Journal of Control, Automation and Systems*, **11** (1), 182–187.

115 Chen, M. and Chen, W.H. (2010) Sliding mode control for a class of uncertain nonlinear system based on disturbance observer. *International Journal of Adaptive Control and Signal Processing*, **24** (1), 51–64.

116 Zhou, Y. and Chen, M. (2013) Sliding mode control for NSVs with input constraint using neural network and disturbance observer. *Mathematical Problems in Engineering*, **2013**, 904830.

117 Chen, W.H. (2003) Nonlinear disturbance observer-enhanced dynamic inversion control of missiles. *Journal of Guidance, Control, and Dynamics*, **26** (1), 161–166.

118 Yang, J., Chen, W.H., and Li, S. (2011) Non-linear disturbance observer-based robust control for systems with mismatched disturbances/uncertainties. *IET Control Theory & Applications*, **5** (18), 2053–2062.

119 Yang, J., Li, S., and Chen, W.H. (2012) Nonlinear disturbance observer-based control for multi-input multi-output nonlinear systems subject to mismatching condition. *International Journal of Control*, **85** (8), 1071–1082.

120 Chen, M., Zhou, Y., and Guo, W.W. (2014) Robust tracking control for uncertain MIMO nonlinear systems with input saturation using RWNNDO. *Neurocomputing*, **144**, 436–447.

121 Luo, Y. and Zhang, H. (2008) Approximate optimal control for a class of nonlinear discrete-time systems with saturating actuators. *Progress in Natural Science*, **18** (8), 1023–1029.

122 Li, Y., Li, T., and Tong, S. (2013) Adaptive fuzzy modular backstepping output feedback control of uncertain nonlinear systems in the presence of input saturation. *International Journal of Machine Learning and Cybernetics*, **4** (5), 527–536.

123 Li, Y., Tong, S., and Li, T. (2014) Adaptive fuzzy output-feedback control for output constrained nonlinear systems in the presence of input saturation. *Fuzzy Sets and Systems*, **248**, 138–155.

124 Cao, Y.Y., Lin, Z., and Hu, T. (2002) Stability analysis of linear time-delay systems subject to input saturation. *IEEE Transactions on Circuits and Systems I: Fundamental Theory and Applications*, **49** (2), 233–240.

125 Wen, C., Zhou, J., Liu, Z., and Su, H. (2011) Robust adaptive control of uncertain nonlinear systems in the presence of input saturation and external disturbance. *IEEE Transactions on Automatic Control*, **56** (7), 1672–1678.

126 Lan, W. and Huang, J. (2003) Semiglobal stabilization and output regulation of singular linear systems with input saturation. *IEEE Transactions on Automatic Control*, **48** (7), 1274–1280.

127 Chen, M., Ge, S.S., and Choo, Y.S. (2009) Neural network tracking control of ocean surface vessels with input saturation. IEEE International Conference on Automation and Logistics, Shenyang, China.

128 Chen, M., Ge, S.S., and Ren, B. (2011) Adaptive tracking control of uncertain MIMO nonlinear systems with input constraints. *Automatica*, **47** (3), 452–465.

129 Yau, H.T. and Chen, C.L. (2007) Chaos control of Lorenz systems using adaptive controller with input saturation. *Chaos, Solitons & Fractals*, **34** (5), 1567–1574.

130 Choi, J. (1999) On the stabilization of linear discrete time systems subject to input saturation. *Systems & Control Letters*, **36** (3), 241–244.

131 Zhou, B., Gao, H., Lin, Z., and Duan, G.R. (2012) Stabilization of linear systems with distributed input delay and input saturation. *Automatica*, **48** (5), 712–724.

132 Li, T., Li, R., and Li, J. (2011) Decentralized adaptive neural control of nonlinear interconnected large-scale systems with unknown time delays and input saturation. *Neurocomputing*, **74** (14), 2277–2283.

133 Huang, J., Wen, C., Wang, W., and Jiang, Z.P. (2013) Adaptive stabilization and tracking control of a nonholonomic mobile robot with input saturation and disturbance. *Systems & Control Letters*, **62** (3), 234–241.

134 Yang, Y., Yue, D., and Xue, Y. (2015) Decentralized adaptive neural output feedback control of a class of large-scale time-delay systems with input saturation. *Journal of the Franklin Institute*, **352** (5), 2129–2151.

135 Iqbal, M., Rehan, M., Hong, K.S., *et al.* (2015) Sector-condition-based results for adaptive control and synchronization of chaotic systems under input saturation. *Chaos, Solitons & Fractals*, **77**, 158–169.

136 Lim, Y.H., Oh, K.K.,and Ahn, H.S. (2013) Stability and stabilization of fractional-order linear systems subject to input saturation. *IEEE Transactions on Automatic Control*, **58** (4), 1062–1067.

137 Luo, J. (2014) State-feedback control for fractional-order nonlinear systems subject to input saturation. *Mathematical Problems in Engineering*, **2014**, 891639.

138 Farina, L. and Rinaldi, S. (2000) *Positive Linear Systems: Theory and Applications*, John Wiley & Sons, Inc., New York.

139 Kaczorek, T. (2002) *Positive 1-D and 2-D Systems*, Springer, New York.

140 Kaczorek, T. (1999) Stabilization of positive linear systems by state feedback. *Pomiary Automatyka Kontrola*, **45** (3), 2–5.

141 Gao, H.J., Lam, J., Wang, C.H., and Xu, S.Y. (2005) Control for stability and positivity: Equivalent conditions and computation. *IEEE Transactions on Circuits and Systems II: Express Briefs*, **52** (9), 540–544.

142 Wang, C.H. and Huang, T.M. (2013) Static output feedback control for positive linear continuous-time systems. *International Journal of Robust and Nonlinear Control*, **23** (14), 1537–1544.

143 Benzaouia, A., Hmamed, A., and El Hajjaji, A. (2010) Stabilization of controlled positive discrete-time T–S fuzzy systems by state feedback control. *International Journal of Adaptive Control and Signal Processing*, **24** (12), 1091–1106.

144 Chen, X.M., Lam, J., Li, P., and Shu, Z. (2013) ℓ_1-induced norm and controller synthesis of positive systems. *Automatica*, **49** (5), 1377–1385.

145 Chen, X.M., Lam, J., Li, P., and Shu, Z. (2014) Output-feedback control for continuous-time interval positive systems under L_1 performance. *Asian Journal of Control*, **16** (6), 1592–1601.

146 Chen, X.M., Lam, J., Li, P., and Shu, Z. (2013) L_1-induced performance analysis and sparse controller synthesis for interval positive systems. International Conference of Applied and Engineering Mathematics, London.

147 Chen, X.M., Lam, J., and Li, P. (2014) Positive filtering for continuous-time positive systems under L_1 performance. *International Journal of Control*, **87** (9), 1906–1913.

148 Chen, X.M., Lam, J., and Lam, H.K. (2015) Positive filtering for positive Takagi Sugeno fuzzy systems under ℓ_1 performance. *Information Sciences*, **299**, 32–41.

149 Shen, J. and Lam, J. (2015) ℓ_∞/L_∞-gain analysis for positive linear systems with unbounded time-varying delays. *IEEE Transactions on Automatic Control*, **60** (3), 857–862.

150 Liu, J.J., Zhang, K.J., Pang, G.C., and Wei, H.K. (2015) Robust stabilisation for constrained discrete-time switched positive linear systems with uncertainties. *IET Control Theory & Applications*, **9** (17), 2598–2605.

151 Liu, J.J. and Zhang, K.J. (2015) State feedback with memory for constrained switched positive linear systems. *Entropy*, **17** (5), 2655–2676.

152 Liu, J.J., Zhang, K.J., Pang, G.C., and Wei H.K. (2016) Controller synthesis for constrained discrete-time switched positive linear systems. *Nonlinear Analysis: Hybrid Systems*, **19**, 1–12.

153 Liu, J.J., Zhang, K.J., Pang, G.C., and Wei, H.K. (2015) Robust stability of positive switched systems with dwell time. *International Journal of Systems Science*, **47** (11), 2553–2562.

154 Podlubny, I. (1999) *Fractional Differential Equations*, Academic Press, San Diego.

155 Kaczorek, T. (2011) *Selected Problems of Fractional System Theory*, Springer, Berlin.

156 Kaczorek, T. (2008) Fractional positive continuous-time linear systems and their reachability. *International Journal of Applied Mathematics and Computer Science*, **18** (2), 223–228.

157 Kaczorek, T. (2010) Stability and stabilization of positive fractional linear systems by state-feedbacks. *Bulletin of the Polish Academy of Sciences: Technical Sciences*, **58** (4), 537–554.

158 Hmamed, A., Mesquine, F., Benzaouia, A., *et al.* (2013) Continuous-time fractional bounded positive systems. 52nd IEEE Conference on Decision and Control, Florence, Italy.

159 Kaczorek, T. (2014) Minimum energy control of fractional positive continuous-time linear systems with bounded inputs. *International Journal of Applied Mathematics and Computer Science*, **24** (2), 335–340.

160 Mesquine, F., Hmamed, A., Benhayoun, M., *et al.* (2015) Robust stabilization of constrained uncertain continuous-time fractional positive systems. *Journal of the Franklin Institute*, **352** (1), 259–270.

161 Benzaouia, A., Hmamed, A., Mesquine, F., *et al.* (2014) Stabilization of continuous-time fractional positive systems by using a Lyapunov function. *IEEE Transactions on Automatic Control*, **59** (8), 2203–2208.

162 Kaczorek, T. (2015) Minimum energy control of fractional positive electrical circuits with bounded inputs. *Circuits, Systems, and Signal Processing*, **35** (6), 1815–1829.

163 Kilbas, A.A., Srivastava, H.M., and Trujillo, J.J. (2006) *Theory and Application of Fractional Differential Equations*, Elsevier, New York.

164 Petráš, I. (2011) *Fractional-Order Nonlinear Systems: Modeling, Analysis and Simulation*, Higher Education Press, Beijing.

165 Pan, I. and Das, S. (2012) *Intelligent Fractional Order Systems and Control: An Introduction*, Springer, Heidelberg.

166 Li, C. and Deng, W. (2007) Remarks on fractional derivatives. *Applied Mathematics and Computation*, **187** (2), 777–784.

167 Aguila-Camacho, N., Duarte-Mermoud, M.A., and Gallegos, J.A. (2014) Lyapunov functions for fractional order systems. *Communications in Nonlinear Science and Numerical Simulation*, **19** (9), 2951–2957.

168 Li, L. and Sun, Y. (2015) Adaptive fuzzy control for nonlinear fractional-order uncertain systems with unknown uncertainties and external disturbance. *Entropy*, **17** (8), 5580–5592.

169 Shao, S., Chen, M., and Yan, X. (2015) Adaptive sliding mode synchronization for a class of fractional-order chaotic systems with disturbance. *Nonlinear Dynamics*, **83** (4), 1855–1866.

170 Ge, S.S., Hang, C.C., Lee, T.H., and Zhang, T. (2001) *Stable Adaptive Neural Network Control*, Kluwer, Norwell, MA.

171 Duarte-Mermoud, M.A., Aguila-Camacho, N., Gallegos, J.A., and Castro-Linares, R. (2015) Using general quadratic Lyapunov functions to prove Lyapunov uniform stability for fractional order systems. *Communications in Nonlinear Science and Numerical Simulation*, **22** (1), 650–659.

172 Wei, Y., Chen, Y., Liang, S., and Wang, Y. (2015) A novel algorithm on adaptive backstepping control of fractional order systems. *Neurocomputing*, **165**, 395–402.

173 Wen, X.J., Wu, Z.M., and Lu, J.G. (2008) Stability analysis of a class of nonlinear fractional-order systems. *IEEE Transactions on Circuits and Systems II: Express Briefs*, **55** (11), 1178–1182.

174 Ye, H., Gao, J., and Ding, Y. (2007) A generalized Gronwall inequality and its application to a fractional differential equation. *Journal of Mathematical Analysis and Applications*, **328** (2), 1075–1081.

175 Wu, Z.B. and Zou, Y.Z. (2014) Global fractional-order projective dynamical systems. *Communications in Nonlinear Science and Numerical Simulation*, **19** (8), 2811–2819.

176 Cao, Y. and Bai, C. (2014) Finite-time stability of fractional-order BAM neural networks with distributed delay. *Abstract and Applied Analysis*, **2014**, 634803.

177 Lim, Y.H. and Ahn, H.S. (2012) Utilization of Gronwall–Bellman lemma in fractional order systems. Technical Report of Distributed Control and Autonomous Systems Lab., Gwangju Institution of Science and Technology, GIST DCASL TR 2012-01.

178 Lorenz, E.N. (1963) Deterministic nonperiodic flow. *Journal of the Atmospheric Sciences*, **20** (2), 130–141.

179 Li, C. and Yan, J. (2007) The synchronization of three fractional differential systems. *Chaos, Solitons & Fractals*, **32** (2), 751–757.

180 Lu, J.G. (2005) Chaotic dynamics and synchronization of fractional-order Genesio–Tesi systems. *Chinese Physics*, **14** (8), 1517.

181 Lu, J.G. (2005) Chaotic dynamics and synchronization of fractional-order Arneodo's systems. *Chaos, Solitons & Fractals*, **26** (4), 1125–1133.

182 Samardzija, N. and Greller, L.D. (1988) Explosive route to chaos through a fractal torus in a generalized Lotka–Volterra model. *Bulletin of Mathematical Biology*, **50** (5), 465–491.

183 Ma, J.H. and Chen, Y.S. (2001) Study for the bifurcation topological structure and the global complicated character of a kind of nonlinear finance system (I). *Applied Mathematics and Mechanics*, **22** (11), 1240–1251.

184 Sheu, L.J., Chen, H.K., Chen, J.H., *et al.* (2008) Chaos in the Newton–Leipnik system with fractional order. *Chaos, Solitons & Fractals*, **36** (1), 98–103.

185 Lü, J., Chen, G., Cheng, D., and Celikovsky, S. (2002) Bridge the gap between the Lorenz system and the Chen system. *International Journal of Bifurcation and Chaos*, **12** (12), 2917–2926.

186 Lü, J. and Chen, G. (2002) A new chaotic attractor coined. *International Journal of Bifurcation and Chaos*, **12** (3), 659–661.

187 Deng, W.H. and Li, C.P. (2005) Chaos synchronization of the fractional Lü system. *Physica A*, **353**, 61–72.

188 Min, F.H., Shao, S.Y., Huang, W.D., and Wang, E.R. (2015) Circuit implementations, bifurcations and chaos of a novel fractional-order dynamical system. *Chinese Physics Letters*, **32** (3), 030503.

189 Ahmad, W.M. (2005) Hyperchaos in fractional order nonlinear systems. *Chaos, Solitons & Fractals*, **26** (5), 1459–1465.

190 Wu, X., Lu, H., and Shen, S. (2009) Synchronization of a new fractional-order hyperchaotic system. *Physics Letters A*, **373** (27), 2329–2337.

191 Huang, X., Zhao, Z., Wang, Z., and Li, Y. (2012) Chaos and hyperchaos in fractional-order cellular neural networks, *Neurocomputing*, **94**, 13–21.

192 Oustaloup, A., Sabatier, J., and Lanusse, P. (1999) From fractal robustness to the CRONE control. *Fractional Calculus and Applied Analysis*, **2** (1), 1–30.

193 Li, T.Z., Yu, W., and Luo, M.K. (2014) Control of fractional chaotic and hyperchaotic systems based on a fractional order controller. *Chinese Physics B*, **23** (8), 80501.

194 Shao, S.Y. and Chen, M. (2016) Fractional-order control for a novel chaotic system without equilibrium. *IEEE/CAA Journal of Automatica Sinica*. doi: 10.1109/JAS.2016.7510124

195 Shao, S.Y., Chen, M., Chen, S.D., and Wu, Q.X. (2016) Adaptive neural control for an uncertain fractional-order rotational mechanical system using disturbance observer. *IET Control Theory & Applications*, **10** (16), 1972–1980.

196 Shao, S.Y., Chen, M., and Yang, Q.Y. (2016) Sliding mode control for a class of fractional-order nonlinear systems based on disturbance observer. IEEE International Conference on Industrial Technology, Taipei, Taiwan.

197 Shao, S.Y., Chen, M., and Wu, Q.X. (2016) Stabilization control of continuous-time fractional positive systems based on disturbance observer. *IEEE Access*, **4**, 3054–3064.

198 Shao, S.Y., Chen, M., and Wu, Q.X. (2016) Tracking control for uncertain fractional-order chaotic systems based on disturbance observer and neural network. *IMA Journal of Mathematical Control and Information*. doi: 10.1093/imamci/dnw024

199 Ziegler, J.G. and Nichols, N.B. (1942) Optimal settings for automatic controllers. *Transactions of the American Society of Mechanical Engineers*, **64**, 759–768.

200 Liang, T.N. (2011) Fractional order PID controllers and analysis of stability region for fractional order systems with uncertain parameters. Doctoral thesis. Xidian University.

201 Luo, Y. and Chen, Y. (2012) *Fractional Order Motion Controls*, John Wiley & Sons, Ltd, Chichester.

202 Xue, D.Y. and Zhao, C.N. (2007) Fractional order PID controller design for fractional order system. *Control Theory & Applications*, **24** (5), 771–776.

203 Wei, L., Fei, M., and Hu, H. (2008) Modeling and stability analysis of grey-fuzzy predictive control. *Neurocomputing*, **72** (1), 197–202.

204 Chen, X., Yang, J., Li, S., and Li, Q. (2009) Disturbance observer based multi-variable control of ball mill grinding circuits. *Journal of Process Control*, **19** (7), 1205–1213.

205 Olivier, L.E., Craig, I.K., and Chen, Y.Q. (2012) Fractional order and BICO disturbance observers for a run-of-mine ore milling circuit. *Journal of Process Control*, **22** (1), 3–10.

206 Chen, Y.Q., Vinagre, B.M., and Podlubny, I. (2003) On fractional order disturbance observer. ASME 2003 International Design Engineering Technical Conferences and Computers and Information in Engineering Conference, Chicago, IL.

207 Lu, J., Xie, W., Zhou, H., and Zhang, A. (2014) Vibration suppression using fractional-order disturbance observer based adaptive grey predictive controller. *Journal of Vibroengineering*, **16** (5), 2205–2215.

208 Son, N.N. and Anh, H.P.H. (2014) Adaptive backstepping self-balancing control of a two-wheel electric scooter. *International Journal of Advanced Robotic Systems*, **11** (10), 165.

209 Li, Y., Chen, Y.Q., and Podlubny, I. (2009) Mittag–Leffler stability of fractional order nonlinear dynamic systems. *Automatica*, **45** (8), 1965–1969.

210 Sadati, S.J., Baleanu, D., Ranjbar, A. *et al.* (2010) Mittag–Leffler stability theorem for fractional nonlinear systems with delay. *Abstract and Applied Analysis*, **2010**, 108651.

211 Liu, S., Li, X.Y., Jiang, W., and Zhou, X.F. (2012) Mittag–Leffler stability of nonlinear fractional neutral singular systems. *Communications in Nonlinear Science and Numerical Simulation*, **17** (10), 3961–3966.

212 Qian, D.L., Li, C.P., Agarwal, R.P., and Wong, P.J.Y. (2010) Stability analysis of fractional differential system with Riemann–Liouville derivative. *Mathematical and Computer Modelling*, **52** (5), 862–874.

213 Zhang, F.R. and Li, C.P. (2011) Stability analysis of fractional differential systems with order lying in (1, 2). *Advances in Difference Equations*, **2011**, 213485.

214 Qin, Z.Q., Wu, R.C., and Lu, Y.F. (2014) Stability analysis of fractional-order systems with the Riemann-Liouville derivative. *Systems Science & Control Engineering*, **2** (1), 727–731.

215 Chen, L.P., He, Y. G., Chai, Y., and Wu, R.C. (2014) New results on stability and stabilization of a class of nonlinear fractional-order systems. *Nonlinear Dynamics*, **75** (4), 633–641.

216 Hua, C.C. and Guan, X.P. (2004) Adaptive control for chaotic systems. *Chaos, Solitons & Fractals*, **22** (1), 55–60.

217 Liu, L.P., Han, Z.Z., and Li, W.L. (2009) Global sliding mode control and application in chaotic systems. *Nonlinear Dynamics*, **56** (1–2), 193–198.

218 Jia, Q. (2007) Hyperchaos generated from the Lorenz chaotic system and its control. *Physics Letters A*, **366** (3), 217–222.

219 Volos, C.K., Kyprianidis, I.M., and Stouboulos, I.N. (2013) Image encryption process based on chaotic synchronization phenomena. *Signal Processing*, **93** (5), 1328–1340.

220 Wang, Z., Huang, X., Li, Y.X., and Song, X.N. (2013) A new image encryption algorithm based on the fractional-order hyperchaotic Lorenz system, *Chinese Physics B*, **22** (1), 010504.

221 Zhou, P. and Bai, R. (2014) One adaptive synchronization approach for fractional-order chaotic system with fractional-order $1 < q < 2$. *The Scientific World Journal*, **2014**, 490364.

222 Chen, L., Chai, Y., Wu, R., and Yang, J. (2012) Stability and stabilization of a class of nonlinear fractional-order systems with Caputo derivative. *IEEE Transactions on Circuits and Systems II: Express Briefs*, **59** (9), 602–606.
223 Lu, J.G. and Chen, G. (2006) A note on the fractional-order Chen system. *Chaos, Solitons & Fractals*, **27** (3), 685–688.
224 Milanovic, V. and Zaghloul, M.E. (1996) Improved masking algorithm for chaotic communications systems. *Electronics Letters*, **32** (1), 11–12.
225 Ishteva, M. (2005) Properties and applications of the Caputo fractional operator. Master thesis. Universität Karlsruhe (TH).
226 Ge, Z.M. and Jhuang, W.R. (2007) Chaos, control and synchronization of a fractional order rotational mechanical system with a centrifugal governor. *Chaos, Solitons & Fractals*, **33** (1), 270–289.
227 Xin, G., and Yu, J.B. (2005) Chaos and chaotic control in a fractional-order electronic oscillator. *Chinese Physics*, **14** (5), 908.
228 Tavazoei, M.S. and Haeri, M. (2008) Limitations of frequency domain approximation for detecting chaos in fractional order systems. *Nonlinear Analysis: Theory, Methods & Applications*, **69** (4), 1299–1320.
229 Ahmad, W.M. and Sprott, J.C. (2003) Chaos in fractional-order autonomous nonlinear systems. *Chaos, Solitons & Fractals*, **16** (2), 339–351.
230 Li, Y., Chen, Y.Q., and Podlubny, I. (2010) Stability of fractional-order nonlinear dynamic systems: Lyapunov direct method and generalized Mittag–Leffler stability. *Computers & Mathematics with Applications*, **59** (5), 1810–1821.
231 Araki, M. and Kondo, B. (1972) Stability and transient behavior of composite nonlinear systems. *IEEE Transactions on Automatic Control*, **17** (4), 537–541.
232 Yu, S.M. (2004) A new type of chaotic generator. *Acta Physica Sinica*, **53** (12), 4111–4119.
233 Daftardar-Gejji, V. and Bhalekar, S. (2010) Chaos in fractional ordered Liu system. *Computers & Mathematics with Applications*, **59** (3), 1117–1127.
234 Jiang, G.P. and Tang, W.K.S. (2002) A global synchronization criterion for coupled chaotic systems via unidirectional linear error feedback approach. *International Journal of Bifurcation and Chaos*, **12** (10), 2239–2253.
235 Yin, Y.Z. (1997) Synchronization of chaos in a modified Chua's circuit using continuous control. *Journal of Electronics*, **19** (6), 824–827.
236 Thamilmaran, K., Lakshmanan, M., and Venkatesan, A. (2004) Hyperchaos in a modified canonical Chua's circuit. *International Journal of Bifurcation and Chaos*, **14** (1), 221–243.
237 Shao, S.Y., Min, F.H., Wu, X.H., and Zhang, X.G. (2014) Implementation of a new chaotic system based on field programmable gate array. *Acta Physica Sinica*, **63** (6), 060501.
238 Suykens, J.A.K., Curran, P.F., Vandewalle, J., and Chua, L.O. (1997) Robust nonlinear H_∞ synchronization of chaotic Lur'e systems. *IEEE Transactions on Circuits and Systems I: Fundamental Theory and Applications*, **44** (10), 891–904.
239 Zhu, H., Zhou, S., and Zhang, J. (2009) Chaos and synchronization of the fractional-order Chua's system. *Chaos, Solitons & Fractals*, **39** (4), 1595–1603.
240 Matignon, D. (1996) Stability results for fractional differential equations with applications to control processing. CESA'96 IMACS Multiconference: Computational Engineering in Systems Applications, Lille, France.

241 Petráš, I. (2002) Control of fractional-order Chua's system. *Journal of Electrical Engineering*, **53** (7–8), 219–222.

242 Jiang, G.P., Chen, G.R., and Tang, W.K. (2003) A new criterion for chaos synchronization using linear state feedback control. *International Journal of Bifurcation and Chaos*, **13** (8), 2343–2351.

243 Monje, C.A., Chen, Y., Vinagre, B. M., *et al.* (2010) *Fractional-order Systems and Controls: Fundamentals and Applications*, Springer, New York.

Index

Robust Adaptive Control for Fractional-Order Systems with Disturbance and Saturation, First Edition.
Mou Chen, Shuyi Shao, and Peng Shi.